Defects in Nanocrystals

Defects in Nanocrystals

Structural and Physico-Chemical Aspects

Sergio Pizzini

CRC Press
Taylor & Francis Group
Boca Raton London New York

CRC Press is an imprint of the
Taylor & Francis Group, an **informa** business

Cover figures are drawn from the Final Technical Report of the Nanophoto Project of the European Community "Modelling Activities." The INFM-CNR team is warmly acknowledged.

First edition published 2020
by CRC Press
6000 Broken Sound Parkway NW, Suite 300, Boca Raton, FL 33487-2742

and by CRC Press
2 Park Square, Milton Park, Abingdon, Oxon, OX14 4RN

© 2020 Taylor & Francis Group, LLC

CRC Press is an imprint of Taylor & Francis Group, LLC

Reasonable efforts have been made to publish reliable data and information, but the author and publisher cannot assume responsibility for the validity of all materials or the consequences of their use. The authors and publishers have attempted to trace the copyright holders of all material reproduced in this publication and apologize to copyright holders if permission to publish in this form has not been obtained. If any copyright material has not been acknowledged please write and let us know so we may rectify in any future reprint.

Except as permitted under U.S. Copyright Law, no part of this book may be reprinted, reproduced, transmitted, or utilized in any form by any electronic, mechanical, or other means, now known or hereafter invented, including photocopying, microfilming, and recording, or in any information storage or retrieval system, without written permission from the publishers.

For permission to photocopy or use material electronically from this work, access www.copyright.com or contact the Copyright Clearance Center, Inc. (CCC), 222 Rosewood Drive, Danvers, MA 01923, 978-750-8400. For works that are not available on CCC please contact mpkbookspermissions@tandf.co.uk

Trademark notice: Product or corporate names may be trademarks or registered trademarks, and are used only for identification and explanation without intent to infringe.

ISBN: 978-0-367-34599-0 (hbk)
ISBN: 978-0-429-32820-6 (ebk)

Typeset in Times
by Deanta Global Publishing Services, Chennai, India

Contents

Contents

Preface

Experience tells us that the physico-chemical properties of the matter depend on the size and shape to which it is brought by mechanical or chemical processes. In fact, melting temperatures of elemental and compound solids decrease with the decrease of the size, the chemical reactivity of gaseous species is enhanced by the use of nano-structured metallic or non-metallic catalysts, and the optical properties of metallic and semiconductor nanoparticles strongly differ from those of their counterpart bulk materials. With the decrease of the size, the mass amount of the atomic species at the surface of a nanocrystal does increase, together with the change of its dimensionality, D, that could lead to 0D dots, 1D nanowires, and 2D sheets, whose physical properties are expected to depend on their dimensionality.

Eventually, the physico-chemical properties of the matter at the nano-size could be finely tuned by the adsorption of chemical species at its surface, and the change of thermodynamic, chemical and physical properties of an elemental or compound solid with the decrease of the size is associated with morphological and structural changes. This could lead to the set-up of core-shell structures, to the thermodynamic stabilization of metastable phases, and to the coexistence of stable and metastable phases in a single nanowire.

Defectivity is, therefore, a fundamental aspect of matter at the nanoscale, if we assume that defects are meant as all possible deviations from the structural characteristics of an ideal crystal, and include point and extended defects, like twins and stacking faults, as well as all the typical morphological features of nanocrystals.

This book is an attempt to treat the growth and properties of semiconductor nano-materials from a purely physico-chemical viewpoint, hoping to demonstrate that physical chemistry is crucial and necessary to fully understand the properties of nanostructures.

The amount of information concerning nanostructures present today in literature is already enormous, and does increase day after day.

I take, therefore, on my own, responsibility the selection of the information included in this book, with the consciousness that the selection made is certainly not exhaustive, and not respectful of the work of a multitude of researchers I could not consider.

I express my gratitude to my colleagues of the Polytechnic School of Milano, of the University of Milano-Bicocca, Bologna, Cagliari, and Como, and of the Universities of Konstanz and Aix-Marseille (Prof. C. Cavallotti, Polytechnic School of Milano [Italy]; Prof. A. Cavallini, University of Bologna [Italy]; Prof. L. Colombo and A. Mattoni, University of Cagliari [Italy]; Prof. Leo Miglio, Prof. M. Guzzi, Prof. M. Acciarri, and Prof. S. Binetti, University of Milano-Bicocca [Italy]; Prof. B. Pichaud, G. Regula and M. Texier, University Aix-Marseille [France]; Prof. Hans von Känel, D. Crastina, and G. Isella, LNESS, University of Como [Italy]; Prof. G. Hahn, Dr. G. Micard , Dr.B. Terheiden and Dr. K. Peter, University of Konstanz [Germany]) for the work we carried out together in the frame of the European Nanophoto Project, addressed at the study of nanocrystalline silicon, as well to a number of colleagues worldwide for their support during the writing of the book and delivery of the material. Among them my particular gratitude goes to Prof. H.E. Ruda, University of Toronto (Canada); Dr. K.H. Chung, Princeton University (U.S.); Dr. A. Irrera and B. Fazio CNR, Messina (Italy); Prof. F. Priolo, Catania University (Italy); Dr. Shailendra K. Saxena, University of Alberta, National Institute for Nanotechnology (NINT) (Canada); Prof. Spyridon Galis, SUNY Polytechnic Institute, Albany, NY (U.S.); Prof. Harri Lipsanen and Dr. Henrik, Mäntynen Aalto University, School of Electrical Engineering (Finland); Prof. S-L.Zhang, Dept. Physics, Beijing University (Popular Republic of China); Prof. A. Fontcuberta i Morral, Ecole Polytechnique Federale, Lausanne (Switzerland); Prof. C.S. Casari and M. Zamani, Polytechnic School, Milano (Italy).

Author Biography

Sergio Pizzini started his career at the Joint Research Centre of the European Commission in Ispra (Italy) and later continued in Petten (Netherlands), committed to studies of a fuel for a molten salt reactor, within a Joint Program with the Oak Ridge Centre in the U.S. After leaving the Commission he joined the University of Milano as Associated Professor, where he started basic studies on solid electrolytes, which also resulted in the realization of a prototype of a solid electrolyte gas sensor and of a solid state sensor for the determination of the stoichiometry of nuclear oxide fuels. Still maintaining his position at the University, he worked for five years as Director of the Materials Department of the Corporate Research Centre of Montedison in Novara, where he launched a number of new R&D activities on advanced materials for electronics. After having left Montedison he founded the Heliosil Company, where, as its CEO, he studied a process for the production of solar-grade silicon and patented a furnace and a process for the directional solidification of silicon in multicrystalline ingots. In 1982 he left all outside duties in order to serve the University of Milano and, later, the University of Milano-Bicocca, as full Professor of Physical Chemistry. In this last period of activity, in addition to his teaching and management duties, he carried out systematic studies on semiconductor silicon, mostly addressed at the understanding of electronic and optical properties of point and extended defects of Czochralski, multicrystalline, and nanocrystalline silicon in the frame of national and European projects.

Sergio Pizzini is author of more than 250 technical papers published in international journals. He has authored or co-authored four books and was chairman or co-chairman of a number of international symposia in the materials science field. Since his retirement, he has served the University of Milano-Bicocca through external cooperation.

List of Abbreviations

AES	Auger electron spectroscopy
AFM	Atomic force microscopy
BCC	Body centered cubic
BS	Black silicon
CSTEM	Cross-sectional scanning tunneling microscopy
CVD	Chemical vapor deposition
DB	Dangling bond
DFT	Density functional theory
DLTS	Deep level transient spectroscopy
DNW	Diamond nanowires
DOS	Electronic density of states
EAM	Embedded atom method
EFTEM	Energy filtered transmission electron
EXD	Energy dispersive X-ray spectroscopy
FTIR	Fourier transform infrared spectroscopy
FWHM	Full width at half maximum
HRTEM	High-resolution TEM
HW-CVD	Hot-wire CVD
LDA	Local density approximation
LED	Light-emitting diode
LEPECVD	Low-energy plasma-enhanced CVD
LID	Light-induced degradation
MACE	Metal-assisted chemical etching
MAEAM	Modified analytics embedded atom method
MBE	Molecular beam epitaxy
MD	Molecular dynamics (simulation)
MOCVD	Metal organic chemical vapor deposition
NC	Nanocrystal
NP	Nanoparticle
NW	Nanowire
PAS	Positrons annihilation spectroscopy
PEB	Perdew–Burke–Ernzernhoff (functional)
PECVD	Plasma-enhanced CVD
PL	Photo luminescence
QC	Quantum confinement
QD	Quantum dots
RBS	Rutherford backscattering
RF	Radiofrequency
SF	Stacking fault
SIMS	Secondary-ion mass spectroscopy
SRO	Silicon-rich oxide
STS	Scanning tunneling spectroscopy

TB	Tight binding (calculations)
TEM	Transmission electron microscopy
UHV	Ultra-high vacuum
VLS	Vapor-liquid-solid (growth)
VSS	Vapor-solid-solid (growth)
XPS	X-ray photoemission spectroscopy
XRD	X-ray diffraction

1 Nanostructured Semiconductors

Basic Physico-Chemical Concepts and Potential Applications of Nanosized Semiconductors*

"There's Plenty of Room at the Bottom"

—*Richard Feynman, December 1959*

1.1 NANOSTRUCTURED SEMICONDUCTORS: BACKGROUND CONCEPTS

A nanostructured semiconductor (NS) is a material with a nanometric size in one or more of its spatial dimensions, whose properties are size- and shape-dependent, and which belongs to a class of systems which could be called "small" systems.

Following Pokropivny and Skorokhod [1] we will consider zero-dimensional (0D) a material for which all dimensions are at the nanoscale, one-dimensional (1D) a material for which two dimensions are at the nanoscale, and 2D or 3D materials those for which one or zero dimensions are at the nanoscale. It is often assumed that the nanoscale range extends below to a size of 100 nm, but varies in function of the material nature, as we will show in the following.

All physical properties of matter are shown to deviate from their bulk properties, where "bulk" means a material with an atomic density equal to the Avogadro number ($N_A = 6.022 \cdot 10^{23}$/mol), as soon as their size decreases [2].

This is particularly true for the thermodynamic properties of a material, whose actual values, in fact, decrease with the decrease of the size, and which do present substantial deviation from their bulk values already in the mesoscale range (>100 nm), but depend, as well, on the shape, which has a significant influence on their actual values, as has been demonstrated, for example, in the case of silicon [3].

For some properties of matter a critical stage is established at the nanoscale, when the radius of the nanoparticle becomes comparable with the Bohr exciton radius, and

---

* https://orcid.org/0000-0002-0542-3219.

TABLE 1.1
Bohr Radii and Energy Gaps of Selected Semiconductors

Elemental or Compound Semiconductor	Bohr Radius of Exciton (nm)	Energy Gap (eV) 300 K
Si	4.5	1.11
Ge	24	0.66
GaAs	14.5	1.43
GaN	2.7	3.39
AlN	1.2	6.1
ZnS	2.5	3.6
ZnSe	4.1	2.70
ZnTe	6.7	2.25
CdSe	5.6	1.74
CdTe	7.3	1.44
PbS	18	0.41

quantum confinement (localization) of carriers in nanocrystalline semiconductors does occur. In turn [4] *Quantum confinement (QC) is defined as a reduction in the degrees of freedom of the carrier particles, implying their reduction in the allowed phase space.**

In an analogous manner, the normal lattice vibrations[†] are also confined (localized) when their wave function is restricted to the volume of the crystallite [5, 6].

Depending on the dimensionality of a specific "small" system, which could be a dot (0D), a wire (1D), or a well (2D), the carriers of the system become quantum-confined within three, two, and one dimension.

Quantum size effects on the opto-electronic properties of semiconductor nanocrystals (NCs) have been discussed extensively by Yoffe [7, 8], with a review of the experimental work available at that time given, together with the theoretical treatment of the quantization effects on the optical properties of low-dimensionality systems.

It is shown that QC might be expected, and is, in fact, found, when the radius R of the NC is of the order of magnitude of the bulk Bohr exciton radius a_B of a particular semiconductor (see Table 1.1)

$$a_B = \frac{h^2 \varepsilon}{e^2} \left(\frac{1}{m_e^*} + \frac{1}{m_h^*} \right) = \frac{h^2 \varepsilon}{e^2 \mu} = a_0 \frac{\varepsilon}{\varepsilon^o} \frac{m_e}{\mu} \tag{1.1}$$

where ε is the dielectric constant of the material, $a_o = \dfrac{\varepsilon^o h^2}{m_e e^2} = 0.0529$ nm is the Bohr radius of hydrogen, ε^o is the permittivity of the free space, m_e is the rest mass of the

---

* Original sentence in [4].
[†] Phonons.

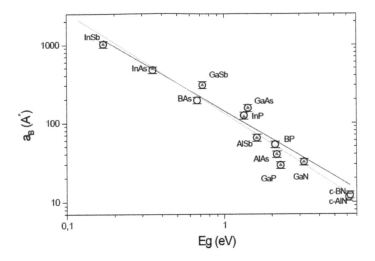

FIGURE 1.1 Bohr radius of the exciton vs energy gap of III–V semiconductors. (*After* N.Korti-Baghdadli, A. E. Merad, Y. Benouaz (2013). Adjusted Adhashi's model of exciton Bohr parameters and new proposed model for optical properties of III–V semiconductors *Amer. J. Mat Sci Technology* **3**, 65–73 *Open access Journal.*)

electron, $\mu = \dfrac{m_e^* m_h^+}{m_e^* + \mu_h^*}$ is the reduced mass of the exciton, and m_e^* and m_h^* are the effective masses of the electron and hole, which depend on the semiconductor nature.

It is important to mention, here, that quantum confinement causes a shift from a continuous density of energy level to discrete, atomic-like, energy levels, an increase of the energy gap E_G, and then a blue shift of the absorption edge and of the luminescence peak.

Therefore, the energy gap E_G of semiconductors is shown to be inversely correlated to the exciton Bohr radius [9], as could be seen for the case of III–V semiconductors in Figure 1.1* [10], but which is common to all semiconductors.

Depending on the relationships between the radius R of a semiconductor NC and its exciton Bohr radius, three main confinement conditions might be foreseen [7].

Weak confinement conditions should be expected when $R \ll a_B$ and moderate or strong confinement conditions when $R < a_B$ or $R \ll a_B$.

In the case of weak[†] or moderate confinement, the Coulomb interaction among carriers predominates and the exciton is a stable species, its motion is quantized, and the energy shift ΔE of the ground state of the exciton as a function of the radius of the NC is given by the equation

$$\Delta E \approx \frac{h^2 \pi^2}{2MR^2} \tag{1.2}$$

where M is the mass of the exciton.

---

* Whose experimental behaviour is fitted by the equation $a_B = \left(\dfrac{A}{E_G}\right)^\alpha$ with $A = 44.594$ eV and $\alpha = 1.288$.

† Which is typical, as, for example, with CuCl NCs.

The case of strong confinement holds when $R \ll a_e$ and $R \ll a_h$, where $a_e = \dfrac{h^2 \varepsilon}{m_e^* e^2}$ and $a_h = \dfrac{h^2 \varepsilon}{m_h^* e^2}$ are the electron and hole Bohr radii, respectively [7], ε as before is the dielectric constant of the semiconductor, and m_e^* and m_h^* are the effective masses of electrons and holes.

Under these conditions, the Coulomb interaction term is negligible, the exciton cannot be formed, electrons and holes are confined as independent entities, the optical spectra should consist of transitions among sub-bands, and the energy shift of the ground state of the exciton as a function of the radius of the NC is given by the equation

$$\Delta E \approx \frac{h^2 \pi^2}{2 \mu R^2} \tag{1.3}$$

Here, the exciton mass M is replaced by the reduced mass μ of electrons and holes, where $\dfrac{1}{\mu} = \dfrac{1}{m_e^*} + \dfrac{1}{m_h^*}$ and m_e^* and m_h^* are the effective masses of electrons and holes, respectively.

A third condition arises when $R \ll a_e$ and $R \gg a_h$, which holds when $a_h \gg a_e$ due to the largest hole mass. Here the sole electron is quantized, and the hole interacts with the electron via the Coulomb potential, $a_B = \dfrac{h^2 \varepsilon}{e^2 \mu}$ and $\Delta E \approx \dfrac{h^2 \pi^2}{2 \mu R^2}$.

Quantum dots (QDs) are a typical family of nanometric objects with a size below 30 to 3 nm, containing from 10^5 to 10^3 atoms, which fall within the last category of quantum-confined systems. QDs are lacking *of translational invariance and may be considered a new kind of condensed matter in between crystalline solids and molecules* [11].*

In small systems, thermodynamics should be used with caution, since classical Boltzmann's thermo-statistics applies only to systems with atomic densities comparable with the Avogadro number ($N_A = 6.022 \cdot 10^{23}$/mol), and with a volume $V \to \infty$, which is not the case of nanocrystals. Extensivity is another, essential condition for thermodynamics to work, i.e. it is a thermodynamic limit. Small systems are "non-extensive" when they are smaller than the extension of their forces, and the entropy of these systems is not the sum of the entropy of the parts in which they could be divided.

Nevertheless, we will see that is common practice to extend the range of applicability of thermodynamics towards the mesoscale, and account thereafter for the considerable amount of experimental knowledge about the dependence of physico-chemical properties of elemental and compound systems from bulk conditions down to a size of hundreds of nanometers ($N \approx 10^8$ atoms).

We will also see that the theoretical work carried out in this field does allow a rational account of thermodynamic properties of multicomponent materials at the nanoscale, which may substantially differ from those at the macroscale, when the number of component atoms N is comparable with the Avogadro number.

---

* Original sentence in the Schulz paper [11].

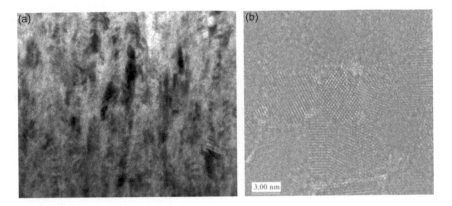

FIGURE 1.2   *Left*: SEM view of a vertical section of an nc-Si film; *right*: HRTEM micrograph of a horizontal section of the same film.

This is the case of certain binary metallic and semiconductor systems, presenting extended solubility gaps at the macro scale, but which turn out to complete solubility behavior at the nanoscale [12].

Materials consisting of heterogeneous mixtures of two or more structurally different phases might, instead, present different structures in more than one length scale, leading to the upset of hierarchical properties, well-known in biological materials like wood, animal bones, nacre [13–15], but also observed in nanocrystalline silicon films, which will be discussed with a large emphasis in Chapter 3.

These films present a disordered structure, consisting of an array of vertically oriented silicon nanocrystalline columns with a section of more than 6 nm dispersed in a matrix of amorphous silicon; see Figure 1.2. The columns, however, consist of an ordered, mosaic structure, where thin shells of amorphous-silicon separate differently oriented silicon nanocrystals, as shown in Figure 1.2 right.

Like in the case biological materials [16], we expect that a structural hierarchy in semiconductor nanomaterials originates from a self-organized mechanism, which plays a major role in determining its structural and physical properties.

According to Prigogine and Nicolis [17], self-organization typically occurs in the frame of dissipative systems, open to mass and energy fluxes, which are non-equilibrium systems, where the set-up of a locally low-entropy state (i.e. of local order) does occur at the expense of the global system in which they are imbedded, where the entropy increases.

Self-organization in a growth process, as is the case of nc-Si films, is, therefore, a way to counterbalance the presence of a local gradient of temperature, pressure, or concentration with the creation of an organized structure.

1.2   DEFECTS IN NANOSTRUCTURED SEMICONDUCTORS*

Like in the case of bulk semiconductors, the physico-chemical properties of semiconductor nanostructures could be supposed to be influenced by a variety of point

---

* A detailed treatment of defects at the nanoscale will be given in Chapter 2.

defects, vacancies, self-interstitials, and impurities, whose actual content would depend on the specific synthetic process used, and on the subsequent chemical or thermal treatments.

Since the surface/volume ratio of nanometric semiconductors is very large, we expect, however, an increasing role of surfaces in their physical properties, as soon as their size decreases, together with that of interfaces in nanocrystalline hetero-structures and of grain boundaries (GB) in polycrystalline nanocrystalline films.

Eventually, also the nanocrystal shape, which depends on the crystalline structure of its surface facets, should play some role in the physical properties of the material.

Unlike extended defects, point defects (vacancies and self-interstitials) in solids are equilibrium species [18], and the equations which account for their formation and for their equilibrium concentration are given using a standard thermodynamic approach and with the Kröger and Vink nomenclature.*

The formation of vacancies does occur with the following formal equilibrium process

$$M_M \rightleftharpoons M_{\text{surf}} + V_M \tag{1.4}$$

where M_{surf} is an atom of M at the surface, and the surface is considered an infinite sink or source of M, and the vacancy concentration $[V_M]$ is given by the following equation

$$[V_M] = \frac{N_V}{N} = B\exp-\frac{\Delta H_f}{kT} \tag{1.5}$$

where N_V is the number of vacancies, N is the total number of atoms of the system, and the preexponential B contains an entropy of formation term, k is the Boltzmann constant and ΔH_f is the enthalpy of formation of a vacancy.

Conversely, the formation of self-interstitials does occur with the following formal equilibrium process

$$M_{\text{surf}} \rightleftharpoons M_i \tag{1.6}$$

and their equilibrium concentration is given by the following equation

$$[M_i] = \frac{N_I}{N} = B\exp-\frac{\Delta H_f}{kT} \tag{1.7}$$

The formation enthalpies of both defects ΔH_f have been experimentally measured and theoretically evaluated for a variety of bulk semiconductors [19]. For the case of silicon, the experimental formation energies of vacancies span from 3.6 to 2.44 eV, while the calculated formation energy values are around 4 eV. For self-interstitials in silicon the experimental values are within 3.18 and 3.8 eV, while the theoretical values, which depend on their local structure, range within 4.16 and 3.31 eV [2]. The geometrical configuration of point defects in bulk semiconductors is, also, known, and calls for four broken bonds and a small atomic displacement for the relaxed vacancy.

Point defects are localized in the lattice, but the associated electronic perturbation is extended to long distances. The dominant characteristic of point defects, including

---

* Which will be used throughout the entire book. A double square bracket will be used for their concentration.

non-dopant impurities, is, however, that of introducing electronic deep levels in the gap.

The same defects are expected to be present in semiconductor nanostructures, with formation energies depending on the size, but their dependence on size and shape has been only theoretically investigated, since the direct experimental measurement of their values is, up to now, still unpracticable.

In fact, the main experimental source of information on the energetics of vacancy and self-interstitial formation comes from self-diffusion measurements,* and self-diffusion measurements on these systems are unpracticable.

Impurities are point defects introduced by external sources *with non-equilibrium reactions* in the semiconductor lattice, where they eventually sit in substitutional or interstitial positions, or in both, in an elemental

$$X + V_A \rightarrow X_A \tag{1.8}$$

$$X \rightarrow X_i \tag{1.9}$$

or in a compound AB lattice

$$X \rightarrow X_A + V_B \tag{1.10}$$

$$X \rightarrow X_B + V_A \tag{1.11}$$

$$X \rightarrow X_i \tag{1.12}$$

where the equilibrium concentration of vacancies and self-interstitials are given by Equations 1.5 and 1.7.

Deviations from stoichiometry of a compound semiconductor, associated with the preferential sublimation of one of the compound components, are also accounted for with the formation of vacancies, though the actual process could involve different types of defects

$$A_A \rightarrow A^g + V_A \tag{1.13}$$

$$B_B \rightarrow B^g + V_B \tag{1.14}$$

Vacancies are involved, as well, in doping processes of compound semiconductors, enabling stoichiometric conditions to be maintained at the expense, however, of the formation of their deep states, and potential degradation of electronic properties, as we will see in Section 2.4 of Chapter 2.

Vacancies are also involved, with the set-up of surface plasmon resonance effects,[†] in non-stoichiometric, nanocrystalline compound semiconductors, as is the case of germanium telluride $[Ge_{1-x}Te]$ [20] and of copper sulfide $[Cu_{1-x}S]$ nanocrystals [21], taken as non-unique examples.

This effect [20] is associated with the presence of a high density of ionized cationic vacancies and, consequently, of a high concentration of holes (10^{21} cm^{-3})

$$V_M \rightleftharpoons V_M^- + h \tag{1.15}$$

which leads to a degenerate semiconductor behavior [20].

---

* On the assumption that impurity self-diffusion is supported by vacancies and self-interstitials.
[†] Typical of metallic nanocrystals.

Also, the electronic properties of films of colloidal quantum dots (3.5 nm) of lead sulfide [PbS] deposited on solid substrates could be finely tuned by varying its stoichiometry from lead -rich to sulfur-rich samples [22], using Rutherford Back scattering (RBS) spectroscopy to measure the stoichiometry of the samples.

It was possible to observe that the energy gap of PbS colloidal dots, measured using photoluminescence (PL) spectroscopy, is strongly depressed by an excess of lead vacancies and enhanced by an excess of sulfur vacancies [22], explained as due to a stoichiometry-dependent quantum confinement of charge carriers.

Impurities are, also, deeply involved in determining or, at least, influencing the physical properties of nanocrystals, although with different qualitative and quantitative effects as compared with the bulk counterparts.

As an example, p-type doping leads to the stabilization of the high-pressure, hexagonal structure of silicon nanowires [23], showing the large impact of the local stress associated with the solution of an impurity with a different atomic radius of silicon in the silicon lattice at the nanoscale.

Hydrogen interaction with defects, like in the case of macroscopic semiconductors, leads to defect passivation, to the saturation of dangling bonds (DB) at the surface, or to its chemical interaction with reducible impurities [2].

The interaction of semiconductor nanoparticles with oxygen and water leads, eventually, to enhanced surface oxidation, hydroxylation, or passivation, with relevant changes of their chemical and optical properties. A cosmological event, consisting of the burst of large stars collapsing in *supernovae*, is considered at the origin of the extended red emission (540–960 nm) visible in the interstellar medium, arising from the photoluminescence of oxygen-passivated silicon nanocrystals [24].

Extended defects (grain boundaries, dislocations, staking faults, twins) are also present in polycrystalline semiconductors at the nanoscale and in nanowires. Extended defects behave, like in bulk semiconductors, as recombination sites for minority carriers and gettering sites* for impurities, but are also responsible for phase changes in semiconductor nanowires.

Gettering might occur with the formation of a localized chemical bond, or with the formation of a new solid phase, which in the case of silicon is a silicide phase. The gettering process is driven by the Gibbs free energy of the gettering reaction, and the heterogeneous equilibrium concentration of the impurity at the gettering phase (g) x_i^g and in solution (s) x_i^s is given by the following equation

$$\frac{x_i^g}{x_i^s} = A\exp-\frac{\Delta G_g}{kT} \tag{1.16}$$

where ΔG_g is the Gibbs free energy of the gettering reaction [2]. Extended defects at the nanoscale might have also a role in biphasic equilibrium transformations. This is the case of silicon nanowires, extensively discussed in Chapter 3, where high densities of stacking faults, and the consequent set-up of surface stress, favor the set-up of a phase transformation from cubic to hexagonal silicon [25].

---

* This is a peculiar characteristic of extended defects.

It is also known that surface stress induces the spontaneous, homogeneous nucleation of dislocations [26] in nanocrystals.

New types of nanometric defects, the skyrmions, were recently discovered in ultra-thin structures. They are nanometric magnetic solitons, apparently stable at room temperatures and in the absence of external magnetic fields*,† [27, 28], thus behaving as equilibrium species in ultra-thin structures.

Size and shape do not affect only the physical properties of a material, but also its chemical reactivity at the nano-level. As already mentioned in the first section of this chapter, metallic and non-metallic nanoparticles have been known, in fact, for years, to behave as excellent heterogeneous catalysts for chemical reactions. Surface defects should influence, also, the chemical reactivity of nanoparticles against the ambient. Therefore, the nanocrystal surfaces, and local inhomogeneities associated with interface defects, should play a role in the properties of nanocrystals supported/embedded‡ on/in different types of substrates and in their compatibility with the environment.

Also, the properties of nanocrystals embedded in a heterogeneous matrix are, or might be, strongly influenced by the presence of inherently highly defective interfaces, as is the case of nc-Si arrays in a SiO_2 matrix [29], or of arrays of organic nanocrystals embedded in inorganic porous media.

Last, but not least, we expect a key role of surface (and interface) defects in the radiative recombination dynamics of minority carriers in nanostructures [30], with the onset, as an example, of light emission induced by surface defect states [31]. These surface defect states are, in fact, the predominant sources of light emission in silicon nanocrystals embedded in a SiO_2 matrix, being localized at a length scale below 1 nm, much beyond the size of the nanocrystals. When surface defects are, however, passivated by hydrogen, core quantum effects are responsible for the light emission [31].

Therefore, attention should be paid to deduce energy gap properties of a semiconductor nanostructure from optical or photoluminescence measurements without considering the role of surface defects, an issue that will considered in detail in Chapter 5.

Surface defects can also induce the preferential surface diffusion of impurities and dopants, as is the case of Au-diffusion in CVD-grown Si and GaAs nanowires [32, 33], and influence, as well, the thermodynamics and kinetics of the nucleation of dislocations in nanostructures [34] and that of a second-phase segregation.

1.3   CHEMICAL AND STRUCTURAL CHARACTERIZATION OF SEMICONDUCTOR NANOSTRUCTURES

The physics and chemistry of processes occurring in the bulk or at the surface of a semiconductor nanoparticle depend on the nature, concentration, and geometrical

---

* Unlike previous literature reports, which mentioned the need for low temperatures and external magnetic fields to stabilize these defects.
† S. Pizzini in these papers is the daughter of the author of this book.
‡ This is the case of nanocrystalline silicon.

arrangement of the atomic species, including that of the contaminants (impurities) present in the material and at its surface.

To detect how, and how far, composition,* structure, size, shape, and defects do affect the properties of semiconductor nanostructures [30], dedicated experimental tools are needed, in view of the difficulty in measuring the local chemical, structural, and physical properties of these nanometric objects.

Chemical characterization of nanostructures might be carried-out with X-ray photoemission spectroscopy (XPS), Auger electron spectroscopies (AES), secondary-ion mass spectroscopy (SIMS), with different physics underneath, and with Rutherford back scattering (RBS).

Modern Auger electron spectrometers, equipped with field electron guns, with probe diameters below 10 nm, are the best tools available for the local analysis of nanostructures [35] leading to a depth resolution from less than 0.1 nm to 4 nm, with electron energies in the range 15–2000 eV.

Unlike bulk solids, where the AES analysis of the surface is biased by the interference of bulk atoms, in an amount that depends on the depth at which the analytical probe operates, with nanostructures the AES analysis is the analysis of the whole, if the size is of the order of the depth resolution. For thicker samples, as was the case of an epitaxial layer of aluminum nitride [AlN], 200 nm thick, grown on a silicon substrate, AES was successfully used to detect the segregation of Si at the surface of the AlN layer [35].

AES and Rutherford back scattering (RBS) spectroscopies might be also used to detect the stoichiometry of nano samples, as was done with zinc oxide [ZnO] nanoparticles as a function of the process steps [36], or with PbS colloidal dots, taken as non-exclusive examples. The case of PbS nanodots is particularly interesting, because it was possible to associate the excess of lead or sulfur vacancies with the shift of the energy gap of the material, measured by photoluminescence spectroscopy [22].

Since a nanoscale resolution was only available for depth analysis, secondary-ion mass spectrometry (SIMS), operated with laser beams of 1 μm, was not considered, until recently, a useful tool for nano-characterization.

Today, in view of the better lateral resolution (50 nm) of the modern SIMS spectrometers joined to their sub-nanometric depth resolution and in view of numerical data processing available, which permit compositional profiles in 2D to be achieved [37], SIMS is on the top list of the characterization tools for nanostructures.

Structural characterization of semiconductor nanostructures, carried out either to follow and optimize a growth process or to establish details of the prepared samples, presents minor problems, when compared to their chemical characterization, due to the already present sub-nanometric resolution of the transmission electron microscopes (TEM) and high-resolution TEMs. It is, however, also common practice to employ conventional XR diffraction techniques on nanocrystal ensembles to get their average structural properties.

Raman spectroscopy, despite not offering a lateral nano-resolution, is intrinsically a nano-probe since the Raman spectra originate from bond vibrations, which

---

* Including deliberate impurity doping.

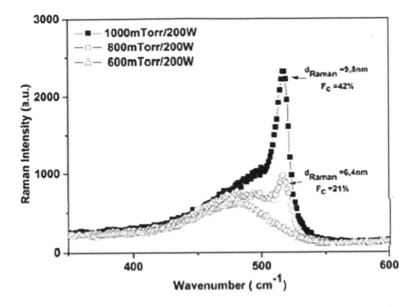

FIGURE 1.3 Raman evidence of the set-up of the amorphous-crystalline transition of a nanocrystalline silicon film as a function of the preparation conditions. (*After* R. Amrani, F. Pichot, L. Chahed, Y.Cuminal (2012) Amorphous-nanocrystalline transition in silicon thin films obtained by argon diluted silane PECVD in *Crystal Structure Theory and Applications* **1**, 57–61 http://dx.doi.org/10.4236/csta.2012.13011 *Published Online December 2012 (http:// www.SciRP.org/journal/csta)).*)

make Raman spectroscopy sensible to short-range structures. It is also systematically used to get structural information [38], since it permits phase identifications based on the Raman peak position of a reference bulk material, but also knowledge about phonon localization effects and of phase transitions. This last case is well-illustrated in Figure 1.3 [39] where one could follow the shift from the spectrum of amorphous silicon, consisting of a broad band peaked at 480 cm^{-1}, toward the spectrum of crystalline silicon, with a sharp peak at 520 cm^{-1}, as a function of the preparation conditions, on which will be discussed deeply in Chapter 3.

1.4 SMALL IS BEAUTIFUL, AND USEFUL

For millennia, much before the invitation of Feynman to enter a new field of physics, which is the *incipit* of this chapter, artwork implied the use of matter at small sizes, metallic colloids and color pigments being the best examples. The development of modern chemistry is, also, indebted to the use of catalysts, micrometric particles of inorganic materials, whose surfaces do influence the reaction rate of a chemical process and might, also, favor the occurrence of a reaction path in a well-defined, also stereotactic, direction. As a typical example we could refer to the use of MgCl$_2$/ TiCl$_4$ micrometric powders, simply prepared by milling, for the industrial production of polypropylene with the Ziegler–Natta process, which earned the Nobel Prize for the inventors.

With the progress of wet- and dry-processes used for the preparation of micro-metric or nanometric powders of metals and semiconductors, the application of nanoparticles to a variety of uses expanded almost exponentially. A few outstanding examples are enough.

Silver nanoparticles applied to water-harvesting processes from the atmosphere and to antibacterial and antitumoral activities represent a modern example of appli-cation of nanotechnology. As well, ultra-thin nanocomposite films composed of sil-ver nanoparticles embedded in a silica glass work as efficient solar light absorbers in a wide range of wavelengths, via the excitation of the plasmonic resonances of the Ag nanoparticles [40]. Their good absorption efficiency is shown to depend on a complex mesoscale and nanoscale structure, relevant to one of the dominant themes of this book.

Size and shape at the nanoscale do improve, also, the mechanical behavior of materials, with impressive applications of metallic nanocomposites, where the inter-faces and their shapes are critical in suppressing the shear bands at ultrahigh yield strength [41]. And it is interesting to remark, here, that the spatial patterns of nano-structured materials, and within them, of nanostructured semiconductors, character-ized by a length scale from a few to several tens of nanometers, often occur as the effect of a self-assembly process, which drives their growth.

Miniaturization is the domain of *small* in one or more dimensions, and the trend toward the miniaturization of electronic devices fabricated on substrates consisting of semiconductor silicon is a key part of the history of technological developments in the last century and in the first decades of the new millennium, which started with the fabrication of the cm-scale first solid state diode, and arrives at the nanometric elements of a microelectronic device.

Without mentioning details of the semiconductor micro- and nano-technologies for microelectronic applications, which are outside the interests of this book, silicon nanostructures are used today in a variety of other applications, which include sen-sors for microfluidic systems, bio-integrated sensors [42], and high-efficiency (>18%) silicon solar cells having on the front surface a film of black silicon, consisting of an array of vertically aligned nanowires, which allow a nearly ideal absorption of the solar spectrum [43, 44]. Miniaturization allowed, also, the development of solid-state optoelectronic devices on substrates consisting of compound semiconductors, and the fabrication of light-emitting diodes (LEDs), which now satisfy the conditions needed for being efficient and low-cost light sources for cars, homes, cities, and a number of battery-powered devices.

New applications are now expected from nanometric objects consisting of semi-conducting dots, sheets, and wires, which are the main subject of study in this book, provided a better knowledge of their fabrication process is developed, enabling con-stant product properties, as well as a better understanding of their properties and of the means of tuning their optoelectronic properties by doping and other suitable post-treatments.

A viable application of semiconductor luminescent dots or nanocrystals as light-emitting sources could be foreseen, provided they are distributed in an optically transparent liquid, or incorporated in a optically transparent nanometric-thick matrix, which behaves as an optical support.

This is the case of silicon nanocrystals dispersed in a SiO_2 matrix, which will be discussed in Chapters 3 and 5, where the nanometric dimension of Si nanocrystals favor quantum confinement conditions, and blue-shifted, reasonably intense light emission, uncommon for an indirect-band semiconductor like silicon. The success of this application is due to the fact that silica is the chemical precursor of silicon, and that nanometric-thick non-stoichiometric silica SiO_x sheets can be thermally decomposed to an ordered distribution of nanometric silicon crystals in a silica matrix [45]

$$SiO_x \rightleftharpoons \left(1-\frac{x}{2}\right)Si + \frac{x}{2}SiO_2 \qquad (1.17)$$

that works as an optically transparent support.

Therefore, the extension of this kind of process to other semiconductors is not trivial at all, since the case of silicon is unique.

Significant difficulties are, in fact, encountered with the synthesis of compound semiconductor nanostructures, where the chemistry of their growth processes is mostly based on empirical concepts, with unknown reasons for a success or a failure.

This is particularly true in the case of compound semiconductor nanowires, discussed in Chapter 4, where, often, but not always, basic thermodynamics concepts are under-evaluated. This makes the optimization of these nanostructures difficult, and the understanding of a crucial question, i.e. of the relationships among the structure of a phase, the nanoscopic organization of the material, and its defectivity with the process parameters (temperature, nature of precursors, surface distribution of catalysts), is very difficult, especially when environmental impurities become dominant.

Despite these problems, common to any technology at its beginning, the field is in a stage of substantial progress.

As an example, axial- or radial-heterostructures could be grown on a single nanowire, as is shown in Figure 1.4 for a GaAs-GaP heterostructure, where the heterostructure formation could be operated by changing the inlet of AsH_3 or PH_3 precursor gases in the growth reactor.

As a further example, Carapezzi and Cavallini in a paper recently appeared in the literature [46], devoted to the study of the surface nanoarchitecture of some sets of Si nanowires produced with the metal assisted chemical etching (MACE) process, that will be discussed in Chapter 3, propose a new model for the analysis of the forces driving the self-organization of the nanowires ensembles and apply fractal analysis to the characterization of their morphology.

Quantum computing is, eventually, at the leading edge of today's nanotechnology, making use of the quantum states of a two-level system, a qbit, suitably incorporated at a length scale of 0.1 nm on a patterned silicon surface, where each element is fully isolated from the environment to avoid decoherence. Ten years after the classic paper of Kane [47], quantum computing begins to be operative with the CES 2019 IBM computer, equipped with 20 qbits, leaving the door open to further, necessary improvements.

As suggested by Feynman, there's plenty of room at the bottom.

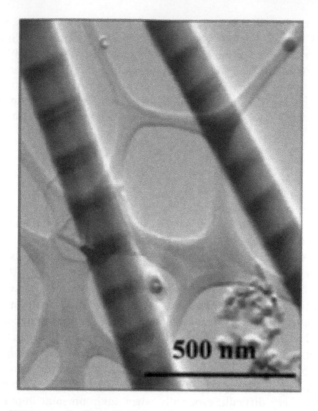

FIGURE 1.4   SEM map of a GaAs-GaP nanowire heterostructure. *(After* www.physics.purdue.edu/academic programs/courses/phys570P/.)

REFERENCES

1. V. V. Pokropivny, V. V. Skorokhod (2008) New dimensionality classifications of nano-structures. *Physica E* **40**, 2521–2525.
2. S. Pizzini (2015) *Physical Chemistry of Semiconductor Materials and Processes,* J.Wiley & Sons, Chichester, UK.
3. M. N. Magomedov (2016) Size dependence of the shape of a silicon nanocrystal during melting. *Tech. Phys.* **42**, 761–764.
4. E. G. Barbagiovanni, D. J. Lockwood, P. J. Simson, L. V. Goncharova (2014) Quantum confinement in Si and Ge nanostructures: Theory and experiments. *Appl. Phys. Rev.* **1**, 011302.
5. H. Richter, Z. P. Wang, L. Ley (1981) The one phonon Raman spectrum in microcrystalline silicon. *Solid State Commun.* **39**, 625–629.
6. I. H. Campbell, P. M. Fauchet (1986) The effects of microcrystal size and shape on the one phonon Raman spectra of crystalline semiconductors. *Solid. State Commun.* **58**, 739–741.
7. A. D. Yoffe (2002) Low-dimensional systems: Quantum size effects and electronic properties of semiconductor microcrystallites (zero-dimensional systems) and some quasi-two-dimensional systems. *Adv. Phys.* **2**, 799–890.
8. A. D. Yoffe (2001) Semiconductor quantum dots and related systems: Electronic, optical, luminescence and related properties of low dimensional systems. *Adv. Phys.* **50**,

1–208. [which is the republication of the original article of 1993 A. D. Yoffe (1993) Low-dimensional systems: quantum size effects and electronic properties of semiconductor microcrystallites (zero-dimensional systems) and some quasi-two-dimensional systems *Adv. Phys.* 42, 173–262].

9. R. Koole, E. Groeneveld, D. Vanmaekelbergh, A. Meijerink, C. de Mello Donegá (2014) Size effects on semiconductor nanoparticles. In: *Nanoparticles*, C. de Mello Donegá Ed., Springer Verlag, Heidelberg.

10. N. Korti-Baghdadli, A. E. Merad, Y. Benouaz (2013) Adjusted Adhashi's model of exciton Bohr parameters and new proposed model for optical properties of III-V semiconductors. *Am. J. Mat. Sci. Technol.* **3**, 65–73.

11. S. Schulz, G. Czycholl (2005) Tight-Binding model for semiconductor nanostructures. *Phys. Rev. B* **72**, 165317.

12. S. Xiao, W. Hu, W. Luo, Y. Wu, X. Li, H. Deng (2006) Size effect on alloying ability and phase stability of immiscible bimetallic nanoparticles. *Eur. Phys. J. B* **54**, 479–484.

13. R. Lakes (1993) Materials with structural hierarchy. *Nature* **361**, 511–515.

14. P. Fratzl, R. Weinkamer (2007) Nature's hierarchical materials. *Prog. Mater. Sci.* **52**, 1263–1334.

15. L. Mishnaevsky, M. Tsapatsis (2016) Hierarchical materials: Background and perspectives. *MRS Bull.* **41**, 661–664.

16. B. D. Witts, P. L. Clode, N. H. Patel, G. E. Schroder-Turk (2019) Nature's functional materials: Growth or self-assembly? *MRS Bull.* **44**, 106–111.

17. G. Nicolis, I. Prigogine (1977) *Self-Organization in Nonequilibrium Systems*, J. Wiley and Sons.

18. J. H. Crawford, L. M. Slifkin (1972) *Point Defect in Solids*, Plenum Press; M. Lannoo, J. Burgoin (1981) Point defects in Semiconductors Springer-Verlag.

19. S. Pizzini (2017) Point defects in group IV semiconductors: Common structural and physico-chemical aspects. *Materials Research Forum LLC*.

20. J. A. Faucheaux, A. L. D. Stanton, P. K. Jain (2014) Plasmon resonances of semiconductor nanocrystals: Physical principles and new opportunities. *J. Phys. Chem. Lett.* **5**, 976–985.

21. Y. Liu, M. Liu, M. T. Swihart (2017) Plasmonic copper sulfide-based materials: A brief introduction to their synthesis, doping, alloying, and applications. *J. Phys. Chem. C* **121**, 13435–13447.

22. D. M. Balazs, K. I. Bijlsma, H.-H. Fang, D. N. Dirin, M. Döbeli, M. V. Kovalenko, M. A. Loi (2017) Stoichiometric control of the density of states in PbS colloidal quantum dot solids. *Sci. Adv.* **3**, eaao1558.

23. F. Fabbri, E. Rotunno, L. Lazzarini, D. Cavalcoli, A. Castaldini, N. Fukata, K. Sato, G. Salviati, A. Cavallini (2013) Preparing the way for doping wurtzite silicon nanowires while retaining the phase. *Nano Lett.* **13**, 5900–5906.

24. Z. Zhou, L. Brus, R. Friesner (2003) Electronic structure and luminescence of 1.1 and 1.4 nm silicon nanocrystals: Oxide shell versus hydrogen passivation. *Nano Lett.* **3**, 163–167.

25. Y. Li, Z. Liu, X. Lu, Z. Su, Y. Wang, R. Liu, D. Wang, J. Jian, J. H. Lee, H. Wang, Q. Yu, J. Bao (2015) Broadband infrared photoluminescence in silicon nanowires with high density stacking faults. *Nanoscale* **7**, 1601–1605.

26. J. Li (2007) The mechanism and physics of defect nucleation. *MRS Bull.* **32**, 151–159.

27. S. Pizzini, J. Vogel, S. Rohart, L. D. Buda-Prejbeanu, E. Jué, O. Boulle, I. M. Miron, C. K. Safeer, S. Auffret, G. Gaudin, A. Thiaville. (2014) Chirality-induced asymmetric magnetic nucleation in Pt/Co/AlO$_x$ ultrathin microstructures. *Phys. Rev. Lett.* **113**, 047203.

28. O. Boulle, J. Vogel, H. Yang, S. Pizzini, D. de Souza Chaves, A. Locatelli, T. Onur Menteş, A. Sala, L. D. Buda-Prejbeanu, O. Klein, M. Belmeguenai, Y. Roussigné, A.

Stashkevich, S. Mourad Chérif, L. Aballe, M. Foerster, M. Chshiev, S. Auffret, I. M. Miron, G. Gaudin (2016) Room-temperature chiral magnetic skyrmions in ultrathin magnetic nanostructures. *Nat. Nanotechnol.* **11**, 449–454.

29. M. Jvanescu, A. Stedsmans, M. Zacharias (2008) Inherent paramagnetic defects in layered Si/SiO$_2$ superstructures with Si nanocrystals. *J. Appl. Phys.* **104**, 103518.

30. P. Kloth, M. Wenderoth (2017) From time-resolved atomic-scale imaging of individual donors to their cooperative dynamics. *Sci. Adv.* **3**, e1601552.

31. U. Gösele (2008) Shedding new light on silicon. *Nat. Nanotechnol.* **3**, 134–135.

32. M. J. Tambe, S. Ren, S. Gradečak (2010) Effects of gold diffusion on n-type doping of gaas nanowires. *Nano Lett.* **10**, 4584–4589.

33. K. Sato, A. Castaldini, N. Fukata, A. Cavallini (2012) Electronic level scheme in boron- and phosphorus-doped silicon nanowires (2012). *Nano Lett.* **12**, 3012–3017.

34. L. Y. Chen, M-R. He, J. Shin, G. Richter, D. S. Gianola (2015) Measuring surface dislocation nucleation in defect-scarce nanostructures. *Nat. Mate.* **14**, 707–713.

35. G. Ecke, V. Cimalla, K. Tonisch, V. Lebedev, H. Romanus, O. Ambacher. J. Liday (2007) Analysis of nanostructures by means of Auger Electron spectroscopy. *J. Electr. Eng.* **58**, 301–306.

36. L. C. Chao, S-H. Yang (2017) Growth and Auger electron spectroscopy characterization of donut-shaped ZnO nanostructures. *Appl. Surf. Sci.* **253**, 7162–7165.

37. E. G. Seebauer, D. E. Barlaz (2016) SIMS for analysis of nanostructures. *Curr. Opin. Chem. Eng.* **12**, 8–13.

38. P. Colomban, G. Gouadec (2007) Raman spectroscopy of nanostructures and nanosized materials. *J. Raman Spectrosc.* **38**, 598–603. with commented references.

39. R. Amrani, F. Pichot, L. Chahed, Y. Cuminal (2012) Amorphous-nanocrystalline transition in silicon thin films obtained by argon diluted silane PECVD. *Cryst. Struct. Theory Appl.* **1**, 57–61.

40. J. Y. Lu, A. Raza, S. Noorulla, A. S. Alketbi, N. X. Fang, G. Chen, T. Zhang (2017) Near-perfect ultrathin nanocomposite absorber with self-formed topping plasmonic nanoparticles. *Adv. Opt. Mater.* **5**, 1700222.

41. M. Demkowixz (2019) Does Shape affect shape change at the nanoscale? *MRS Bull.* **44**, 31–39.

42. J. A. Rogers (2019) Materials for biointegrated electronic and microfluidic systems. *MRS Bull.* **44**, 195–202.

43. H.-C. Yuan, V. E. Yost, M. R. Page, P. Stradins, D. L. Meier, H. M. Branz (2009) Efficient black silicon solar cell with a density-graded nanoporous surface: Optical properties, performance limitations, and design rules. *Appl. Phys. Lett.* **95**, 23501.

44. J. Oh, H-C. Yuan, H. M. Branz (2012) An 18.2%-efficient black-silicon solar cell achieved through control of carrier recombination in nanostructures. *Nat. Nanotechnol.* **7**, 743–748.

45. L. X. Yi, J. Heitmann, R. Scholz, M. Zacharias (2002) Si rings, Si clusters, and Si nanocrystals—different states of ultrathin SiO$_x$ layers. *Appl. Phys. Lett.* **81**, 4248–4266.

46. S. Carapezzi, A. Cavallini (2019) The importance of design in nanoarchitectonics: Multifractality in MACE silicon nanowires. *Beilstein J. Nanotechnol.* **10**, 2094–2102.

47. B. E. Kane (1998) A silicon-based nuclear spin quantum computer. *Nature* **393**, 133–137.

2 Physics and Thermodynamics of Nanostructures*

A strong need exists for a microscopic description of the melting process of metal clusters, especially in view of the growing interest in their properties and applications.

It has long been known that the melting temperature TM decreases with decreasing diameter d. Several phenomenological models have been proposed to account for this fact. Almost all consider the cluster as consisting of "bulk" and "surface" atoms, and the drop of TM results from the fact that, at the bulk melting temperature, the surface free energy of the liquid is lower than that of the crystal.

Although qualitatively correct, it is doubtful whether such simple approaches should remain quantitatively valid down to clusters with a number of atoms N < 1000.[†]

2.1 THERMODYNAMIC PROPERTIES OF NANOSTRUCTURES: NANOTHERMODYNAMICS

Unlike conventional solids, that are extended systems, with a number of atoms N comparable with the Avogadro number ($N_A = 6.022.10^{23}$ at/mol), and whose equilibrium thermodynamic properties are independent of the system's size, colloidal semiconductors, nanocrystals and nanowires, are finite and, possibly, inhomogeneous systems with ($N \ll N_A$), for which the experimental evidence tells us that their thermodynamic properties are size-dependent.

This means that the intensive variables (μ, T, p), which are independent of each other in a macroscopic system, in a "small" system are interrelated and determined by its size [1]. Therefore, thermodynamics could be applied to samples of nanoparticles provided they are treated as *ensembles* of independent or *quasi*-independent particles [1]. Furthermore, for finite systems, surface and interface effects, not only volume effects, on entropy, and on other thermodynamic functions, should be considered. Neglecting these constraints could lead, in fact, to apparent violations of the second law of thermodynamics [2]. Eventually, while in extended systems the

---

* https://orcid.org/0000-0002-0542-3219.
† F. Ercolessi, W. Andreoni, and E. Tosatti (1961) Melting of small gold particles: Mechanism and size effects. *Phys. Rev. Lett.* **66** 911.

entropy function S(E) is concave, and its derivative is always positive, in finite systems S(E) might be concave or convex, and the equilibrium properties of the system are defined not by the temperature, but by the Boltzmann entropy $S=k\ln W$ of the whole system.

Due to their large surface to volume (S/V) ratio, nanostructures should be dominated by the energetics of their surfaces, and the most simple, and also qualitatively correct, approach to their description should consider them as consisting of "core" and of "surface" atoms [3].

Using molecular dynamics (MD) modeling, Ercolessi et al. [3] succeeded in confirming the experimental decrease of the melting temperature of gold particles with the decrease of size, and, also, in demonstrating the occurrence of sharp melting conditions for clusters with $N \approx 100–1000$.

Since MD calculations apply well only to systems with $N \leq 10^5$, i.e. up to a size of tens of nm, complementary, although phenomenological, methods have been developed to model nanostructures also in the mesoscopic range.

It is, in fact, experimentally demonstrated that nanostructures (nanocrystals, nanofilms, and nanowires) of different nature and composition (ionic, metallic, semiconductor, organic) present substantial deviations from their bulk thermodynamic properties already in the mesoscopic range. These deviations do increase with the decrease of the size and lead to a smearing of the solid/liquid transition below a critical size less than a few nm,* or according to the Alivisatos definition [4], to the size at which "a nanocrystal contains just enough atoms to have an identifiable interior."[†]

An example of these methods is given by the application of the Kelvin $p = p^o \exp\left(\dfrac{2\sigma_{l,g}V_{m,l}}{rRT}\right)$ and Gibbs–Thomson $\Delta H_m^o = T\Delta S_m^o + \dfrac{2\sigma_{sl}V_{s,m}}{r}$[‡] equations to the discussion of the equilibrium conditions of nanostructures, which show immediate similarities with the mesoscopic liquid droplets to which those equations has been successfully applied [5]. In these equations the surface energy at the liquid/gas interface $\sigma_{l,g}$ plays the major role in determining the equilibrium pressure p at a curved surface of radius r with respect to the pressure at a flat surface p^o of a droplet of a material of molar volume $V_{m,l}$.

As an answer to the Ercolessi et al. [3] concern on the use of (too) simple assumptions to account for the effect of surfaces on the properties of a nanostructure,[§] Buffat and Borel [6, 7] argued that we can expect "a thermodynamical size effect due to the surface free energy which induces a size-dependence of the chemical potential μ of the particle."[¶]

$$\mu = \mu^o + \frac{2\sigma}{\rho r} \tag{2.1}$$

where μ^o is the chemical potential of a macroscopic particle, σ is the surface tension, ρ is the density, and r is the radius of the particle.

---

* The absolute value depends on the nature of the nanostructure, as will be seen later on.
† Original sentence of Alivisatos.
‡ Both of which lead to a phenomenological $1/r$ scaling law.
§ See top of first page.
¶ Original sentence of Buffat.

TABLE 2.1

Exciton Bohr Radius (nm) for Various Semiconductors [12]

CuCl	Diamond	GaP	CdTe	CsS	GaN	Si	CdSe	GaAS	Ge
0.7	0.85	1.17	2.8	2.9	3.6	4.9	5.6	12	17.7

This equation leads to a $1/r$ scaling law, valid for a droplet of an incompressible liquid, but which could be extended to a solid particle obeying the Curie–Wulf equation [8, 9], that represents the equilibrium shape conditions for a solid particle

$$\frac{\sigma_i}{h_i} = \text{const} \tag{2.2}$$

where σ_i is the surface tension of a face $\{hkl\}$ of the solid and h_i is the distance of the face $\{hkl\}$ from the center of the particle.

This behavior is supposed to be associated with the presence of a skin of a few surface and sub-surface layers of undercoordinated atoms,* that could lead to the features of a true non-homogeneous phase, when the skin represents the major part of the entire mass of the particle.

As we will see later in detail, this physical inhomogeneity could be accompanied by a structural inhomogeneity, i.e. by the presence of a new surface phase, as occurs in the case of the surface oxidation of silicon nanopowders, or by deviations from the stoichiometry typical of compound semiconductors, or by chemical inhomogeneities due to unwanted chemical contaminations.

According to Rowlison [11], who discusses the case of nanometric molecular phases, all nanometric systems are, at least, microscopically inhomogeneous, but they could be treated satisfactorily well using classical equilibrium thermodynamics, provided their size is larger than a critical size corresponding to a correlation length ξ, i.e. a length that depends on the system considered, and that in molecular systems has an order of magnitude comparable with the interaction length of intermolecular forces, which amounts to less than a nm in simple liquids.

In semiconductor nanostructures, the correlation length ξ corresponds to the exciton Bohr radius, which we have already discussed in Chapter 1, which varies within 2 and 20 nm, depending on the nature of the material considered (see Table 2.1). This is the condition at which quantum confinement does occur for both electrons and holes, associated with the increase of the energy gap $E_G^{nc}(D)$ as a function of the diameter D of a spherical particle

$$E_G^{nc}(D) = E_G^o + E^{\text{conf}}(D) = E_G^o + \frac{A}{m_e^* D^2} + \frac{B}{m_h^* D^2} \tag{2.3}$$

where $E^{\text{conf}}(D)$ is the confinement energy, A and B are constants, and m_e^* and m_h^* are the effective masses of electrons and holes [13].

---

* Associated with the presence of dangling bonds, point defects, and structural parameters different from the bulk ones [10].

Quantum confinement effects should decay, however, for sizes larger than two to three times the Bohr exciton radius [13], leading to a transition to mesoscopic conditions.

For metallic nanoparticles the critical size is the size at which statistical energy fluctuations are of the order of the enthalpy of melting. Below this size it is impossible to distinguish the very nature of the phase, and the system can be considered as a superposition of two phases [14].

For metallic dots of Pt, the critical size ranges around 6 nm. We could however use for the critical size $\delta*$ of any kind of material the value suggested by Shi [15] $\delta* = 6$ h, where h is the thickness of a monolayer of the surface, i.e. below a size of 50–500pm, depending on the material.

Equilibrium conditions which hold for the local temperatures $T(r)$ and for the local chemical potentials $\mu_i(r)$, at a point r within an inhomogeneous phase, are given by the following equations [11]

$$\nabla T(r) = 0 \qquad (2.4)$$

$$\nabla \mu_i(r) = 0 \qquad (2.5)$$

Thus, both the temperature and the chemical potential are constant throughout the inhomogeneous phase, since T and μ_i are scalar quantities and the gradient of a scalar is constant.

Therefore, for a nanostructure of size comparable with its correlation length ξ, also the specific heat C and the melting temperature $T_m(r)$, despite being different from those of the bulk phase C° and T_m^o, are well-defined equilibrium properties.

Instead, the condition that holds for the local pressure $p(r)$

$$\nabla p(r) = 0 \qquad (2.6)$$

does not lead to constant pressure conditions within the system, since $p(r)$ is a vectoral quantity.

Therefore, in systems of size lower than ξ we could expect local deviations from mechanical equilibrium and the set-up of local strain.

The question that remains open is whether the chemical potentials of the components of a nanophase of defined size do change with the depth along the surface region, as is theoretically forecasted for the formation energy of vacancies on Al surfaces [16].

And, also, it is not easy to account for how an ionic, semiconducting or metallic behavior "develops with increasing particle size, starting with the atomic state"* and how the thermodynamic properties of a nanoparticle could be theoretically modeled in the intermediate size range representing nascent ionic/semiconducting/metallic condition, where the electronic properties are determined by the statistics of the electron-level distribution, in addition to the density of states [17].

Our interest in the next two sections will be mainly addressed to systems presenting particle sizes above, or well above their correlation lengths, to which the

---

* Original Halperin's [17] statement.

conventional thermodynamic approach could be applied, though some hints will be given also to quantum dots.

Most of the examples taken into consideration do refer to metallic nanoparticles, due to the limited knowledge available on semiconductor nanocrystals, but the conclusions could be safely extrapolated to the behavior of semiconductors.

2.2 THERMODYNAMICS OF SMALL SYSTEMS: PHENOMENOLOGICAL EVIDENCE

Following the Buffat and Borel [6, 7] arguments presented in the previous section, a size dependence could be foreseen for all the thermodynamic properties of nanostructures.

The phenomenological evidence that size does affect the thermodynamic properties of a special family of colloidal nanoparticles, consisting of ligated clusters of Pd metal, was deduced by Volokitin et al. [18] on the basis of the deviation of the specific heat of colloidal Pd particles from that of bulk Pd, at temperatures lower than 1 K.

According to the Debye–Einstein theory (in the absence of surface contributions), the specific heat of a metal can be decomposed in two terms

$$C_{\mathrm{met}} = C_{\mathrm{elect}} + C_{\mathrm{vibr}} = \frac{\pi^2 N_A k^2}{2E_F} T + \frac{12\pi^4 N_A k}{5T_D^3} T^3 \qquad (2.7)$$

where N_A is the Avogadro number, T_D is the Debye temperature, k is the Boltzmann constant, and E_F is the Fermi energy, of which the first term C_{elect} is due to conduction electrons, and is strongly size-dependent, while C_{vibr}, with a cubic law dependence on the temperature, is due to lattice vibrations.

Apparently, for two classes of colloidal particles the electronic contributions are shown to dominate the specific heat, since for these clusters an experimental $C \propto T$ relationship is experimentally observed, and the actual C value is definitively lower than that of bulk Pd, and could be attributed to a size effect.

Due to the chemical complexity of these nanoparticles, consisting of a metallic core and of an organic shell, it would be difficult to extrapolate their behavior to conventional *naked* metallic or semiconducting nanoparticles, where intrinsic surface effects should deeply influence their physical properties, and for which, instead, a rich amount of information on the size dependence of their thermodynamic properties exists in the literature.

As an example, specific heat measurements in the 1.5–15 K temperature range on small lead (22, 37, and 60 Å) and of indium (22 Å) particles embedded in porous glass were carried out by Novotny and Meinke [19]. In the absence, or in the case of negligible contribution of electronic contributions, the theoretical temperature dependence of the low-temperature specific heat of bulk crystals should be described by the equation

$$C = C_{\mathrm{surf}} + C_{\mathrm{vibr}} = BT^2 + \frac{12\pi^4 Nk}{5T_D^3} T^3 \qquad (2.8)$$

where the first (quadratic) term accounts for surface (surf) contribution and the second is the usual Debye term associated with thermal vibrations, and B is a constant.

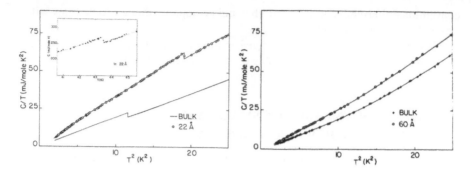

FIGURE 2.1 Experimental enhancement of the specific heat of small (22 Å) In particles (left panel) and of a 60 Å Pb particle (right panel) in the superconducting state below 5 K. (*After* V. Novotny and P.P.M. Meinke (1973) Thermodynamic lattice and electronic properties of small particles *Phys.Rev. B* **8** 4186 *Reproduced with permission of American Physical Society NP/19/DEC/021014 License date: Dec 08, 2019.*)

Therefore, in the presence of substantial surface contributions to the specific heat, we would expect either a quadratic dependence on temperature, at least in a limited interval, or an overall enhancement of their heat capacity with respect to the bulk values.

The experimental behavior of In and Pb spherical nanoparticles (see Figure 2.1) embedded in the pores* of a porous glass [19] actually shows only a systematic enhancement of the heat capacity compared to that of the parent bulk metals, which could be interpreted as due to a contribution of the quadratic (surface) term.

There is, also, vast experimental evidence that the melting temperature T_m of inorganic and organic nanostructures strongly decreases with the decrease of the particle size, indicating a dependence of T_m on $1/r$, where r is the radius of the particle [6, 7, 15, 20–36].

Unlike some earlier authors, who found a close to linear dependence of the melting temperature of spherical metallic particles on $1/r$, Wronsky [20], see Figure 2.2, experimentally found that the melting temperature T_m of small particles of tin on carbon- or silicon monoxide-substrates decreases in a non-linear fashion and that this behavior could be fitted by the following phenomenological relationship

$$T_m = 232 - 410\left(\frac{\sigma_{sl}}{17.1(r-t_0)} - \frac{1}{r}\right) \tag{2.9}$$

where σ_{sl} is the interfacial tension between liquid and solid tin and t_0 is "the smallest thickness of a region of liquid tin surrounding the solid to have the physical significance of a liquid."†

---

* Of almost constant size, assuming the absence of interactions of glass with the metallic particles.
† Original sentence of Wronski.

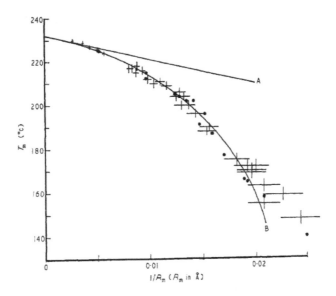

FIGURE 2.2 Size dependence of the melting point of small particles of tin. (*After* C.R.M. Wronski (1967) The size dependence of the melting point of small particles of tin *British J. Appl. Phys.*, **18** 1731–1737 © *IOP Publishing. Reproduced with permission. All rights reserved.*)

Equation 2.9 shows the presence of a scaling coefficient $\alpha = \dfrac{\sigma_{sl}}{17.1(r - t_0)}$, which depends on the surface tension and on a critical size t_0, whose value is empirically assigned.

Using Equation 2.9, the theoretical size dependence of the melting temperature is plotted in Figure 2.2, taking $t_0 = 0$ for the line A and $t_0 = 3.2$ nm for the curve B, and one could remark that the experimental values of the melting temperatures of the samples fit well with curve B.

The concept underneath this equation is the Reiss and Wilson criterion [37], that states that the conditions of thermodynamic equilibrium during the melting of a nanoparticle are satisfied provided the volume of the liquid phase which embeds the solid particle "is large enough for the atomic structure of matter to be ignored,"* independently of the size of the particle.

The Reiss and Wilson criterion comes from the straightforward application of the two-phase equilibrium thermodynamics to the melting of a nanoparticle, for which

$$\mu_i^s = \mu_i^l \quad \left(T_m = \text{cost}\right) \tag{2.10}$$

where μ_i is the chemical potential of a chemical species i, component of an elemental or multinary phase, and l and s are the symbols of a liquid and solid phase.

---

* Original sentence of Reiss and Wilson.

An implementation of the Wronski model [20] was developed by Peters et al. [25], who confirmed that the melting temperature varies inversely with the crystallite size and showed as well that the surface melting model is favored over the homogeneous melting model, leading to a reasonable qualitative insight of the melting process of nanoparticles.

Some other examples are reported in Figures 2.3 and 2.4, holding for a metal (Au) [6] and a compound semiconductor (CdS) [21], respectively, which show a melting point depression following a 1/r law, which start well above a critical radius of 3.2 nm. In both cases the predicted behavior is well-fitted by experimental data. Incidentally, this behavior is also presented by organic nanocrystals [26].

Unlike Wronski [20], Buffat and Borel [6] fitted the experimental curve displayed in Figure 2.3 with the following equation

$$\frac{T}{T^\circ} = 1 - \frac{2}{\rho_s \Delta H_m r_s}\left[\gamma_s - \gamma_l\left(\frac{\rho_s}{\rho_l}\right)^{2/3}\right] \qquad (2.11)$$

which brings in cause not only the surface tension γ of the particle, but also its density ρ_s, but considers the enthalpy of fusion ΔH_m independent of size.

For Buffat and Morel [6], like for most of the early authors, the theoretical discussion about the thermodynamics of melting is based on the assumption of the practical validity of Lindemann's law of melting, despite it accounting only for the presence of a solid phase, while melting is a cooperative phenomenon among a solid and a liquid phase [38]. For Lindemann's law, or Lindemann's criterion, melting might be

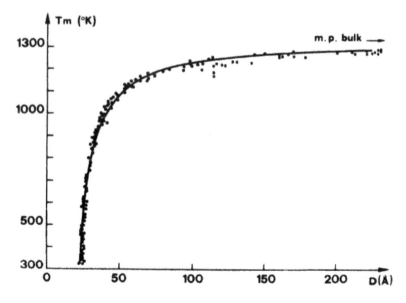

FIGURE 2.3 Experimental and theoretical dependence of the melting temperature of Au nanocrystals supported on amorphous carbon substrate. (*After* Ph. Buffat and J.P. Borel (1976) Size effect on melting temperature of gold particles *Phys. Rev. A* 13, 2287–2298, *Reproduced with permission of APS, License number RNP/19/DEC/021015 License date Dec 08, 2019.*)

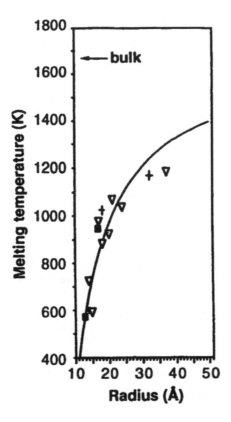

FIGURE 2.4   Size dependence of the melting point of CdS nanocrystals. (*After* A.N. Goldstein, C.M. Echer, A.P. Alivisatos (1992) Melting in semiconductor nanocrystals *Science* **256** 1425–1427 *Reproduced with permission of Science License Number 4724321103455 License date Dec 08,2019.*)

expected when the root mean vibration amplitude $\sqrt{\langle u^2 \rangle}$ exceeds a certain threshold value, approximated to 10% of the distance to the nearest neighbor.

Under these conditions, the melting temperature T_m could be expressed by the following equation

$$T_m = 4\pi^2 mc_L a^2 / k_B \tag{2.12}$$

where m is the atomic mass, a is the interatomic distance, c_L is the Lindemann constant, and k_B is the Boltzmann constant.

Since Lindemann's law is well-followed by many systems with body centered cubic (BCC) and open structures and could be considered topical of the nascent melting state, when the equilibrium liquid is not yet present, its application was beneficial for the earliest information on the thermodynamics of melting of nanostructures.

As a first example, a simple model for a size-dependent melting temperature of nanocrystals on the basis of Lindemann's criterion, and then, on the size-dependent thermal vibrations in a crystal, was developed by Shi [15], who starts by remarking that the assumption of a simple $1/r$ dependence of the melting temperature T_m cannot be applied at the nanometric size, where the amount of surface atoms is a substantial contribution to the total number of atoms of the particles.

The model is developed in terms of the parameters of the corresponding bulk crystal, i.e. the ratio between the amplitude of thermal vibrations for surface atoms and that for the bulk ones, and leads to an exponential dependence of melting temperature

$$T_m = T_m^o \exp-\left[(\alpha-1)\left(\frac{r}{3h}-1\right)^{-1}\right] \tag{2.13}$$

where r is the radius of the particle, h is the thickness of a monolayer, and α is taken as the ratio

$\alpha = \dfrac{\sigma_s^o}{\sigma_v^0}$, with σ_s^o and σ_v^o as the average amplitudes of thermal vibrations at the surface and in the volume of the bulk crystals, respectively. Although it is reasonable to suppose that α should vary with the size of the particle, it is maintained constant as a first approximation.

This model fits well with the size dependence of melting temperature of CdS nanocrystals displayed in Figure 2.4 and with the experimental results of Buffat and Morel [6].

The model of Shi [15] was used by Jiang et al. [26, 39] and by Zhao et al. [28] to fit the monotonical decrease of the melting temperature T_m, the melting enthalpy ΔH_m, and the melting entropy ΔS_m* of different organic and inorganic nanocrystals with the following equations

$$\frac{T_m(D)}{T_m^o} = \exp-\left(\frac{2\Delta S_m^o}{3R}\frac{1}{D/D^o-1}\right) \tag{2.14}$$

$$\frac{\Delta H_m(D)}{\Delta H_m^o} = \exp-\left(\frac{2\Delta S_m^o}{3R}\frac{1}{D/D^o-1}\right)\left[1-\frac{1}{D/D^o-1}\right] \tag{2.15}$$

$$\frac{\Delta S_m(D)}{\Delta S_m^0} = 1-\frac{1}{D/D^o-1} \tag{2.16}$$

where D is the diameter of the nanocrystal, R is the constant of gases, and T_m^o, ΔH_m^o, and ΔS_m^o are the corresponding bulk values.

A simple model, free of adjustable parameters, based on the Mott equations [40] for the enthalpy and temperature of melting

$$\frac{\nu_L}{\nu_S} = \exp-\frac{40\Delta H_m}{T_m} \tag{2.17}$$

where ν_L and ν_S are the vibrational frequencies of the liquid and solid at equilibrium, has been, instead, developed by Zang et al. [27] to simulate the size dependence of the melting enthalpy and entropy.

---

* With major concern for metallic or organic particles embedded in porous media.

The equation for the melting enthalpy

$$\frac{\Delta H_m(r)}{\Delta H_m^o} = \frac{T_m(r)}{T_m^o}\left[\frac{1}{r(r^\circ)^{-1}-1}\right] \tag{2.18}$$

is expressed as a function of the melting enthalpy and temperature of the bulk solid and of a critical radius r° at which all the atoms of the nanocrystals are located at the surface.

The model has been applied to Al and Sn nanocrystals with excellent results.

Also the nano-calorimetric measurements of Lai et al. [41], which allowed the determination of the size dependence of the melting temperature T_m and of the melting enthalpy ΔH_m of Sn nanoparticles, see Figure 2.5, are well-fitted by the following equations

$$\Delta H_m = \Delta H_o\left(1 - \frac{t_o}{r}\right)^3 \tag{2.19}$$

$$T_m = T_m^o - A\left[\frac{\sigma_s}{B(r-t_o)} - \frac{1}{r}\right] \tag{2.20}$$

which have been slightly improved in a later work [42]

$$T_m - T_0 = \frac{2T_o}{\Delta H_0}\left[\frac{\sigma_{sl}}{\rho_s(r-t_0)} + \left(\frac{\sigma_{lv}}{r} + \frac{\Delta P}{2}\right)\left(\frac{1}{\rho_s} - \frac{1}{\rho_l}\right)\right] \tag{2.21}$$

under the assumption of the presence of a two-phase system [38] at melting, and therefore, that of a layer of liquid of critical thickness t_0 embeds the melting solid and that the melting process depends on the interfacial tension σ_{sl} between the solid and the liquid phases.

In these equations T_o and ΔH_0 are the melting temperature and enthalpy of the bulk metal, A is a constant, σ_{sl} and σ_{lv} are the solid–liquid and liquid–vapor interfacial energies, r and t_0 are the actual and critical radii of the particles, and ρ_s and ρ_l are the densities of the solid and of the liquid.

The Gibbs–Thomson equation has also been used to describe the melting point depression and the equilibrium conditions of nanomaterials [5, 31].

For a spherical particle the Gibbs–Thomson (G–T) equation could be written

$$\frac{T_m}{T_m^0} = 1 - \frac{2\sigma_{sl}}{\Delta H_m^0 \rho_s r} \tag{2.22}$$

where T_m^o is the melting point of the bulk material, σ_{sl} is the solid–liquid interfacial energy, ΔH_m^o is the melting enthalpy of the bulk material, ρ_s is the density of the material, and r is the radius of the particle.

In this equation the scaling coefficient takes the form $\alpha = \dfrac{2\sigma_{sl}}{\Delta H_m^o \rho_s}$, and includes, unlike the previous cases, all the macroscopic thermodynamic signatures of the nanophase involved.

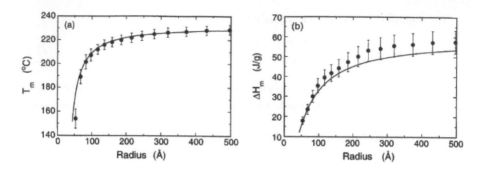

FIGURE 2.5 Size dependence of the melting temperature and melting enthalpy of Sn particles experimental points solid lines fittings using Equations 2.15 and 2.16. (*After* S.L. Lai, J.Y. Guo, V. Petrova, G. Ramanath, L.H. Allen (1996) Size-dependent melting properties of small tin particles: Nanocalorimetric measurements *Phys. Rev. Lett.* **77** 99 *Reproduced with permission of APS License number RNP/19/DEC/021016 License date: Dec 08, 2019.*)

Although the G–T equation fits well most of the experimental results found in the literature, this is not the case of oxidized Al nanoparticles, since the presence of an oxide layer on their surface results in a compressive pressure on the Al core, with the consequence of a higher melting point and melting enthalpy [31] with respect to the calculated values using Equation 2.22 and to the experimental results of un-oxidized Al nanoparticles.

In the case of Al nanoparticles it is also experimentally demonstrated that the depression of the melting enthalpy ΔH_m is considerably larger than that calculated with the relationship

$$\Delta H_m = \Delta H_m^o - \frac{2\sigma_{sl}}{\rho_s r} \tag{2.23}$$

used in the framework of derivations of the G–T equation [31].

Sun and Simon [31] postulate, therefore, that the larger enthalpy depression observed with Al nanoparticles is due to the presence of defects, whose concentration depends as well on size, and suggest that a defect-depending enthalpy term $\Delta H_{\text{defect}}(r)$ should be added to Equation 2.23

$$\Delta H_m = \Delta H_m^o - \frac{2\sigma_{sl}}{\rho_s r} - \Delta H_{\text{defect}}(r) \tag{2.23 bis}$$

to make the G–T equation fully applicable to nanostructures.

We will discuss the size dependence of the enthalpy of formation of point defects in Section 2.6 to show whether this postulate is realistic, or if it depends on the easy oxidation of Al which shall unavoidably lead to the formation of an oxide layer.

In this context, a physical weakness of all phenomenological models so far considered is to neglect the effect of a compressive deformation of the particle core, arising from surface tension effects, which does increase with the decrease of the particle radius. It should lead, in fact, to a different pressure dependence of the melting temperature of metallic and semiconducting particles, since the melting temperature of metals increases with the increase of the pressure while it decreases in the case of semiconductors.

However, the pressure influence on the melting point is very modest, as it looks for the case of an oxide cap on Al nanoparticles, for which the calculated* increase of the melting point for an estimated pressure of 52 and 110 MPa is only 6.5 and 19.2 K.

A comparison among different phenomenological melting models has been eventually given by Zhu et al. [36], who considered the details of the Reiss and Wilson [37], Pawlow [43], and Rie [44] melting models, and derived accurate equations for describing the melting process at the nanosize, as well as the decrease of the melting temperature during the melting process associated with the decrease of the particle size.

The Reiss model relies on an assembly of dispersed spherical particles, where the effects of reciprocal interactions are negligible. Each solid particle of radius r_s is embedded in a liquid layer of thickness t when the system is at the melting temperature. At the melting temperature, therefore, the radius of the particle $r = r_s + t$.

Assuming that the density of the liquid ρ_l is equal to that of the solid ρ_s at the melting temperature, which is however never the case in nature, the size dependence of the melting temperature is given by the following approximate equation

$$\frac{T_m}{T_m^o} = 1 - \frac{2\sigma_{sl}V_s}{\Delta H_m(r-t)} \tag{2.24}$$

where ΔH_m is the melting enthalpy, V_s is the molar volume of the solid, and T_m^o is the melting temperature of the bulk phase.

The melting law derived from Pawlow's model is approximated by the following equation

$$\frac{T}{T^o} = 1 - \frac{2V_s}{\Delta H_m r_s}\left[\sigma_{sv} - \sigma_{lv}\left(\frac{\rho_s}{\rho_l}\right)^{2/3}\right] \tag{2.25}$$

that corresponds to Equation 2.11 of the Wronski model, while the melting law derived from the Rie's model is approximated by the following equation if $|T - T^o| \ll T^o$

$$\frac{T}{T^o} = 1 - \frac{2\sigma_{sl}V_s}{\Delta H_m r_s} \tag{2.26}$$

All these approximate equations are unable, however, to fit the experimental values when the nanocrystal size is lower than 5 nm. For these critical sizes the accurate equations derived from the three models should be used, which permit the experimental results to be fit down to less than 1 nm.

The fact that the main interest in the phenomenological aspects of melting has been addressed to noble nanometal samples is trivially due to the ease of their fabrication.

---

* With the Clausius Clapeyron equation.

A similar $1/r$ behavior was, however, observed in the case of inorganic oxides by Guenther et al. [45] and by Magomedov [46] in the case of silicon nanoparticles, who took also in account the influence of a shape factor, formerly unrewarded.

The results concerning the behavior of Si nanoparticles [47–49] show that the experimental points strongly deviate from a linear $1/r$ behavior, above $1/r$ values larger than 0.3 (less than 3.3 nm) in good agreement with our previous arguments and with the value of Bohr radius for silicon (see Table 2.1).

The size dependence of the cohesive energy E_C of nanoparticles also deserved attention, since many other properties, like the melting enthalpy and the formation energy of defects, could be deduced from it.

Like for the other thermodynamic properties, an exponential dependence on $1/D$ was demonstrated to hold for the cohesive energy of metallic (Mo and W) nanocrystals [50], which is well-described by the equation

$$\frac{E_C(D)}{E_C^o} = \exp\left[-\frac{2S_{ch}}{3R}\frac{1}{D/D^o-1}\right]\left[1-\frac{1}{D/D^o-1}\right] \tag{2.27}$$

where E_C^o is the bulk value of the cohesive energy, $S_{ch} = \dfrac{E_c(T_{sv})}{T_{sv}}$ is the sublimation entropy, T_{sv} is the actual sublimation temperature, R is the gas constant, and D is the diameter of the crystal.

The important consequence of these results is that the decrease of the cohesive energy implies a decrease of the thermodynamic stability of a nanocrystal with respect to that of the bulk crystal. This effect seems to be very pronounced below 20 nm.

W. H. Qi [30] extended its former model developed for nanoparticles [51] to the case of nanowires and nanofilms, which are systems of large technological importance.

The model accounts for the size and the coherence of nanowires and thin films with their embedding matrixes in terms of the ratio N_s/N (where N_s is the number of surface atoms and N is the total number of atoms of the system, which is also proportional to the diameter of the wire or to the thickness of the film) and of a constant p which accounts for the coherence and is used for the calculation of the cohesion energy, of the formation energy of vacancies, and of the melting temperature and enthalpy.

Under simplified assumptions concerning the bond energy and coordination of the atoms at the surface and the surface relaxation, the following equation holds for the melting temperature of a nanowire

$$\frac{T_m}{T_m^o} = \left[1-\frac{3}{8}\frac{N_s}{N}\left[2-p-p\frac{E_m}{E_b}\right]\right] = 1-\alpha\frac{N_s}{N} \tag{2.28}$$

where E_m is the bond energy of the atoms of the matrix, E_b is the bond energy of an atom in the interior of the nanostructure, and $p=0$ for a free surface, $p=1$ for a coherent interface, and $p=\frac{1}{2}$ for a semi-coherent interface. Again $\alpha = \dfrac{3}{8}\left(2-p-p\dfrac{E_m}{E_b}\right)$ is the scaling constant, which depends on the energy ratio E_m/E_b.

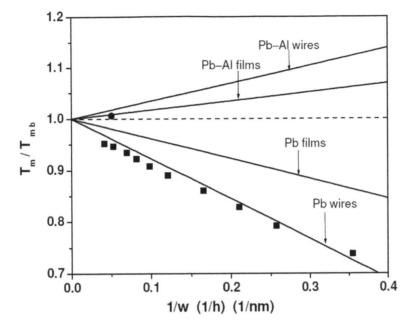

FIGURE 2.6 Dependence of the melting temperatures of freestanding and embedded metallic nanowires and films as a function of the diameter w or of the thickness h of the films. (*After* W.H. Qi (2006) Size- and coherence-dependent thermodynamic properties of metallic nanowires and nanofilms *Modern Physics Letters B* **20** 1943–1951 *Open Access Journal.*)

The model accounts well for the experimental values of the melting temperature of free-standing (isolated) Pb wires ($p=0$), and thin films, as seen in Figure 2.6 where a linear w^{-1} or h^{-1} relationship is followed, with w and h as the diameter of the wire and the thickness of the film, respectively.

A different approach was used by Wautelet [10, 22] to model the influence of the shape on the size dependence of the physical properties of nanoparticles.

He argues that because the $1/r$ scaling law for melting depends on the values of the surface tension γ_s^* of the solid particle and of the surface tension γ_l of the liquid in equilibrium with it, the shape of the particle and the nature of its faces should influence the melting temperatures and the other thermodynamic properties.

The equation that satisfies these requirements takes the following form for the melting temperature

$$T_m = T_{m,\infty} + f(\gamma_l - \gamma_s)/BN^{1/3} = T_{m,\infty}\left(1 - \frac{\alpha}{2r}\right) \quad (2.29)$$

where r is the radius of the particle, f is a shape constant, B is a constant, N is the number of atoms in the particle, and α is the scaling constant.

The synergistic effect of size and shape on the thermodynamics of nanocrystals has been also investigated by numerous authors.

---

* Attention should be given to the different dimensionalities of surface tension γ and surface energy σ.

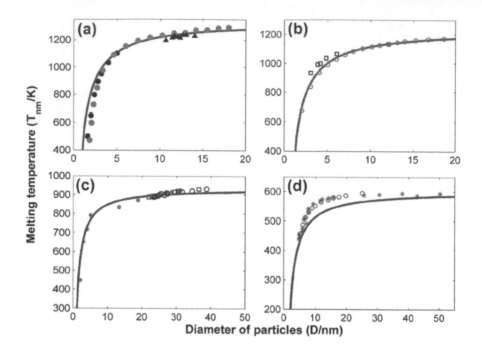

FIGURE 2.7 Size dependence of the melting temperature of Au (a), Ag (b), Al (c), and Pb (d). (*After* S. Xiong, Y. Cheng, M. Wang, Y. Li (2011) Universal relation for size dependent thermodynamic properties of metallic nanoparticles *Phys.Chem.Chem.Phys.* **13** 10652–10660; solid curves:simulation results. *Open Access Journal*.)

Xiong et al. [33, 52] studied the effect of size and shape on the melting temperature, enthalpy, entropy, as well of the Curie and Debye temperature and specific heat of nanoparticles, including the case of embedded nanoparticles, for which superheating effects are experimentally demonstrated. By extending the previous model to the surface free energy at the nanoscale, they demonstrated that the equation

$$\frac{\Upsilon}{\Upsilon^o} = 1 - \frac{K}{D} \tag{2.30}$$

(with $D=2r$) is an universal relation which holds for the size dependence of a generic thermodynamic function Υ (melting temperature, enthalpy, entropy, as well the Curie temperature, the Debye temperature, and the specific heat) of metallic nanoparticles (Ag, Al, Cu, Ni), taking for Υ^o the bulk value of Υ and for K the ratio $\dfrac{\pi N_A d_n^4 (\gamma_s - \gamma_l)}{\Delta H_m^o}$, where N_A is the Avogadro number, d_n is the atomic radius of the constituent, and ΔH_m^o is, again, the bulk melting enthalpy. K has, therefore, the role of a universal constant.

One can observe in Figure 2.7 the excellent fit of the experimental and simulation results for Au, Ag, and Al, while important deviations are observed for Pb, that are explained considering that Pb is much softer than Au, Ag, and Al, thus allowing an easier relaxation of the surface atoms.

Furthermore, they calculated also the effect of size and shape on the phase transformations of Ti and Zr nanoparticles [53] which follow an HCP to FCC transformation, unlike bulk phases that show an HCP to BBC transformation.

The method followed is based on the calculation of the temperature dependence of the Gibbs free energy G of the different nanophases, which is the sum of a surface and a bulk free energy, and the phase transformation temperature is obtained when $G_{HCP} = G_{FCC}$. The calculation is carried out also for the bulk phases, with excellent fits with the experimental transformation temperatures.

An equation similar to Equation 2.30, which describes both the size and shape dependence of a thermodynamic or physical property ξ of a nanoparticle, including the melting and the Curie temperature, the melting enthalpy, the enthalpy of a phase transformation, the cohesive energy and the formation energy of a vacancy, was independently proposed by Guisbiers* [54–57]

$$\frac{\xi}{\xi^o} = \left[1 - \frac{\alpha_{\text{shape}}}{2r}\right]^{1/2S} \tag{2.31}$$

where ξ^o is the bulk value of the thermodynamic property ξ, r is the radius of the particle, α_{shape} is a scaling constant, and S equals 1/2 or 1 if the elemental entities involved in the considered physical phenomenon follow the Fermi–Dirac or the Bose–Einstein statistics.

As an example, the elemental entities[†] involved in the cohesive energy and in the formation energy of vacancies are electrons, for which the Fermi–Dirac statistics hold.

The equation which describes the size and shape dependence of melting temperature is given here below

$$\frac{T_m}{T_m^o} = 1 - \frac{AD*(\sigma_s - \sigma_l)}{2rV\Delta H_m^o} = 1 - \frac{\alpha_{\text{shape}}}{2r} \tag{2.32}$$

where

$$\alpha_{\text{shape}} = \frac{D^*(\sigma_s - \sigma_l)}{\Delta H_m^o} A/V \tag{2.33}$$

is a size-independent but shape-dependent scaling coefficient, $D*$ is an adjustable parameter (in nm), σ_s and σ_l are the surface energies of the solid and liquid phases, ΔH_m^o is the melting enthalpy of the bulk phase, and the ratio A/V (in nm^{-1}) indirectly accounts for the particle shape.

Once the α_{shape} values can be calculated, the size and shape dependence of the thermodynamic functions of metallic or semiconductor nanoparticles can be easily calculated. For this reason, α_{shape} is a universal constant.

As a first example of the application of the Guisbiers model, Figure 2.8 displays the ξ/ξ^o ratios vs. the α_{shape} parameter for properties which are determined by fermions ($S = 1/2$) or bosons ($S = 1$), for different values of the diameter D (nm) of the

---

* S. Pizzini is strongly indebted to Gregory Guisbiers for a detailed discussion on the α coefficient.
† Called by Guisbiers "particles."

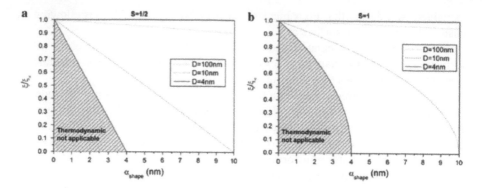

FIGURE 2.8 Shape dependence of the ξ/ξ_∞ ratio for a system following the Fermi–Dirac statistic (*left panel*) or the Bose–Einstein statistics (*right panel*). (*After* G. Guisbiers (2010) Size-Dependent Materials Properties: Toward a Universal Equation *Nanoscale Res. Lett.* **5** 1132–1136: *Open Access Journal.*)

particle. When $\alpha_{shape}=0$ (vertical red line) or $\xi/\xi_\infty=1$ (horizontal red line) the system experiences no size effects and its properties are the bulk ones.

It is possible to see that if the Fermi–Dirac statistics is followed, a linear decrease of ξ/ξ^o with the decrease of the shape factor could be forecasted, that is not the case when the Bose–Einstein statistics is followed.

A different approach has been followed by Kumar [34], who formulated the following equation for the size dependence of the Debye temperature θ_{nm} of spherical crystallites

$$\frac{\theta_{nm}}{\theta_b}=\left(1-\frac{N}{2n}\right)^{1/2} \tag{2.34}$$

and for the Debye frequency ν_{nm} of nanomaterials

$$\frac{\nu_{nm}}{\nu_b}=\left(1-\frac{N}{2n}\right)^{1/2} \tag{2.35}$$

where θ_b is the bulk Debye temperature, ν_b is the bulk Debye frequency, N is the number of surface atoms, and n is the total number of atoms [58]. Thus, considering that in metallic crystals the melting process is approximately vibrational in nature, the melting entropy corresponds to the vibrational entropy and the melting enthalpy corresponds to the vibrational enthalpy.

Since the bulk vibrational entropy S^o_{vibr} is related to the bulk melting temperature T^o_m

$$S^o_{vibr}=\frac{3R}{2}\ln\left(\frac{T^o_m}{C}\right) \tag{2.36}$$

where C is a constant, the melting entropy of bulk materials S^o_m could be written

$$S^o_m=\frac{3R}{2}\ln\left(\frac{T^o_m}{C}\right) \tag{2.37}$$

and that of nanomaterials S_{mn}

$$S_{mn} = \frac{3R}{2} \ln\left(\frac{T_{mn}}{C}\right) \qquad (2.38)$$

as a function of the corresponding melting temperatures T_{mn}.

Thus

$$S_{mn} = S_{mb} + \frac{3R}{2} \ln\left(\frac{T_{mn}}{T_{mb}}\right) \qquad (2.39)$$

Considering that Liang and Li [59] show that

$$\frac{\theta_{nm}}{\theta_b} = \left(\frac{T_{mn}}{T_m^o}\right)^{1/2} \qquad (2.40)$$

it turns out that

$$S_{mn} = S_m^o + \frac{3R}{2} \ln\left(\frac{\theta_n}{\theta_b}\right)^2 \qquad (2.41)$$

and for Equation 2.34

$$S_{mn} = S_m^o + \frac{3R}{2} \ln\left(1 - \frac{N}{2n}\right) \qquad (2.42)$$

which holds for spherical dots.*

At the melting temperature, since the Gibbs free energy of melting is zero, we have $H_m = TS_m$ and the melting enthalpy of a spherical nanodot results to be

$$H_{mn} = \left[H_m^o + \frac{3RT_m^o}{2} \ln\left(1 - \frac{N}{2n}\right)\right]\left[1 - \frac{N}{2n}\right]$$

$$= \left[H_{mb} + 3RT_{mb} \ln\frac{v_{nm}}{v_b}\right]\left[1 - \frac{N}{2n}\right] \qquad (2.43)$$

and

$$\frac{H_{mn}}{H_m^o} = \left[1 + \frac{3RT_m^o}{H_m^o} \ln\frac{v_{nm}}{v_b}\right]\left(1 - \frac{N}{2n}\right) = \alpha\left(1 - \frac{N}{2n}\right) \qquad (2.44)$$

where $\alpha = 1 + \frac{3RT_{mb}}{H_{mb}} \ln\frac{v_{nm}}{v_b}$ is a dimensionless scaling coefficient,† which turns out to be a function of the same thermodynamic parameters of the scaling coefficient α_{shape} as Guisbers in Equations 2.31 and 2.32.

The results for the melting enthalpy of spherical Sn dots are reported in Figure 2.9, which shows the excellent fit of the calculated values (continuous dark line) and the experimental results.

---

* The authors display also the relationships which hold for non-spherical nanodots and for wires.
† Different from $\alpha_{shape} f$; Equations 2.31 and 2.32.

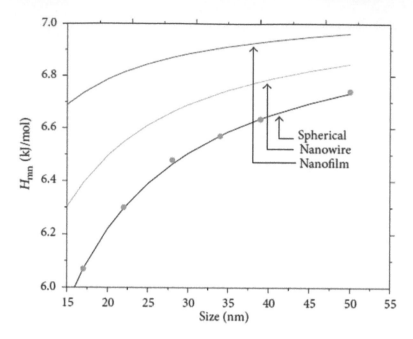

FIGURE 2.9 Size dependence of the melting enthalpy of spherical Sn dots. (*After* R. Kumar, G. Sharma, M. Kumar (2013) Effect of size and shape on the vibrational and thermodynamic properties of nanomaterials *J. Thermodynamics* art.ID 328051 DOI: 10.1155/2013/328051 Sep 16, 2014 *open access paper under Creative Common Attribution License.*)

The most recent studies on this topic carried out by Xiao et al. [60] confirm the significant effects of shape on the thermodynamic stability of non-spherical quantum dots with a size lower than 10 nm.

The arguments followed by the authors are based on the application of the Zhang and Jiang model [61], which predicts the size dependence of the melting temperature of a spherical dot with the following equation

$$\frac{T_m(D)}{T_m^o} = \exp-\frac{2S_{vibr}}{3R(\delta-1)} \tag{2.45}$$

where S_{vibr} is the vibration entropy of the crystal, R is the ideal gas constant, T_m^o is the bulk melting temperature, $\delta = \dfrac{D}{D^o}$, D is the diameter of a *ideally spherical* particle, and D^o is the critical size at which all of the atoms are placed on the surface.

For non-spherical dots Equation 2.45 might be modified by adding a λ term

$$\frac{T_m(D)}{T_m^o} = \exp-\frac{2S_{vibr}}{3R\lambda(\delta-1)} \tag{2.46}$$

where $\lambda = \dfrac{\delta}{\delta^o}$ and δ^o is the δ value for a spherical dot.

Since R is a constant and S_{vibr} is considered by the authors also a constant, the size and shape dependence of the melting temperature could be calculated with simple geometrical concepts, for a variety of macroscopic shapes, as is shown in Figure 2.10.

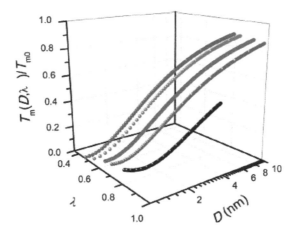

FIGURE 2.10 Size and shape dependence of melting temperature $T_m(D,\lambda)$ of icosahedral (red line), dodecahedral (pink line), cuboctahedral (olive line), octahedral (blue line), decahedral (dark-yellow line), and tetrahedral (black line) Ag quantum dots. (*Reprinted with permission from* 1 H.-J. Xiao, T.-S. Zhu, H.-C. Xuan, and M. Li (2015) Size Consideration on Shape Factor and Its Determination Role on the Thermodynamic Stability of Quantum Dots *J. Phys. Chem. C* **119** 12002–12007 *Copyright {2015} American Chemical Society.*)

The problem inherent to the conclusions of this work is that also S_{vibr} will change with λ, following the changes of the surface energy $F = \sum_s s\gamma_s$ of the nanocrystal at their equilibrium shapes, where γ_s is the surface energy of the facet involved (see Wulff criterion, Chapter 3, Section 3.3.5) and its changes could not be manipulated with simple geometrical concepts.

These arguments could not be straightforwardly applied to the energetics of solid phase transformations of semiconductor dots down to a size of a few nm, as could be inferred from the work of Alivisatos [4] and, later, of Chen et al. [62], who enlighten, also, new aspects of defectivity in nanostructures.

A scaling law applies, in fact, also to solid phase transformations, as is the case of the wurtzite to rock salt transformation of CdSe and of other solid phase transformations of CdSe, InP, and Si nanometric clusters presenting a well-developed structural shape. There, however, the mechanical driving force (the transformation pressure) at constant temperature increases with the decrease of the size (see Figure 2.11), just the opposite of what is found for melting, as the consequence of local deviations from mechanical equilibrium and the set-up of local strain (see Equation 2.6).

The thermodynamics of this process could be qualitatively understood by considering that the structural disorder at the surface of the crystal during the transformation from the tetrahedral to the octahedral structure of the high-pressure phase increases with the decrease of the size and inhibits the establishment of the structure of the thermodynamic stable phase.

In a subsequent work, carried out in cooperation with Chen et al. [62], Alivisatos extended the analysis to the kinetics of the phase transformation process and demonstrated that the kinetic barrier (the activation energy E_{act}) for the wurtzite–rock salt transformation of CdSe nanocrystals, at a constant pressure of 4.9 GPa, increases

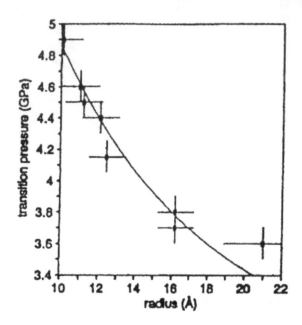

FIGURE 2.11   Effect of size on the transformation pressure from wurtzite to rock salt in CdSe. (*Reproduced with permission from* A.P. Alivisatos (1996) Perspectives on the physical chemistry of semiconductor nanocrystals. *J. Phys. Chem.* **100**, 13226–13239 *Copyright 1996 American Chemical Society.*)

with the number of unit cells and thus with the size, following the conventional scaling trend observed in melting processes.

The mechanism of phase transformation of nanometric clusters is, however, very different from that occurring in larger crystals, since in these small crystals the transformation* "involves the coherent deformation of the entire cluster in one single step."

And it is also forecasted that if the temperature is sufficiently high, at the equilibrium transformation pressure, *the crystals will fluctuate between the two structures*, recalling the conditions of the equivalent equilibrium state among molten and solid states, discussed at the beginning of this chapter.

The last issue considered in this section concerns the analysis of size dependence of the physico-chemical properties of Au and Ag nanoparticles, which exhibit a 100% increase of their elastic modulus and a 70% decrease of the melting temperatures with respect to the bulk values, carried out by Sun's group [32] in the framework of the bond order-length-strength (BOLS) correlation mechanism [63–65] which is addressed directly at the physics of the process underneath, with a *first principle* approach, not at its phenomenology.

---

* Original sentence in Chen's paper.

The basic assumption, here, is that the size dependence of the physico-chemical properties of the matter, at the nano-size, depends entirely on the undercoordination of the atoms at the outer surface of the sample and at a few [3] sub-surface layers,* which shortens and strengthens the remaining bonds [66], while the relative magnitude of these changes depends on the surface to volume ratio.

Using a sum of the core-shell property terms and the BOLS correlation mechanism, the size-dependence function $Q(R)$ of a generic property Q could be derived, that is uniquely a function of the number of the undercoordinated atoms at the shell structure

$$Q(R) = Q_b + \sum_{i \leq 3} N_i(q_i - q_b) \tag{2.47}$$

where $Q_b = Nq_b$ and the q terms are mass densities of the Q property.

Thus

$$\Delta Q = \frac{Q(R) - Q_b}{Q_b} = \sum_{i \leq 3} \frac{N_i(q_i - q_b)}{Nq_b} = \sum_{i \leq 3} \beta_i \frac{(q_i - q_b)}{q_b} \tag{2.48}$$

and β_i is the surface to volume ratio of the ith layer

$$\beta_i = \frac{S_i}{V} = \frac{N_i}{N} = \frac{\tau C_i d_b}{R} = \frac{\tau C_i}{K} \tag{2.49}$$

where τ is the dimensionality of the nanostructure ($\tau = 3$ for a spherical nanoparticle of a radius R), C_i is the bond contraction coefficient, d_b is the bond length, and $K \approx R/d_b$ is the number of atoms along the radius of the sphere [67].

Thus

$$\Delta Q = \sum_{i \leq 3} \frac{\tau C_i}{K} \frac{(q_i - q_b)}{q_b} = \sum_{i \leq 3} \frac{\tau C_i d_b}{R} \frac{(q_i - q_b)}{q_b} \tag{2.50}$$

which shows that the relative change ΔQ of the thermodynamic function Q scales with $1/R$ and that the scaling constant depends on fundamental properties of the matter at the atomic scale.

The theory has been applied to the melting temperatures of Ag and Au nanospheres, with an excellent agreement between the experimental and theoretical results, that demonstrates the validity of the BOLS approach in evaluating the effect of surface undercoordination on the thermodynamic properties of nanostructures.

Using the conventional language of defect physics, it seems, therefore, demonstrated that defects at the surface of nanostructures, consisting of undercoordinated and dangling bonds and, possibly, of point defects, dominate the overall physico-chemical properties of the material.

In conclusion, the phenomenological evidence that nanoparticles are less stable than their bulk counterparts and that the particle size influences the thermodynamics

---

* The atoms in the core region maintain their bulk properties, as will be discussed in the last section of this chapter.

of melting and of other phase transformations of nanomaterials might be understood by considering that the properties of a nanomaterial depend on the structure and energetics of the interfaces between the equilibrium phases. This conclusion addresses further interest in the basic understanding of the thermodynamics of multicomponent systems at the nanoscale, of the size dependence of the surface composition and shape, and of the interfacial energetics of multi-component phases, which are the topics of the next two sections.

2.3 PHASE DIAGRAMS OF NANO-SYSTEMS

Two main issues can be discussed in this context. The first deals with the effect of size or shape, or of both, on the onset of a phase transformation $M(\alpha) = M(\beta)$ of an elemental or compound nano object (a crystallite or a nanowire, as an example) induced by self-strain, where α and β are two stable phases of the element M.

The second deals with the effect of size and shape on the thermodynamic equilibria of binary or multinary systems, for which, as an example, the temperature dependence is described by a eutectic diagram.

Since self-strain effects will be deeply discussed in Chapter 3, Section 3.4.5 when looking to the thermodynamic stability of hexagonal silicon nanowires, here we will be only concerned with the size and shape effects on the thermodynamics of phase transformations of multinary systems.

2.3.1 THERMODYNAMIC EQUILIBRIA IN MULTINARY ALLOYS IN THE ABSENCE OF SEGREGATION

Multinary semiconductor nanophases consist of binary or multinary systems, presenting full or partial solubility conditions, for which it is customary to look at the role of size in their temperature and composition properties, i.e. in their phase diagrams.

As could be expected, the features of phase diagrams of nano-systems should, in fact, depend on the system's size, because not only are the melting enthalpies and the melting temperatures size-dependent, but also the mixing enthalpies, which might deviate from the values of their bulk counterparts, even in the case where bulk solid solutions present an ideal behavior, due to chemical interactions induced by strain or disorder.

The experimental study of the size dependence of even the simplest phase diagrams is, however, very difficult, if not impossible [68], but this gap is filled by several theoretical models concerning the solution theory of nanoalloys.

Phase transformations of "small systems" were investigated by Hill [1, 69, 70], Gross [71], Wautelet [72], and Pohl et al. [73].

Their work gives a valuable insight into the characteristics of "small systems"* and into the basic features of phase transformations and phase stability of

---

* Which have linear dimensions comparable with the interaction distances between the particles according to Gross [71].

nano-systems, which could be evaluated provided their size is sufficiently large to allow a proper use of the thermodynamic variables* [72].

As an example, the phase diagram of the Ge–Si system at the nanoscale was simulated by Wautelet et al. [23] by assuming that the solid and liquid solutions are nearly ideal (mixing enthalpy $\Delta H_{mix}=0$ like in the bulk case), and that the melting temperatures depend on size, and making the approximation that the melting enthalpy ΔH_m does not.

Under these conditions, the solid–liquid equilibria could be calculated by equating the chemical potentials of the solvent and of the solute (called A and B) in the two phases and writing [74]

$$kT \ln \left(\frac{x_{\text{sol}}}{x_{\text{liq}}} \right)_A = \Delta H_{m,A} \left(1 - \frac{T}{T_{m,A}} \right) \qquad (2.51)$$

$$kT \ln \left(\frac{1 - x_{\text{sol}}}{1 - x_{\text{liq}}} \right)_B = \Delta H_{m,B} \left(1 - \frac{T}{T_{m,B}} \right) \qquad (2.52)$$

where $\Delta H_{m,A}$ and $\Delta H_{m,B}$ are the melting enthalpies of the pure substances in the bulk state.

If this process is carried out at any temperature T between the melting temperature of pure A ($T_{m,A}$) and of pure B ($T_{m,B}$), normalized to the particle size

$$T = T_m^o \left[1 - \frac{\alpha}{2R} \right] \qquad (2.53)$$

where α is the scaling constant considered in the previous section [10], the calculated phase diagram (see Figure 2.12) for spherical nanoparticles of an ideal

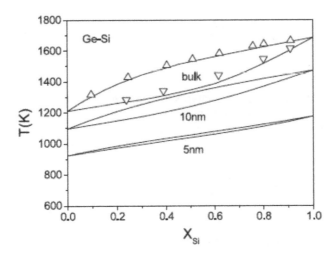

FIGURE 2.12 Phase diagram of the Ge–Si systems for bulk and nanometric conditions. (*After* L.H. Liang, D. Liu, Q. Jang (2003) Size-dependent continuous binary solution phase diagram *Nanotechnology* *14* 438–442 Reproduced with permission of IOP Nov 24, 2018.)

---

* We will see that a size of 4–5 nm represents an empirical limit.

Ge–Si solution, containing $2\cdot10^7$ atoms (about 30 nm in diameter) shows, as could be expected, only a systematic deviation of the whole diagram towards lower temperatures as compared to that of the bulk system, without substantial changes in the composition of the liquid and solid phases, since only the temperature dependence on size is accounted for.

Similar arguments were used by Vallee et al. [24], who extended the discussion to the issue concerning regular solutions.

In this case

- The mixing enthalpy $\Delta H_{\mathrm{mix}} = x(1-x)\Omega$ of the liquid and solid solutions in mutual equilibrium is ruled by two interaction coefficients Ω_l and Ω_s, given by $\Omega = \dfrac{H_{AB}}{H_{AA} + H_{BB}}$ and H_{AA}, H_{BB}, and H_{AB} are the bond energies of the AA, BB, and AB bonds.
- The Gibbs free energy of mixing is given by the following equation*

$$\Delta G_{\mathrm{mix}} = x_A x_b \Omega - T\Delta S_{\mathrm{conf}} - x_A x_b \eta \qquad (2.54)$$

where the third term is a non-configurational entropy term which has a negligible role in bulk systems, but could be important at the nanoscale.

On this basis, the equilibrium conditions are obtained by equating the chemical potentials of A and B in the two phases

$$kT \ln\left(\frac{x_{A,\mathrm{sol}}}{x_{A,\mathrm{liq}}}\right) = \Delta H_{m,A}\left(1 - \frac{T}{T_{m,A}}\right) + \Omega_l\left(1-x_{A,l}\right)^2 - \Omega_s\left(1-x_{A,s}\right)^2 \qquad (2.55)$$

$$kT \ln\left(\frac{x_{\mathrm{sol}}}{x_{\mathrm{liq}}}\right)_B = \Delta H_{m,B}\left(1 - \frac{T}{T_{m,B}}\right) + \Omega_l\left(x_{A,l}\right)^2 - \Omega_s\left(x_{A,s}\right)^2 \qquad (2.56)$$

where the second and third terms of both equations are mixing enthalpy terms, and as before, only the melting temperatures are supposed to depend on size, while the melting enthalpies are considered unaffected by size.

These equations have been applied to the binary Au–Cu system, that in bulk conditions presents a continuous series of solid solutions, with a congruence point at $x_{\mathrm{Cu}} = 0.48$, and solution demixing at low temperatures.† Taking $\Delta H_{m,A}$, $\Delta H_{m,B}$, Ω_l, and Ω_s independent of size and $\Omega_s < 0$, $\Omega_l < 0$ as for the bulk phases, the results obtained for particles consisting of 10^6 atoms indicate almost negligible size effects on the equilibrium phase compositions, but a systematic shift toward lower temperatures.

Different results are obtained if the equilibrium conditions are calculated using for the size, shape, and composition dependence of the enthalpy of formation of a

---

* For the thermodynamics of solutions see S. Pizzini (2015) *Physical Chemistry of Semiconductor Materials and Processes*, Wiley.
† A solution demixing at low temperatures is expected for positive or small, negative values of the enthalpy of mixing.

nanoalloy the following equation formulated by Xiao et al. [75], Qi et al. [76], and Xiong et al. [77]

$$\Delta H_{NP}^{f} = \Delta H_{b}^{f}\left(1-\alpha N^{1/3}\right)-\alpha N^{-1/3}\left[x_{A}E_{b}^{A}\left(1-x_{A}^{-1/3}\right)+x_{B}E_{b}^{B}\left(1-x_{B}^{-1/3}\right)\right] \quad (2.57)$$

where ΔH_{NP}^{f} and ΔH_{b}^{f} are the enthalpy of formation of the (ideal) nanostructured alloy and of the bulk alloy respectively, α is a shape factor, N is the total number of atoms, and E_{b}^{A} and E_{b}^{B} are the bulk cohesive energies of A and B.

The ratio of this equation is that the formation enthalpy ΔH_{f}^{AB} of a bulk ideal binary $A_{x}B_{1-x}$ alloy is the difference between the cohesive energy of the solution E_{b}^{AB} and the sum of the fractional cohesive energies of the components E_{b}^{A} and E_{b}^{B}

$$\Delta H_{f}^{AB} = E_{b}^{AB} -\left[xE_{c}^{A}+(1-x)E_{b}^{B}\right] \quad (2.58)$$

and that the cohesive energy of the ideal solution E_{b}^{AB} could be calculated with the generalized embedded atom method (EAM) potential [78] as

$$E_{b}^{AB} = \frac{1}{2}\phi^{A}(r)+F^{A}(\rho)x+\frac{1}{2}\phi^{B}(r)+F^{B}(\rho)(1-x) \quad (2.59)$$

where the $\phi(r)$ and $F(\rho)$ terms are pair-potential functions and embedding energy terms, with $\rho(r)$ as the electron density.*

As an example, if the size dependence of the formation enthalpy (the heat of mixing) is calculated with the EAM method for the Au–Cu solutions, one sees that the negative values of the enthalpy of mixing increase with the decrease of the size, leading to an increase of the thermodynamic stability of the solid solution, and to the decomposition of the low-temperature $Au_{n}Cu_{m}$ phases, stable in the case of bulk conditions below a size of 2–3 nm ($N=321$).

Similar results are reported by Xiao et al. [75] concerning the phase diagram of Au–Pt alloys, which demonstrate that the calculated heat of solution ΔH_{mix} of these alloys, which is positive in the bulk case and induces a wide solubility gap centered at 60% of Au, becomes negative when the number of atoms in the NP is less than 7,000 (about 6 nm in diameter), leading to the thermodynamic stability of the Au–Pt solution in the whole composition range.

Analogous results are obtained for the Ag–Cu system, which presents an extended insolubility range in bulk conditions, while complete solubility conditions are shown to occur at the nanoscale, below a size of 2–3 nm.

Details on the thermodynamics of nanometric powders of binary alloys with a size of the order of a few nm and lower, where melting temperature and thermo-dynamic functions show the largest deviations from bulk conditions, were deeply discussed by Liang et al. [79] and M. Cui et al. [68]. The Liang et al. [79] study con-cerned Cu–Ni and Si–Ge alloys, assumed to behave as regular solutions in both the liquid and solid state, with size-dependent interaction coefficients Ω_{s} and Ω_{l} which simply scale with a $1/D$ relationship [80]

---

* See the papers in question for major details.

$$\frac{\Omega(D)}{\Omega^\circ} = 1 - \frac{2D^\circ}{D} \tag{2.60}$$

where Ω° is the interaction coefficient in the bulk phase and D° is the sample size.

On that basis, the evolution of the phase diagram of the Si–Ge system, of the Cu–Ni alloys, and of several other binary alloys was calculated for bulk conditions and for $D = 10$ and $D = 4$–5 nm, all showing an excellent fit with the available experimental results concerning the bulk systems.

It has been also found that when the size of the components decreases to a critical value, that correspond to 4 or 5 nm, the composition of the liquid and solid phase are close or equal and, therefore, the segregation coefficient $k_{segr} = \dfrac{x_j^s}{x_j^l}$ is very close or equal to 1 in the entire range of concentrations for these systems.

Apparently, in the case of the Si–Ge system, see Figure 2.12, the critical size value (5 nm) fits with a value of the correlation length close to the exciton Bohr radius for Si (see Table 2.1).

Also in the case of the binary Cu–Ni system the critical size is very close to that shown to hold for metallic nanoparticles, where the critical size is the size at which statistical energy fluctuations are of the order of the enthalpy of melting (see Section 2.1) and is of the order of a few nm (6 nm in the case of Pt NPs). Below this size the system can be considered as a superposition of two phases [14].

In their work addressed at the study of binary metallic alloys Cui et al. [68] start by mentioning that the simulation of a binary phase diagram at the nanoscale should account not only for the ideality or regularity of the solutions involved, but for size and shape (λ) effects on the melting temperature, melting enthalpy ΔH_m, and interaction coefficient Ω.

They assume, additionally, that the surface composition might be different from that of the core, and would depend on the enthalpy of surface segregation ΔH_{segr} [81] according to the following empirical equation

$$x_{surf} = \frac{x_{core} \exp \dfrac{\Delta H_{segr}}{RT}}{1 - x_{core} + x_{core} \exp \dfrac{\Delta H_{segr}}{RT}} \tag{2.61}$$

where the segregation enthalpy is the energy involved in the exchange of bulk atoms with surface atoms, driven by a difference in their respective chemical potentials.

Eventually, the following equation, derived according to Lindemann's criterion, describes the size and shape λ dependence of a generic thermodynamic function $X(D,\lambda)*$ [82]

$$\frac{X(D,\lambda)}{X} \approx \left[1 - \frac{1}{12(D/D_o) - 1} \right] \exp - \left[\frac{2\lambda S_o}{3R} \frac{1}{12(D/D_o) - 1} \right] \tag{2.62}$$

---

* Including the segregation enthalpy.

where D_o is the critical thickness of the particle at which all the atoms are at the surface, and S_o is the bulk solid–vapor transition entropy.

In absence of segregation phenomena, the phase diagram of a binary system behaving as a regular solution could be calculated from the melting temperature $T_{m,A}$ of the component A to the melting temperature $T_{m,B}$ of the component B, suitably scaled to a specific size and shape via the Equation 2.62, using the following equations [79]

$$\ln\left(\frac{x_B^s}{x_B^l}\right) = \frac{\Delta H_{m,B}\left(1 - T/T_{m,B}\right) - \Omega^s\left(1 - x_B^s\right)^2 + \Omega^l\left(1 - x_B^l\right)^2}{RT} \tag{2.63}$$

$$\ln\left(\frac{x_A^s}{x_A^l}\right) = \frac{\Delta H_{m,A}\left(1 - T/T_{m,A}\right) - \Omega^s\left(x_B^s\right)^2 + \Omega^l\left(x_B^l\right)^2}{RT} \tag{2.64}$$

where also the ΔH_m^A, ΔH_m^B, and Ω terms are scaled to a specific size and shape.

The phase diagrams of binary Ni–Cu alloys calculated following this procedure for the bulk case and for a size of 10 and 4 nm are displayed in Figure 2.13 for different particle shapes.

It could be observed that the shape has a key role in the temperature shift of the diagrams toward lower temperatures, that keeps a maximum for the tetrahedral structure.

It is also remarkable that when the size of the particles decreases toward a critical value, that corresponds to 4 nm in the present case, the composition of the liquid and solid phases in equilibrium are very close for all shapes or even equal in the case of tetrahedral structure. Therefore the segregation coefficient $k_{segr} = \frac{x_j^s}{x_j^l}$ is very close to 1 in the entire range of concentrations, making the solid and liquid phase chemically indistinguishable when $k_{segr} \equiv 1$.

The rationale of these results could be deduced by a simple analysis of the conditions at which the composition of the two phases in thermodynamic equilibrium coincides.

Since

$$\ln k_{segr} = \frac{\mu_j^{o,s} - \mu_j^{o,l}}{RT} - \ln\frac{\gamma_j^s}{\gamma_j^l} = \frac{\Delta G_j^o}{RT} - \ln\frac{\gamma_j^s}{\gamma_j^l} \tag{2.65}$$

where γ_j^s and γ_j^l are the activity coefficients of the component j in the solid and the liquid phase and ΔG_j^o is the Gibbs free energy of melting, $k_{segr} = \frac{x_j^s}{x_j^l} = 1$ when

$$\ln k_{segr} = \frac{\Delta G_j^o}{RT} - \ln\frac{\gamma_j^s}{\gamma_j^l} = 0 \tag{2.66}$$

Considering that the heat of fusion depends on size and tends to zero at values of i lower than the critical size, the condition of equal composition for the two phases in equilibrium requires that the ratio of the activity coefficients $\frac{\gamma_j^l}{\gamma_j^s} = 1$.

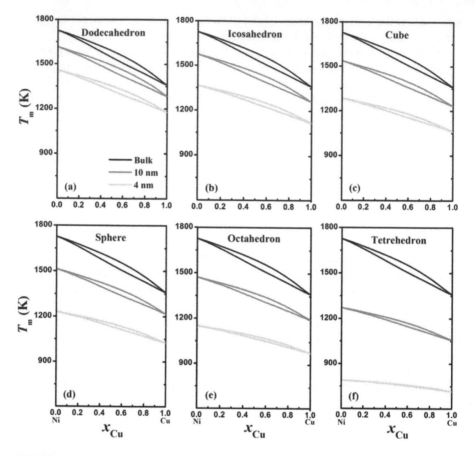

FIGURE 2.13 Size and shape dependence of the phase diagram of binary Cu–Ni alloys. (*After* M. Cui, H. Lu, H. Jiang, Z. Cao, X. Meng (2017) Phase diagram of continuous binary nanoalloys: Size, shape, and segregation effects *Sci Rep.* **7** 41990 *This work is licensed under a Creative Commons Attribution 4.0 International License.*)

The fact that the liquid and the solid phase are chemically indistinguishable at the critical size is in good agreement with studies on other model systems, which showed that nanometric materials at their critical sizes simultaneously display solid-like and liquid-like characteristics [3].

The effect of size and surface tension effects on binary phase diagrams has been, eventually, investigated for the Ag–Au system at the nanoscale by Park and Lee [83] and Monij and Jabbareh [84], showing that the effect is significant, again, at the limit of 5 nm, when the majority of the atoms reside at the surface.

The effect of the surface energy in the simulation of phase diagrams at the nanoscale, where the size effects are entirely associated with surface energies, has been also considered by Ghasemi et al. [85] in their work on the In–Sb phase diagram.

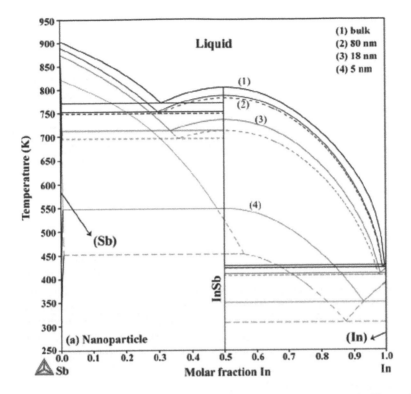

FIGURE 2.14 Calculated phase diagram of the In–Sb system. (*After* M. Ghasemi, Z. Zanolli, M Stankovski and J. Johansson (2015) Size- and shape-dependent phase diagram of In–Sb nano-alloys *Nanoscale* **7**, 17387–17396] *This article is licensed under a Creative Commons Attribution 3.0 Unported License.*)

Under the assumption of non-ideality of the solution, the total Gibbs free energy of the binary $A_x B_{1-x}$ solution at the nanoscale is given by the equation

$$\Delta G_{\text{nano}}^{\text{total}} = \Delta G_{\text{bulk}} + \Delta G_{\text{surf}}$$

$$= x_A \Delta G_A^{o,\text{nano}} + x_B \Delta G_b^{o,\text{nano}} + RT(x_A \ln x_A + x_b \ln x_B) + \Delta G^{\text{exc}}$$

(2.67)

where the excess Gibbs free energy ΔG^{exc} is given as the sum of a $x_A x_B \Omega$ term due to the non- ideality of the solution and of an excess Gibbs free energy term due to the surface free energy $\Delta G_{\text{surf}}^{\text{exc}}(T, x_\gamma^{\text{surf}})$, where γ is A or B.

The phase diagram of the In–Sb calculated under this approach is displayed in Figure 2.14 for different particle sizes and compared to the phase diagram of the bulk system. It is again apparent that the effect of particle size is almost negligible down to 18 nm, while it is critical at 5 nm, where the intermediate In–Sb phase disappears and the eutectic is shifted towards high In contents.

2.3.2   THERMODYNAMIC EQUILIBRIA IN MULTINARY ALLOYS
IN THE PRESENCE OF SEGREGATION

Different thermodynamic equilibrium conditions in nanostructures might be induced by surface segregation, which leads to significant surface composition changes, observed both in the case of ideal and regular binary A_xB_{1-x} solutions [86].

Surface segregation effects could be evaluated following the Langmuir–Mc Lean theory [87, 88] that shows that the ratio of the surface compositions x_A^{surf} and x_B^{surf} of the components of an A_xB_{1-x} alloy depends on the Gibbs free energy ΔG_{segr} of the surface segregation process* (that is, the driving force of the process), and on the composition ratios in the core of the NP

$$\frac{x_A^{surf}}{x_B^{surf}} = \frac{x_A^{core}}{x_B^{core}} \exp - \frac{\Delta G_{segr}}{kT} = B \frac{x_A^{core}}{x_B^{core}} \exp - \frac{\Delta H_{segr}}{kT} \qquad (2.68)$$

where B is an entropic term.

It is very disappointing that the Gibbs free energy ΔG_{segr} or the enthalpy of segregation ΔH_{segr} are quantities that are difficult to calculate in the case of nanostructures, where segregation depends also on the size and on the shape of the particle.

However, there are several parameters which offer a semiempirical background to a driving force to segregation [88] common to the bulk case and to that of NPs.

The first is the relative strength H_{AB} of the A–B bond with respect to the A–A and B–B bonds. When $H_{AB} > H_{AA}$, H_{BB} the A_xB_{1-x} solution is stable and surface segregation does not occur. Conversely, if $H_{AB} < H_{AA}$, H_{BB} the species with the lower homonuclear bond energy tends to segregate at the surface.

A second criterion operates in the case of structural mismatch, arising from the difference in size of the atoms A and B, with the set-up of local strain, which could be partially released if the species with the largest radius segregates at the surface.

As an example, surface segregation driven by lattice mismatch has been theoretically modeled using the density functional theory (DFT) method in the case of CoPt nanoparticles, considering that surface Pt segregation could be predicted, in view of the difference in atomic radii (Pt 177 pm, Co 152 pm) and was experimentally proven for CoPt bulk crystals [89]. The calculated Pt segregation energy ranges around 5.5 eV in the case of cuboidal particles and 2.8 eV for cuboctahedric CoPt particles.

This condition does however not hold if the mismatch strain could be accommodated by changes of both the bond length and bond angle within the alloy, as is the case of Ge–Si solutions, where, in spite of the substantial difference of the atomic radii of the components atoms (Si 111pm, Ge 125 ppm) most of the atomic strain is accommodated by changes of both the bond length and bond angle within the alloy [90].

A third conditions arises when the A and B species present different values of surface tension σ_{surf} (or sublimation enthalpies ΔH_{sub}), which favor the surface segregation of the species with higher σ_{surf} or ΔH_{sub}.

---

* The Gibbs free energy term ΔG_{segr} is the difference between the free energy of the surface region and of the bulk. When $\Delta G_{segr} = 0$ no segregation occurs.

Segregation effects driven by surface tension differences in a bulk binary alloy could be quantitatively calculated using the equation of Miedema [91]

$$\frac{x_{A,s}}{x_{A,b}} = \frac{f\Delta H_{sol}(AB) - g(J_A - J_B)V_A^{2/3}}{RT} \qquad (2.69)$$

where $\Delta H_{sol}(AB)$ is the heat of solution of A in B (A is the solute), $J_A - J_B$ is the difference of surface enthalpies of A and B, V_A is the molar volume of A, and f and g are constants.

Positive values of $\Delta H_{sol}(AB)$ lead to the thermodynamic instability of the solution and thus to surface segregation of A, if the process is not compensated by the second term in Equation 2.68.

The same equation could be used to calculate surface enrichment of NPs, provided the size and shape dependence of $\Delta H_{sol}(AB)$ and of the J_A and J_B terms could be accounted for.

The last condition for segregation comes from different chemical reactivities of A and B with respect to chemical contaminants (water, oxygen) in the gas phase in equilibrium with the NP. The species with higher reactivity migrates towards the surface where they are irreversibly bound as oxides or hydrides.

Formation of core-shell structures in NP has been discussed by Shirinyan and Wautelet [92] who mention that the necessary condition for the formation of core-shell structures is that the two subsystems wet each other, with the surface energies of the two subsystems that drive the system to the thermodynamic equilibrium.

Surface segregation is known to play a key role in semiconductor alloys, as in the case of Si–Ge alloys and of Si–Ge superlattices, which have been an hot subject of research for many years [93–98].

As is well-known, the silicon–germanium bulk alloys do behave as almost ideal solid solutions in the entire range of compositions, but Ge tends to segregate at the surface in view of its lower surface energy in comparison to Si [Ge 1835 erg/cm^2 <111> 1300 erg/cm^2 <110>; Si 2130 erg/cm2<100> 1510 erg/cm2 <110>] and lower bond strength [Ge–Ge 42,6 kcal/mole; Si–Si 45.5 kcal/mole] [99].

Segregation effects on Si–Ge NPs were investigated by Vallee et al. [24] based on the theoretical background of a former study dealing with the size dependence of thermodynamic properties of Si–Ge NPs in the absence of segregation [10].

There, the hypothesis was that the driving force for segregation originates from the different sublimation enthalpies of pure Ge (334 kJ/mol) and Si (359 kJ/mole), that the overall size dependence is determined by the size dependence of the melting temperatures $T_m = T_m^b\left[1 - \dfrac{\alpha}{2R}\right]$, that Si–Ge nanocrystalline ($N = 10^6$ atoms) solid solutions do behave as regular solutions, and that the interaction coefficient Ω ranges between two limiting values ($0.09kT < \Omega < -0.1kT$).

As expected, the deviations of the surface composition from that of the bulk are shown to depend on the value assumed to hold for the interaction coefficient Ω, that rules the mixing enthalpy of the solid phase [$\Delta H_{mix} = x(1 - x)\Omega$], that is larger for positive values of Ω.

Due to the number of arbitrary assumptions made, the phase diagrams elaborated by the authors should be regarded, only, as a qualitative example of surface segregation driven by selective sublimation effects.

Eventually, the case of Si–Ge nanowires was considered by Abudukelimu et al. [100], who started from the known phase diagram of spherical particles previously modeled, accounting for the different surface segregation effects on the faces of the nanowires via an α_{shape} coefficient.

On that basis, the calculated phase diagram of Si–Ge nanowires with an average radius $R = 16.8$ nm is very close to that calculated for spherical nanoparticles of the same size.

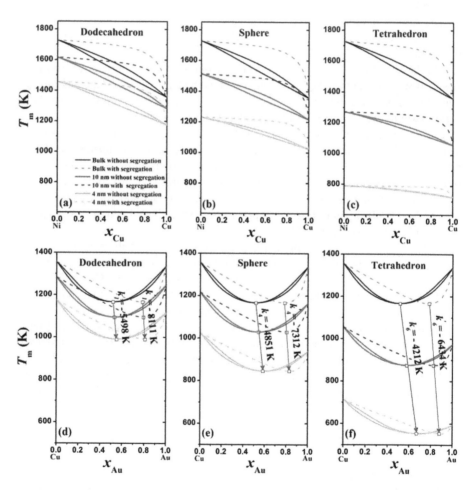

FIGURE 2.15 Phase diagrams of Cu–Ni alloys in absence (solid lines) and in presence of surface segregation (dotted lines solid surface composition). (*After* M. Cui, H. Lu, H. Jiang, Z. Cao, X. Meng (2017) Phase diagram of continuous binary nanoalloys: Size, shape, and segregation effects *Sci Rep.* **7** 41990 *licensed under Creative Commons Attribution 4.0 International License.*)

Surface segregation effects in nanometric clusters with sizes of the order of a few nm have been also discussed by Cui et al. [68] for the case of metallic solid solutions, on which we have already discussed in the previous section.

Assuming that the surface concentration depends on the enthalpy of surface segregation ΔH_{segr} according to Equation 2.60 [81] and that Equation 2.61 describes the size and shape (λ) dependence of ΔH_{segr}, ΔH_m, and Ω, the calculated surface compositions of binary Ni–Cu alloys, in the absence and in presence of surface segregation, are displayed in Figure 2.15.

It could be observed that the surface composition of the solid NPs is largely in excess of that of the core and mostly depends on the shape of the NP, which also strongly influences the thermodynamics of this binary system in the absence of segregation.

2.4 THERMODYNAMICS OF DOPING OF NANOCRYSTALLINE SEMICONDUCTORS

Doping is the basic process used to control the carrier concentration of bulk semiconductors [101–103], which allowed the development of the modern microelectronic technologies.

It is carried out by solid phase diffusion, ion implantation, or molecular beam epitaxy (MBE) of heterovalent, substitutional impurities, which form a solid solution in the parent lattice and create thermally ionizable defect states in the energy gap of the semiconductor.

Dopant impurities are also used to create luminescence or magnetic centers in the semiconductor lattice, as is the case of Er-doping or of several transition metal impurities.

Similarly, it is imperative to develop and understand doping processes in semiconductor nanostructures if these materials should positively evolve into practical applications in novel optoelectronic technologies [104].

To allow a better understanding of the basic issues concerning doping processes at the nano-level, a short discussion of the problems involving doping at the bulk level will be preliminarily given, considering both elemental and compound semiconductors.

2.4.1 DOPING THERMODYNAMICS OF BULK SEMICONDUCTORS

In the case of silicon and germanium doping, B and P are the acceptor and donor impurities more frequently used to control the electron n and hole h concentration, whose concentration would depend on the B or P dopants content, according to the following equations

$$B(s) + V_{Si} \rightleftharpoons B_{Si}$$
$$Si_{Si} \rightleftharpoons Si_{surf} + V_{Si}$$

$$B(s) + Si_{Si} \rightleftharpoons Si_{surf} + B_{Si} \tag{2.70}$$

$$B_{Si} \rightleftharpoons B_{Si}^- + h \qquad (2.71)$$

$$P(s) + V_{Si} \rightleftharpoons P_{Si}$$

$$Si_{Si} \rightleftharpoons Si_{surf} + V_{Si}$$

$$P(s) + Si_{Si} \rightleftharpoons Si_{surf} + P_{Si} \qquad (2.72)$$

$$P_{Si} \rightleftharpoons P_{Si}^+ + e \qquad (2.73)$$

where Equations 2.70 and 2.72 account for the formation of a substitutional impurity in the semiconductor lattice from a solid dopant source, and imply that the enthalpy of formation of a substitutional impurity $\Delta H_f = \Delta H_{sol} + \Delta H_{exch}$ is the sum of a heat of solution and of a second term accounting for the energy transfer of a bulk silicon atom to a surface reservoir of infinite capacity, and Equations 2.71 and 2.73 represent the impurity ionization equilibria, with $h = [B_{Si}^-]$ and $e = [P_{Si}^+]$. We will see that the calculation of the energy of formation of a dopant in a semiconductor nanocrystal follows the same procedure.

A defect-assisted thermal diffusion ensures the establishment of constant dopant concentration in the system or in a part of the system. The limiting dopant concentration corresponds to the dopant solubility at the process temperature (at 900°C, [B] = 10^{20} cm^{-3}, [P] = $6 \cdot 10^{20}$ cm^{-3}),* which, in the case of B, might be enhanced by an excess of vacancies above their equilibrium concentration [105].

As expected from our previous considerations about the role of vacancies on the dopant dissolution, the solubility of dopants in silicon stays in the order of magnitude of the concentration of silicon vacancies, and the expected exponential dependence of the dopant concentration vs. temperature is experimentally demonstrated.

Since the electron and hole product is constant, $nh = K_T^*$, we have also, in the presence of B- or P-doping, $k_T = n[B_{Si}^-]$ and $k_T = h[P_{Si}^+]$, respectively.

The ionization (binding) energies of donor and acceptor impurities in Si and Ge in their lower lying states are reported in Table 2.2.

Doping of II–VI and III–V compound semiconductors is a less trivial task, and limiting issues are self-compensation effects by amphoteric doping or by point defects, point defect–impurity pairs, solubility of dopants, deactivation of dopants by lattice relaxation, and finally, by second phase formation [106] when the solubility limits are overcome.

TABLE 2.2
Ionization (Binding) Energies (meV) of Donor and Acceptor Energies in Si and Ge [101]

Semiconductor	P	As	Sb	B	Al	Ga	In
Si	45.5	53.7	42.7	45	67	72	100
Ge	12.76	14.04	10.19	10.3	10	10.8	11.4

---

* $3.3.10^{20}$/cm^3 at 1100°C and 1.2×10^{21}/cm^3 for phosphorus at the same temperature.

TABLE 2.3

Experimental Values of the Binding Energies (in meV) of Donor Impurities in Compound Semiconductors [101]

	Si	Ge	S	Sn	Te	In	Al	Cl
GaAs	5.84	6.0	5.87	6.0				
InSb					0.6			
CdTe						14		
ZnSe							26.3	26.9

TABLE 2.4

Experimental Values of the Binding Energies (in meV) of Acceptor Impurities in Compound Semiconductors (*) GaP Is an Indirect Gap Semiconductor [101]

	C	Be	Mg	Si	Zn	Cd	Li
GaAs	26	28	28	35	31.0		
GaP (*)					64.0		
InSb						~ 10	
ZnSe							114

In the case of II–VI and III–V compound semiconductors, many of which are direct gap materials, doping involves lattice sites of one of the two sublattices or of both, when amphoteric doping does occur. Unlike elemental semiconductors, where the sp^3 binding is fully covalent and the energy gap is indirect, bonds of compound semiconductors present an ionicity fraction different from zero, which depends on the electronegativity of the elements.

The ionization (binding) energies (meV) of donor and acceptor dopants in some, important, compound semiconductors are reported in Tables 2.3 and 2.4

Taking GaAs as a typical example of compound semiconductors, its doping with Si from a silicon solid source, under full equilibrium conditions for stoichiometry, native defects concentration, and impurity ionization, could be described by extending the thermodynamic analysis of the doping process from a liquid phase given by Teramoto [107, 108] to doping from a solid phase

$$Si(s) + V_{Ga} \rightleftharpoons Si_{Ga}$$

$$Ga_{Ga} \rightleftharpoons Ga_{surf} + V_{Ga}$$

$$Si(s) + Ga_{Ga} \rightleftharpoons Ga_{surf} + Si_{Ga} \qquad (2.74)$$

with the formation of a non-stoichiometric GaAs phase, with the Ga excess sitting at its surface. Stoichiometry conditions are obtained if

$$Ga_{surf} \rightleftharpoons Ga_{Ga} + V_{As} \tag{2.75}$$

with the formation of a vacancy in the As-sublattice, leading to the final equilibrium conditions for Si-doping of stoichiometric GaAs

$$Si(s) \rightleftharpoons Si_{Ga} + V_{As} \tag{2.74 bis}$$

and

$$Si_{Ga} \rightleftharpoons Si_{Ga}^+ + e \tag{2.76}$$

where Si_{Ga} behaves as a donor.

Similarly, the Si-doping in the As-sublattice and its ionization is described by the following equations

$$Si(s) + As_{As} \rightleftharpoons As_{surf} + Si_{As} \tag{2.77}$$

$$As_{surf} \rightleftharpoons As_{As} + V_{Ga} \tag{2.78}$$

$$Si(s) \rightleftharpoons Si_{As} + V_{Ga} \tag{2.77 bis}$$

and

$$Si_{As} \rightleftharpoons Si_{As}^- + h \tag{2.79}$$

where Si_{As} behaves as an acceptor.

It could be supposed that the processes running with Equations 2.74 bis and 2.77 bis do occur simultaneously, with relative rates which should depend on temperature.

It has been, in fact, shown by Vazquez-Cortas et al. [109] that in thermally annealed MBE-grown GaAs samples Si is mainly incorporated in Ga sites at 450°C, mainly in As sites at 500°C, and above 500°C in both Ga and As sites, leading to self-compensated samples with different degrees of self-compensation.

More information on the temperature dependence of the concentration of vacancies in GaAs, and on the correlated diffusion of Si in GaAs, comes from the work of Ahlgren et al. [110], who show that only above 800°C, where Si sits in As positions, the solubility C_{Si} (at cm^{-3}) of Si in substitutional positions, deduced from diffusion experiments, follows an exponential relationship

$$C_{Si} = \left(A + B \times 10^7 \exp - \frac{C}{kT} \right) 10^{18} \tag{2.80}$$

where $A = 5.5$, $B = 5.7$, and $C = -1.44$ eV (= 138.9 kJ/mol), which is the enthalpy of solution of silicon in substitutional As sites of GaAs.

This equation leads to Si solubilities in As sites of 10^{19} cm^{-3} at 800°C and 10^{20} cm^{-3} at 1000°C.

The experimental access at the direct observation of dopant impurities and defects is difficult, due their low concentration, though cross-sectional scanning tunneling microscopy (CSTEM) has been successfully applied to observe vacancies in several bulk III–V compounds, after exposure of core defects to the surface by cleavage and STM imagining [111]. AFM image analysis and secondary-ion mass spectroscopy (SIMS) were also used in order to look at the Si incorporation in Si-doped GaAs films [112].

2.4.2 THERMODYNAMICS OF DOPING OF NANOSTRUCTURED SEMICONDUCTORS

Understanding doping effects in semiconductor nanostructures is still a major challenge, since dopants might be incorporated both at their surfaces and in their cores, with a coupling of doping effects at the surface and quantum confinement in the core [113]. This would lead to inhomogeneous dopant distribution, and to self-purification effects, due to the irreversible migration of the impurities toward the surface, as shown by Dalpian and Chelikowski [114] and by Yuan et al. [115] for the case of p-type doped Ge NCs, where the B atoms preferentially sit at the surface of the NCs, rather than at the NC core.

We expect, also, that in addition to an inhomogeneous dopant distribution, the dopant at the surface would present different physical and electronic properties in comparison to those of the core, due a relaxed bond configuration and to the presence of dangling bonds [104].

Last but not least, there is a limited possibility of the experimental localization of the dopants within the core and the surface, due to intrinsic limitations of the experimental approaches available [116] as well as an almost total absence of experimental evidence about the role of point defects and of the size- and shape-dependent equilibria between impurities and point defects in semiconductor nanostructures.

We are, therefore, almost totally indebted to theoretical computation methods for our present knowledge on these issues.

As a first example, B- and P-doping thermodynamics of Si NCs has been theoretically discussed by Cantele et al. [117], who show that a B atom in a substitutional position is slightly displaced along the <111> direction when it is placed in the center of a Si cluster, thus having only three Si–B bonds of equal length, and that the bond lengths depend on the cluster size, as is shown in Figure 2.16, with the three shorter bonds (squares in the figure) which do decrease with the decrease of the size.

They calculated, as well, the size dependence of the formation energies ΔE_f of a neutral impurity X in substitutional positions of a silicon matrix

$$\Delta E_f = E(\text{Si}_{n-1}X) - E(\text{Si}_n) + \mu_{Si} - \mu_X \qquad (2.81)$$

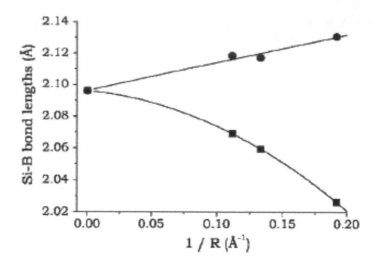

FIGURE 2.16 Dependence of the B–Si bond lengths with the decrease of the cluster size. The bond length at $1/R=0$ corresponds to the B–Si bond length in bulk silicon ($R=2.367$ Å). (*Reproduced with the permission of American Physical Society, after* G. Cantele, E. Degoli, E. Luppi, R. Magri, D. Ninno, G. Iadonisi and S.Ossicini (2005) First-principles study of n- and p-doped silicon nanoclusters *Phys. Rev.* B **72** 113303 RNP/19/APR/014373 *License date: Apr 29, 2019.*)

Following the scheme of Equations 2.70 and 2.72 for B- and P-doping, the formation energy is defined and calculated as the energy needed to insert a dopant atom in a silicon cluster* after the removal of a silicon atom (i.e. formation of a vacancy) transferred to a chemical reservoir, assumed to be the cluster surface, with μ_{Si} and μ_X as the chemical potentials of Si and of the impurity X.

The results of this calculation are reported in Figure 2.17 for B and P dopants in silicon.

It is apparent that the formation energies do increase with the decrease of the size in both the case of B (left size of the figure) and of P and that relaxation effects are very important in the case of B, for which the relaxed configuration implies that all four Si–B are equal in length.

In both cases the conditions of the minimum energy of the cluster are obtained when the impurities segregate at the surface of the cluster, with calculated energy variations (0.35 eV) as a B impurity moves to a Si layer just below the surface, associated with negligible variations of the core energy. This result indicates that atomic relaxation around the impurities is easier at the surface, as could be also qualitatively assumed.

Similar results should hold for the formation energies of a neutral impurity in both the sublattices of a compound semiconductor, with their increase with the decrease of the size of the nanocrystal.

---

* Which is hydrogen saturated at the surface in the paper of reference [117].

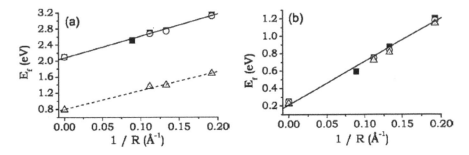

FIGURE 2.17 Size dependence of the formation energy of a substitutional B (a) and P (b) impurity in a silicon cluster. Empty triangles Δ display the formation energies when relaxation effects have been accounted for. (*Reproduced with permission* of the *American Physical Society*, after *et* G. Cantele, E. Degoli, E. Luppi, R. Magri, D. Ninno, G. Iadonisi and S. Ossicini (2005) First-principles study of n- and p-doped silicon nanoclusters *Phys. Rev. B* **72** 113303 *RNP/19/APR/014373 License date: Apr 29, 2019.*)

For the case of an impurity X (Si as an example) in the Ga-sublattice or in the As-sublattice of GaAs, the formation energy should be given by the sum of the formation energy of the impurity and of the formation energy of the vacancy

$$\Delta E_f = E(Ga_{n-1}X) - E(Ga_n) + \mu_{Ga} - \mu_X + \Delta E_f^{V,As} \qquad (2.82)$$

$$\Delta E_f = E(As_{n-1}X) - E(As_n) + \mu_{As} - \mu_X + \Delta E_f^{V,Ga} \qquad (2.83)$$

For a non-stoichiometric compound, the formation energy of a neutral impurity should be, instead, calculated without accounting for the presence of vacancies.

This is the case discussed by Dalpian and Chelikowski [114] who calculated the formation energy of a neutral impurity (Mn) in a Cd-rich, CdSe lattice using the following equation

$$\Delta E_f = E_T^{Mn} - E_T + (E_{Cd} + \mu_{Cd}) - (E_{Mn} + \mu_{Mn}) \qquad (2.84)$$

where E_T^{Mn} is the total energy of the nanocrystal with the impurity, E_T is the total energy of the nanocrystal without the impurity, E_{Cd} and E_{Mn} are the energies of elemental Cd and Mn, and μ_{Cd} and μ_{Mn} account for the specific thermodynamic conditions of the impurities in the CdSe lattice.

The results of this calculation are displayed in Figure 2.18, which shows that the formation energy increases dramatically with the decrease of size toward a few nm.

To conclude, it is important to anticipate that a common trend exists between the calculated size dependence of the formation energies of substitutional impurities and the calculated size dependence of the formation energy of native point defects, an issue which will be discussed in Section 2.6, and foresees the decrease of the role of point defects in nanostructures

Achieving damage-free, uniform, abrupt, ultra-shallow junctions while simultaneously controlling the doping concentration on the nanoscale is an ongoing challenge to the scaling down of electronic device dimensions.

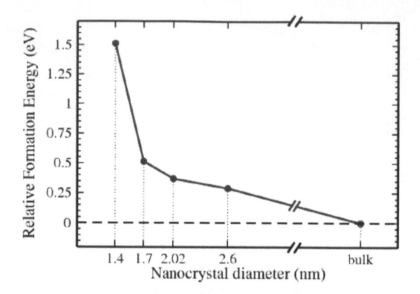

FIGURE 2.18   Variation of the formation energy of Mn impurities in CdSe with the decrease of the NC diameter. (*After* G.M. Dalpian and J.R. Chelikowsky (2006) Self-purification in semiconductor nanocrystals *Phys.Rev.Lett.* **96,** 226802 Reproduced with permission of American Physical Society *RNP/19/DEC/021046. License date: Dec 09, 2019.)*

As an example, 30% of the atoms of an undoped CdSe nanocrystal of 5 nm in diameter, which hosts $\simeq$2,400 atoms, sit at the outer surface, where they are exposed to interaction with the atmosphere or with the solvent embedding them, with a relaxed bonding configuration in comparison to the core atoms. It is well-understandable that dopants at the surface will have different physical and electronic properties of core dopants, leading to substantial dopant inhomogeneity [104].

2.5   NANOTHERMODYNAMICS

The availability of a substantial amount of experimental knowledge about the thermodynamic properties of the matter at the nano-range stimulated, in very recent years, increasing interest in the theoretical modeling of thermodynamic properties of nanoparticles and on their size dependence, limited to an average size of a few hundreds of atoms, i.e. in the range of nanometric dots, with DFT studies and applied to ensembles of 10^6 atoms (i.e. a size of several nanometers) with semiempirical tight binding (TB) calculations.

As an example, semiempirical TB models have been applied to calculate the electronic properties of surface-passivated and uncapped nanometric semiconductors [118] while total energy or Monte Carlo calculations have been used to forecast the thermodynamic properties of nanometric semiconductors.

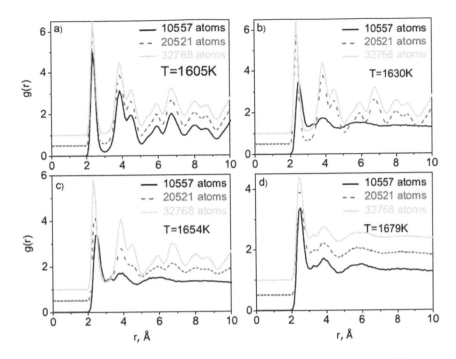

FIGURE 2.19 Radial distribution functions for three silicon samples at different temperatures. (*After* N.T.T. Hang (2014) Size dependent melting of silicon nanoparticles *Communications in Physics,* **24** (3) 207–215 *Free E-Journal.*)

Instead, Hang [119] carried out total energy calculations using the Stillinger–Weber potential to account for the atomic interactions in three silicon crystal samples consisting of 10,557, 20,521 and 32,768 atoms, corresponding to an average size of 7, 6, and 5 nm.

The calculated radial distribution functions, see Figure 2.19, show that a sudden change of the heat capacity is observed between 1608 and 1670 K, corresponding to the solid–liquid transition, i.e. to the melting of the nanoparticle.

It shows, in addition, a limited, but defined dependence of the transition temperature on the size of the sample, which decreases, in fact, with the decrease of the sample size, although with relatively small differences.

There is, however, the evidence that the computed melting temperature of these silicon nanoparticles having a size very close to the critical radius (3.9 nm for Si, see Table 2.1) lies very close to the melting point of bulk silicon, while it should range around 800 K (see Figure 1.1).

It is, therefore, apparent that the structures of these samples have not been properly relaxed to assume their equilibrium configurations.

The work of Dalgik and Domenikeli [120], who carried out MD simulations of the size dependence of the melting temperature of tin particles, gives more reliable results, which fit well with the known experimental values, as can be seen in

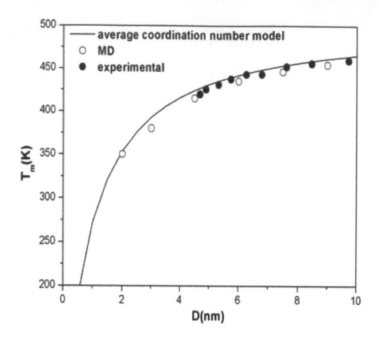

FIGURE 2.20 Size dependence of the melting point of tin. (*After* S.S. Dalgic, U. Domenikeli (2009) Melting properties of tin nanoparticles by molecular dynamics simulation *J. Optoelectron. Advanced Mater.* **11**, 2126–2132 *Free E-Journal.*)

Figure 2.20, showing the potentiality of MD calculations extended to the nanostructures, provided the system is properly relaxed toward the equilibrium configuration.

S. Xiao et al. [29] applied MD simulations to compare the melting temperature of polycrystalline (PC)–Ag, that consists of *infinite** ensembles of nanoparticles with a mean grain size ranging from 3.03 to 12.12 nm, with the melting temperatures of isolated Ag nanocrystals. The simulation has been carried out with the modified analytics embedded atom method (MAEAM), already successfully used for other nanoparticles and bulk materials.

Here, the equilibrium configuration of the NC–Ag system at 0 K is preliminarily obtained by an annealing process at 300 K, followed by a cooling step down to 0 K.

The melting conditions are eventually obtained in correspondence with the sudden change of the mean atomic configuration energy during a heating cycle, observed when the system turns to the liquid state.

They found that melting starts at grain boundaries and that the calculated melting temperatures of Ag nanocrystals, Figure 2.21,[†] are significantly in excess of the experimental melting temperatures of isolated silver nanoparticles of 4 nm (T = 723 K) measured by Asoro et al. [121], or 780 K measured by Nanda et al. [122], still decreasing with their mean size.

* Original term of the Xiao paper.
† The melting temperature of bulk Ag is 1,234 K.

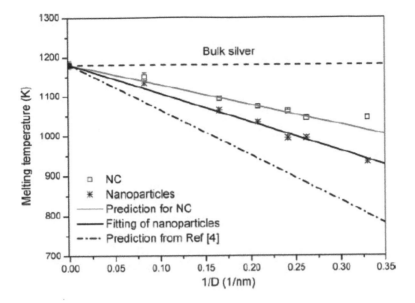

FIGURE 2.21　MD simulated size dependence of the melting temperature of NC-AG and nc-Ag (Ref.4 in the Figure corresponds to Ref.107 in this chapter). (*Reprinted with permission from* S. Xiao, W. Hu, J. Yang (2005) Melting behaviors of nanocrystalline Ag, *J. Phys. Chem. B* **109** 20339–20342 *Copyright (2005) American Chemical Society.*)

The qualitative explanation given is that the coordination number of atoms at GBs is higher than at the surface of the isolated nanoparticles and the interfacial energy σ_{GB} of GBs is, then, less than the surface energy of nanoparticles, a feature of considerable interest for the defect-induced behavior of nanometric crystalline arrays, with which will be dealt in the next chapter.

J.A. Purton et al. [123] used a Monte Carlo exchange technique to simulate the thermodynamic properties of the ternary $Mg_xMn_{1-x}O$ system with samples from 1,780 to 21,953 atoms.

The enthalpy of formation of the solid solution ΔH_f

$$xMgO + (1-x)MnO \rightleftharpoons Mg_xMn_{1-x}O \qquad (2.85)$$

was calculated and shown to depend on the size of the system; see Figure 2.22.

Also, the consolute temperature at which the ternary systems split in two phases depends on the size and decreases with the particle size as is shown in Figure 2.23.

Pohl et al. [73] developed a Monte Carlo model capable of accounting for the segregation phenomena occurring at the nanoparticle surface and of calculating the equilibrium thermodynamics of particles of binary alloys.

The MC method was applied to metallic Pt–Rh particles consisting of 9,201, 2,075, and 807 atoms, corresponding to diameters of 7.8, 4.3, and 3.1 nm, fixing the cluster shape during the simulation, and the results were compared to the calculated diagram of bulk Pt–Rh alloys, showing a broadening of the concentration range of ordered phases, attributed to the role of the surface of the nanoparticles which work as a reservoir for excess atoms.

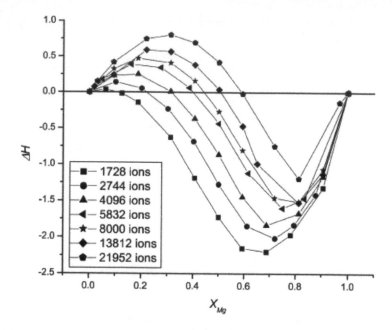

FIGURE 2.22 Calculated values of the enthalpy of formation of the solid solution $Mg_xMn_{1-x}O$. (*After* J.A. Purton, C.S. Parker and N.L. Allan (2013) Monte Carlo simulation and free energies of mixed oxide nanoparticles *Phys. Chem. Chem. Phys.*, **15** 6219–6225 *Open Access Journal.*)

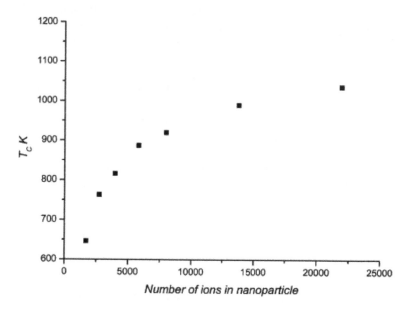

FIGURE 2.23 Size dependence of the consolute temperature. (*After* J.A. Purton, C.S. Parker and N.L. Allan (2013) Monte Carlo simulation and free energies of mixed oxide nanoparticles *Phys. Chem. Chem. Phys.*, **15** 6219–6225 *Open Access Journal.*)

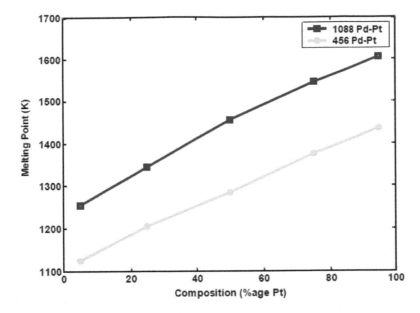

FIGURE 2.24 Calculated size- and composition-dependence of the melting temperature of Pd–Pt alloys. (*Reproduced with permission of APS after* KRSS. Subramanian, R.B. Venkat and J. Babu (2005) A molecular dynamics simulation study on melting of Pd-Pt nanoclusters *Phys,Rev.B* **71** 195415 *License Number: RNP/19/DEC/021047 License date: Dec 09, 2019.*)

An *ab initio* method has been, instead, developed and used to calculate the thermodynamic properties of spherical Cu and Cu_xAg_{1-x} alloys with a size of 20 Å by I. Golovnev et al. [124, 125].

The results show that the calculated thermodynamic properties of the core of the copper nanoparticles coincide with those of the corresponding bulk materials and that the surface atoms are responsible for the significant calculated excess of the total energy of nanoparticles with respect to bulk material.

Eventually, Subramanian et al. [126] used a molecular dynamics model to study the melting properties of Pd–Pt nanoclusters,* consisting of 456 and 1,088 atoms. Monte Carlo simulations with the Quantum Sutton–Chen potential were used to generate minimum energy configurations, with a calculated initial configuration of the Pd–Pt clusters corresponding to a Pd-rich surface with a Pt-rich core.

The thermodynamic stability of this configuration is associated with a lower solid surface energy [2.027 J/m² vs. 2.482 J/m²] and a lower cohesive energy of Pd compared to Pt [127].

In agreement with a previous study of Lewis [128] melting begins at the Pd-rich surface followed by melting of the Pt-rich core.

Therefore, the calculated melting temperatures correspond to the melting temperatures of the Pt-rich core and depend on the alloy composition and on the size of the clusters, as is shown in Figure 2.24. The calculated melting temperatures of

---

* Bulk Pd–Si alloys form a continuous series of ideal solutions.

Pd–Pt NPs are well in the range of the experimental and theoretical values of melting temperatures of Pt–Pd nanoparticles given by Guisbiers et al. [127], and the almost linear variation with the composition can be explained with the almost linear change of size of the Pt core with composition.

This behavior is expected to represent a general feature of multinary alloys at the nanoscale, and a new type of equilibrium defectivity [129].

2.6 THERMODYNAMICS OF POINT DEFECTS AT THE NANOSCALE

2.6.1 INTRODUCTION

It is matter of consolidated experimental and theoretical knowledge that point defects (vacancies and self-interstitials) are thermodynamically stable species in ionic, metallic, semiconducting, and organic solids, different from extended defects (dislocations, stacking faults, grain boundaries), that are non-equilibrium species.

It is also well-known that point defects interact with impurities with the formation of defect–impurity complexes, thus reducing the concentration of the unreacted species.

While we could expect that point and extended defects would be present also in nano-structures, their very nature (configuration and concentration, density in the case of extended defects) should depend on the type, structure,* and composition of the nanostructure involved (nanocrystals and colloidal crystals, nanowires, nano-crystalline films, nanocrystals embedded in a porous matrix) and on their growth procedure and temperature.

Furthermore, the large amount of "surface" material, with undercoordinated and strained bonds, that, possibly, could also be environmentally reacted,† would lead to a predominant role of surface defects with respect to "core"‡ point defects in all kind of elemental nanostructures.

Since in multinary alloys at the nano-size, surface segregation would lead to core-shell structures, with various mass amounts and different compositions, both the core and shell material would also present different amounts of point defects and impurities in lattice sites.

Similar considerations hold for the case of nanostructured elemental semiconductor films, where two different structural phases could coexist in metastable local equilibrium, each saturated with a different concentration of point defects (and impurities).

Unlike the case of bulk materials, where experimental results concerning the concentration§ of point defects are available for a number of metals and semiconductors,¶ in the case of nanostructured materials there is still a practical absence of experimental

---

* We will see that nanowires can present different equilibrium phases.
† We can expect surface oxidation or hydrogenation phenomena, depending on the growth/operating atmosphere.
‡ "Core" point defects take here the meaning of defects present in the core of a nanoparticle.
§ And therefore of their formation enthalpies.
¶ The present author has recently reviewed the state-of-the-art knowledge of defect thermodynamics in semiconductors of the fourth group [130].

results concerning the thermodynamic properties of thermal defects, due to a true characterization bottleneck [131].

In absence of consolidated knowledge, the use of a pure phenomenological approach could bring some preliminary light to few general issues concerning defects in nanostructures.

2.6.2 PHENOMENOLOGICAL ASPECTS

Already in the 1950s, the low concentration of point defects at the surface of small crystals was considered by Turnbull [132] at the origin of their growth rates in a liquid medium, which apparently was orders of magnitude lower than that of large crystals.

Although Turnbull ignored that extended defects would also influence the growth rate, Chen et al. [133] and Richter et al. [134] were able to demonstrate by their TEM characterization that Pd-, and Cu-, Au-Al, and Si-nanowiskers with a diameter down to a few nm could be grown dislocation-free.

The defect scarcity seems to be, therefore, a peculiar property of nanophases, almost independently of their nature.

The hypothesis that nanostructures are virtually point defect-free could also be proven under the assumption that the formation enthalpies of vacancies and self-interstitials in NC would have values comparable with those known for the corresponding macroscopic systems.

In fact, for the case of silicon, if we take for the equilibrium concentration of self-interstitials at 800°C the largest value quoted in literature (10^{12} cm^{-3}), and for that of vacancies a value of 10^{15} cm^{-3} [135], we should find an unmeasurable defect concentration of self-interstitials and few vacancies in the core of a silicon nanocrystal with a size less than 100 nm and a volume of ca 4×10^{-15} cm^{-3} ($N = 2 \times 10^{8}$).

Therefore, the scarcity of thermal defects in nanostructures depends on the high values of their formation energies and on the limited space where are confined, and statistically distributed. The formation energies of point defects in NCs, and, thus, also their concentration, could be evaluated from the values of the bulk counterparts if a law describing their size dependence could be theoretically foreseen, in analogy with the size dependence of other thermodynamic properties of nanostructures, which we have discussed in previous sections.

We should, however, not ignore that the formation energy of "core" defects would possibly differ from that of the surface ones [136], following the stimulus of Knauth and Tuller [137] who argue that the lower stability of nanocrystalline TiO_2 in a reducing atmosphere, and thus the lower Gibbs free energy of reduction, might be due to a lower formation energy of surface defects.

Although there is no theoretical reason to suppose that the formation enthalpies of defects in thermal equilibrium at the nanoscale would not be size-dependent, we will see that an unambiguous, theoretical answer to this question [16, 138, 139] and about the influence of quantum-size effects on these properties [18] is still lacking in the literature.

We will discuss details of this matter in the last sections of the chapter.

2.6.3   THERMODYNAMICS OF DEFECTS AT THE NANOSCALE

Although the application of thermodynamics to point defects at the nano-size deserves some concern, we have already seen that it has given consolidated results for many thermodynamic properties down to a size of several nanometers.

The basic assumption made when defect thermodynamics has been developed at the bulk level is that defects are diluted,* and that the ideal mass law is followed. Under these conditions it is assumed that in the presence of defects, both the configurational ΔS_{conf} and the vibrational entropy ΔS_{vibr} of a solid as well as its enthalpy do increase, and that the increase of the Gibbs free energy due to disorder, at a fixed temperature, is given by the equation [140]

$$\Delta G_f = \Delta H_f - T\Delta S_{\text{conf}} - T\Delta S_{\text{vibr}} \qquad (2.86)$$

The configurational entropy due to disorder $\Delta S_{\text{conf}} = k\ln W$, where W is the complexion number, could be calculated with the Stirling formula

$$\Delta S_{\text{conf}} = k\left\{N\left(\ln N - 1\right) - \left(N - n\right)\left[\ln(N-n) - 1\right] - n\left(\ln n - 1\right)\right\} \qquad (2.87)$$

and the corresponding change of free energy associated with the formation of n defects over N lattice positions could then be obtained by minimization of the equation

$$\frac{\delta\Delta G}{\delta n} = \Delta G_d - T\frac{\delta\Delta S_{\text{conf}}}{\delta n} \qquad (2.88)$$

where $\Delta G_d = \Delta H - T\Delta S_{\text{vibr}}$ is the molar Gibbs free energy of formation of a defect and

$$\frac{\delta\Delta S_{\text{conf}}}{\delta n} = k\left(\frac{\delta N}{\delta n}\ln\frac{N}{N-n} + \ln\frac{N-n}{n}\right) \qquad (2.89)$$

If $N \gg n$ (i.e. the system is a dilute solution of defects)

$$\frac{\delta\Delta S}{\delta n} = k\ln\frac{N}{n} \qquad (2.90)$$

$$\frac{\delta\Delta G}{\delta n} = \Delta G_d + kT\ln\frac{n}{N} \qquad (2.91)$$

and

$$x_D = \frac{n}{N} = \exp-\frac{\Delta G_D}{kT} = B\exp-\frac{\Delta H_f}{kT} \qquad (2.92)$$

where x_D is the molar fraction of defects (taking N as the number of lattice positions mol^{-1}), B is a constant which accounts for the entropy contribution to the free energy,

---

* No interaction among them.

and ΔH_f is the formation enthalpy of the defect in bulk conditions, which in the physical literature is generally called the energy of formation of the defect.

The corresponding value of defect concentration c_D in terms of atoms cm^{-3} could be easily derived from their molar fraction, accounting for the atomic density of the material, that for silicon holds $5 \cdot 10^{22}$ cm^{-3}.

In the case of vacancies, for which $\dfrac{\delta N}{\delta n} = 1$ because we have one excess atom (at the surface) for each vacancy formed, and $N = N^o + n$

$$\frac{\delta \Delta S}{\delta n} = k\left(\ln \frac{N^o + n}{N^o} + \ln \frac{N^o}{n} \right) = k\left(\ln \frac{N^o + n}{n} \right) \qquad (2.93)$$

and

$$x_D = \frac{n}{N^o + n} = B \exp - \frac{\Delta H_f}{kT} \qquad (2.94)$$

The enthalpy of formation of the defects in Equation 2.94 should be a function of the nature and size of the nanoparticle.

According to the Guisbiers model [56, 141] the enthalpy of formation of the vacancies, as an example, does follow a general phenomenological $1/D$ law given by the equation*

$$\frac{\Delta H_{f,V}}{\Delta H_{f,V,\infty}} = \left[1 - \frac{\alpha_{\text{shape}}}{D} \right] \qquad (2.95)$$

where $\Delta H_{f,V,\infty}$ is the formation enthalpy of vacancies in the bulk, D is the size of the dot, $\alpha_{\text{shape}} = \left[\dfrac{D^*(\sigma_s - \sigma_l)}{\Delta H_{m,\infty}} \right] A/V$, D^* is an adjustable parameter (in nm), σ_s and σ_l are the surface energies of the solid and liquid phases, ΔH_m^∞ is the melting enthalpy of the bulk phase, and the ratio A/V (in nm^{-1}) accounts for the particle shape and for the structural inhomogeneity of the particle.

According to this model, the formation energy decreases with the size, and Figure 2.25 displays the calculated size dependences of the energy of formation of vacancies in gold nanoparticles, according different models, details of which are given in reference 57.

It forecasts a severe decrease of the formation energy of vacancies with respect to its bulk value [0.95 eV], in good agreement with the results of other models, all of which are based on a scaling law, with which we have already dealt.

Furthermore, from Equations 2.93 and 2.94 we obtain the size-dependent average vacancy concentration

$$x_V = x_{V,\infty} B \exp \frac{\Delta H_{f,V,\infty} \alpha_{\text{shape}}}{kTD} \qquad (2.96)$$

that increases with the decrease of the size.

---

* An analogous one should hold for self-interstitials.

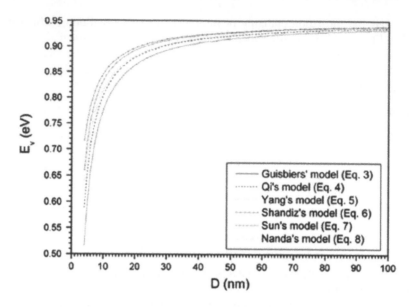

FIGURE 2.25 Size dependence of the energy of formation of vacancies in a gold nanoparticle. (*After* G. Guisbiers (2010) Size-dependent materials properties: Toward a universal equation *Nanoscale Res. Lett.* **5** 1132–1136. *Open Access Journal.*)

Figure 2.26 displays the c_v/c_V^o ratios calculated for several metallic and semiconductor nanoparticles at 298 K, which, in fact, increase up to a factor 10^6 with the decrease of size and diverge for D values smaller than 20 nm

A different approach is followed by Müller and Albe [142] who note that the automatic extension of the models adopted for the evaluation of the size dependence of the melting temperature or the melting enthalpy, to the evaluation of the size dependence of the enthalpy of formation of vacancies might lead to erroneous results, since these models ignore the physical role of the surface in the formation process of a vacancy.

The process of formation of vacancies,* in fact, involves both the core of the nanoparticle where the vacancy is formed and its surface, where an atom is transferred

$$A_A^c \rightleftharpoons A_A^s + V_A^c \tag{2.97}$$

with an increase of the amount of surface (and radius) at the expense of surface energy, which is, however, negligible in the case of macrosystems.

Therefore, the energy of formation E_{fV} of a thermal vacancy in the core of a metallic[†] NC is given by the following equation

$$E_{f,V} = E^* - E_b + \mu = E^* - E_c + \mu = E_{f,V}^o + \mu \tag{2.98}$$

where E^* is the energy of a volume of the core of the NC containing a vacancy, E_b $\approx E_c$ is the energy of a volume of the crystal entirely filled by component atoms,

---

* Not that of self-interstitials.
[†] Where vacancies are the prevailing defects.

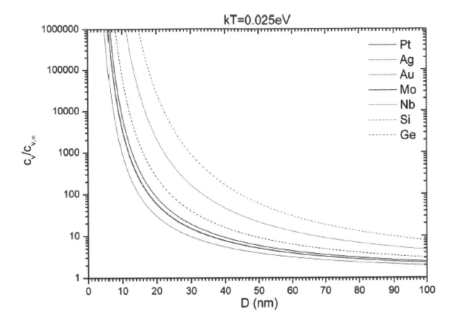

FIGURE 2.26 Calculated dependence of the relative vacancy ratios of several metallic and semiconductor nanoparticles at 298 K. (*Reprinted with permission from G. Guisbiers (2011) Schottky defects in nanoparticles J.Phys.Chem. C 115 2616–2621 Copyright {2011} American Chemical Society.*)

E_c is the cohesive energy, and $\mu = \dfrac{\delta E(N)}{\delta N}$ is the chemical potential of N atoms, that accounts for the work done to insert the atom removed from the bulk to create the vacancy in a reservoir, best represented by the surface.

In turn, the work done to bring the atom removed from the bulk to the surface of a sphere of a radius R is given by the sum

- Of a term $dw = \sigma\,dA = \dfrac{2\sigma\Omega}{R}$, accounting for the creation of an additional surface area $dA = \dfrac{2\Omega}{R}$, where σ is the surface free energy per unit area and Ω is the atomic volume, which results in an additional increase of the dot volume, and therefore of the radius R.
- Of a term that accounts for the work associated with the expansion of the volume against the pressure, associated with the creation of the additional surface area and therefore of radius, which holds $dw = \dfrac{2f\Delta V}{R}$, where f is the area-weighted mean surface stress and ΔV is the volume expansion.

Therefore, the formation energy of vacancies and their thermal equilibrium concentration is given by the following equations

$$E_{f,V} = E^o_{f,V} + \frac{2(\sigma\Omega + f\Delta V)}{R} \qquad (2.99)$$

$$c_V = c^o_V \exp-\left[\frac{2(\sigma\Omega + f\Delta V)}{kTR}\right] \qquad (2.100)$$

where $E^o_{f,V}$ is the bulk formation energy of vacancies and their thermal equilibrium concentration is c^o_V.

In a nanocrystal, the weight of the term $2(\sigma\Omega + f\Delta V)R^{-1}$ of Equations 2.99 and 2.100 might be large and the calculated values of the vacancy concentration decrease with the decrease of the size, see Figure 2.27, while the calculated energy of formation of a vacancy linearly increases with the decrease of R (increase of R^{-1}), as it is shown in just the opposite to the Guisbiers model, but in excellent agreement with calculated values of vacancy concentrations in Cu NPs, see Figure 2.28, using an algorithm developed by Bortz et al. [143].

This work has been recently revisited by Salis et al. [144], who accounted also for the formation of vacancies at the surface of a nanoparticle, whose formation energy is given by the equation

$$E^S_{f,V} = \eta - \frac{\sigma\Omega}{a} + \frac{2\sigma\Omega}{R} = \chi_{\text{surf}} + \frac{2\sigma\Omega}{R} \qquad (2.101)$$

where η is work done to create a vacancy, $\sigma\Omega/a$ is the energy gained in the removal of a surface atom from a surface layer of thickness a, $\chi_{\text{surf}} = \eta - \dfrac{\sigma\Omega}{a}$, and the term

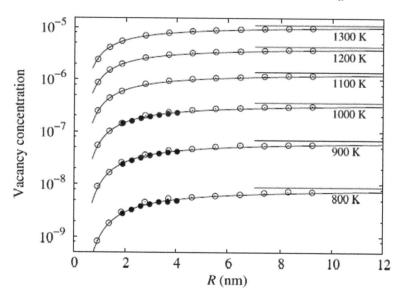

FIGURE 2.27 Calculated size dependence of the concentration of vacancies in Cu NPs. (*After* M. Müller, K. Albe (2007) Concentration of thermal vacancies in metallic nanoparticles *Acta Materialia* **55** 3237–3244 *Reproduced with the permission of Elsevier License Number 4447730700980, License date Oct 14, 2018.*)

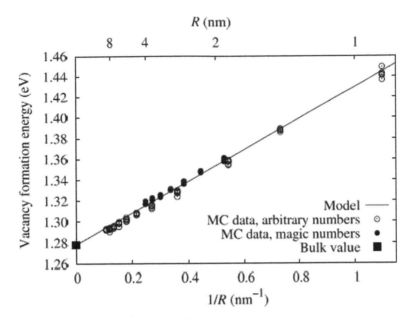

FIGURE 2.28 Calculated size dependence of the energy of formation of vacancies in Cu NPs. (*After* M. Müller, K. Albe (2007) Concentration of thermal vacancies in metallic nanoparticles *Acta Materialia* **55** 3237–3244 *Reproduced with the permission of Elsevier, License number 4447730700980. License date Oct. 14, 2018.*)

$2\sigma\Omega R^{-1}$ is the energy spent for the creation of an additional surface area $dA = 2\Omega/R$, where σ is the surface free energy per unit area and Ω is the atomic volume, which results in an additional increase of the particle volume and of the radius R.

Therefore, the vacancy concentration at the surface c_V^S is given by the following equation

$$c_V^S(R) = c_V^{OS} \exp{-\frac{\chi_{\mathrm{surf}} + 2\sigma\Omega R^{-1}}{kT}} \qquad (2.102)$$

where c_V^{OS} is the surface concentration of vacancies in the bulk crystal,* while the core concentration of vacancies is again given by Equation 2.100.

Both the surface and core concentration of thermal vacancies decrease with the decrease of the radius, but their fractional concentration does increase, with a peak at 5 nm, as is shown in Figure 2.29, with a strong dependence on the term χ_{surf}.

The crucial problem with the discussion of these results is the absence of experimental data which would permit a quantitative support, although X-ray diffraction and positrons annihilation spectroscopy (PAS), which is well-known to be selectively suited for the detection of negative and neutral vacancies and of negatively charged ion-type defects, are expected to give, at least, some limited hints.

---

* A quantity, however, which has never been taken into consideration in the literature and is experimentally unavailable.

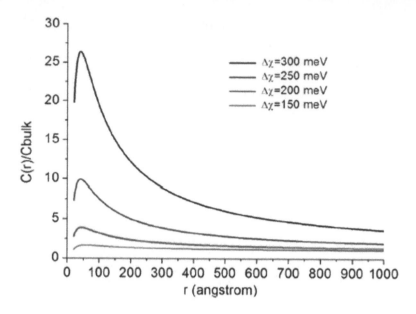

FIGURE 2.29   Calculated size dependence of the fractional concentration of thermal surface vacancies vs. bulk vacancies at 600 K, for various values of χ_{surf}. (*After* M. Salis, C.M.
Carbonaro, M. Marceddu, P.C. Ricci (2013) Statistical thermodynamics of Schottky defects
in metal nanoparticles *Nanoscience and Nanotechnology* **3** 27–33, *Reproduced under a
Creative Common Attribution 4.0 International License.*)

As an example, the effect of vacancies on the <111> and <200> XR diffraction peak intensities of a face-centered cubic crystal was modeled by Makinson et
al. [145] who showed that a significant decrease of both peak intensities could be
observed above a vacancy concentration of 4%.

Since this range of vacancy concentrations is well above the equilibrium concentration of thermal vacancies in metal and semiconductor nanocrystals, the effect is
of little interest for defects in thermodynamic equilibrium conditions.

Also the few experimental results dealing with the determination of defects in
silicon nanocrystals with positrons annihilation spectroscopy (PAS) [146, 147] show
that PAS coupled with photoluminescence spectroscopy measurements reveal only
the presence of dangling bonds in amorphous and crystalline silicon dots embedded
in an SiO_2 matrix, located at the Si/SiO_2 interface.

The presence of Zn vacancies (V_{Zn}) and their concentration decrease by high-
temperature annealing was, instead, put in evidence by PAS on nanometric (30 nm)
ZnO crystallites in polycrystalline aggregates by Wang et al. [148]. It was also shown
that V_{Zn} are not bulk species, but probably reside at the surface of the NCs or at the
GBs of the nanocrystalline aggregates.

Eventually, the presence of vacancy-type defects in silicon nanocrystals has been
indirectly demonstrated by Mokry et al. [136] with PAS and photoluminescence (PL)
measurements. Results of these measurements suggest, in fact, that the diffusion process yielding the coarsening of Si-nc depends on the concentration of vacancy-type

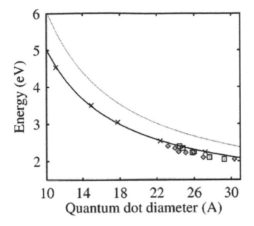

FIGURE 2.30 Experimental size-dependence of the optical gap of silicon nanocrystals (solid line) and calculated quasiparticle gap (dotted line). (*Reproduced with permission after S. Öğüt, J.R. Chelikowsky, S.G. Louie (1997) Quantum confinement and optical gaps in Si nanocrystals Phys. Rev. Lett. **79** 1770 Copyright 1997 by American Physical Society, License nr. RNP/18/NOV/009618 License date: Nov 20, 2018.*)

defects, since when the majority of these defects have been removed by annealing, the growth of the Si-nc slows down or totally stops.

2.6.4 First-Principle Studies of Point Defects at the Nanoscale

We have seen in the previous section that the formation energies of point defects in nanocrystalline semiconductors were obtained using semiempirical calculations, which postulate the transferability of the bulk physical properties to a nanocrystal environment.

While this assumption seems to be rather viable in the mesoscopic range, it is not reliable at the true nanoscopic limit, from 10 nm and below, since neglecting the changes in self-energy induced by quantum confinement would cause, as an example, the underestimation of the size dependence of the optical gap of semiconductor nanocrystals [149–151].

Instead, *ab initio* calculations show that the experimental size-dependence of the optical gap of Si nanocrystals could be described quite well, using as the model a spherical quasiparticle silicon cluster, surface passivated with hydrogen [151], although the calculated optical gap values are about 0.6 eV larger than the experimental values (see Figure 2.30).

Despite the limits, the often-excellent agreement between the phenomenological size dependence of the physical properties of metallic and semiconductor nanocrystals and the available experimental results favorably supports the use of the empirical methods* for the evaluation of the formation enthalpy of point defects and of

---

* Which formally accounts for an increasing surface contribution to the formation energy of the defects with the decrease of the size.

TABLE 2.5

Formation Energy $E_f^{q=0}$ of the Neutral Silicon Monovacancy at 0 K (Unless Specified) from ab Initio Studies

$E_f^{q=0}$ (eV)	Computational Method	Cell Size/Symmetry	References
3.529 (relaxed)	DFT/LDA	216	Wright [152]
3.486	DFT/GGA	512	
3.473		1,000	
3.97 (4.11)	TB/MD	216-(511)	Tang [153]
3.69 (relaxed)	DFT/LDA	256	Centoni [154]
3.53 (relaxed)	DFT/LDA	68/cubic	Al-Mushadani [155]
3.46 (T=1,685 K; N=1,728)	DFT/GGA)	64, 216, 512, 1,000, and 1,728	Sholihun [156]
4.13 (unrelaxed)	DFT/LDA	256	Corsetti [157]
3.4 (relaxed) N=1,000			

their size dependence from the mesoscopic to the beginning of the nanoscopic range, as we have done in the last section.

When the size of the particle under investigation is, instead, a few nanometers or less than a nanometer, *first-principles studies* should be necessarily used, but also this approach is not free of problems. One of these concerns is the size of the simulation cell used, which should be sufficiently large to avoid the effects of lattice disruption caused by the defect center, due to lattice relaxation around the defect, which is long range, and to maintain the ideal dilution conditions, necessary to calculate the formation energy of the unrelaxed and relaxed defect.

It would be, therefore, necessary to investigate on the finite size convergence of the property to be studied, as was done, as an example, by Wright [152] for the silicon vacancy in its different charge q states, using three supercells of 215, 511, and 999 atoms, whose formation energy E_f^q is conventionally calculated as the difference of the energy of a cell with the vacancy and the energy of the cell without a vacancy, i.e.

$$E_f^q = E_{N-1} - \frac{N-1}{N} E_N + q\mu_e \qquad (2.103)$$

where E_{N-1} is the energy of the cell with a vacancy y and E_N is the energy of the cell without the vacancy, μ_e is the chemical potential of the electron, and N is the number of atoms of the cell.

The results of the simulation for the relaxed neutral vacancy in the D_{2d} configuration,* which preliminarily requires the evaluation of the physical properties of the material under study (silicon in this case) and of their convergence with experimental results, are reported in Table 2.5 [152–157].

It is possible to see that according to Wright [152] the calculated formation energy of the relaxed neutral vacancy in the D_{2d} configuration decreases with the increase of N and converges to a value of 3.473 eV with a supercell of N $\geq$ 1000 atoms.

---

* In Wright's paper results for the charged vacancies are also reported.

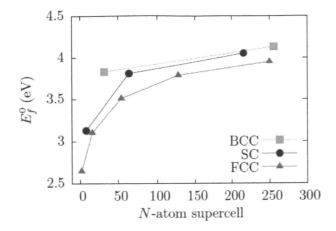

FIGURE 2.31 Convergence of the unrelaxed defect formation energy with the size of the simulation cell. (*After* F. Corsetti (2012) On the properties of point defects in silicon nanostructures from ab initio calculations *PhD Thesis, Imperial College London. February 2012 License date Nov 25, 2019.*)

This formation energy value fits reasonably well with the experimental values of Watkins (3.6 ± 0.5 eV) [158] arising from EPR experiments, of Danneafer (3.6 ± 0.5 eV) [159] from PAS experiments, and of Shimizu (3.6 eV) [160] from Raman shift experiments.

Similar results were obtained by Sholihun et al. [157] who obtained convergence conditions with a larger supercell of 1,728 atomic sites, leading to a value of the formation energy of the relaxed vacancy of 3.46 eV and a concentration of vacancies of 7.4×10^{16} cm^{-3} at the melting point of silicon, in good agreement with the work of Wright previously cited.

A value of 3.5 eV for the relaxed vacancy in the D_{2d} configuration was also obtained from an analogous work carried out by Corsetti [157–161] using a 1,000-atom supercell (see Table 2.5).

They also show (see Figure 2.31) that convergence of the formation energy of the unrelaxed vacancy is already obtained with a supercell of 256 atoms.

One can see, in fact, that with 256-atom supercells of BCC, SC, and FCC symmetry, the formation energies of the unrelaxed neutral vacancy (q=0) converge to a value of 4.13 eV, which, incidentally, fits with a recent experimental literature value of 4 eV in high-purity bulk silicon with a RT carrier concentration of $4 \cdot 10^{12}$ cm^{-3},* given by Fukata et al. [162].

Some results on the formation energies of point defects in compound semiconductors are also available. As an example, Li et al. [163] carried out *first-principles* DFT studies on point defects in lead chalcogenides PbX, where X is S, Se, and Te.

In PbX, like in all compound semiconductors, both cation and anion vacancies, and cation and anion interstitials are potentially stable

---

* The samples used were sealed in quartz capsules in an atmosphere of hydrogen.

$$Pb_{Pb} \rightleftharpoons Pb_{surf} + V_{Pb} \qquad (2.104)$$

$$Pb_{Pb} \rightleftharpoons Pb_i + V_{Pb} \qquad (2.105)$$

$$X_X \rightleftharpoons X_{surf} + V_X \qquad (2.106)$$

$$X_X \rightleftharpoons X_i + V_X \qquad (2.107)$$

as well as Frenkel pairs $V_{Pb} - V_X$, according to the reaction

$$V_{Pb} + V_X \rightleftharpoons V_{Pb} - V_X \qquad (2.108)$$

In addition, deviations from stoichiometry are common, which could be accommodated either at the surface or in the bulk of the Nc.

DFT studies were carried out on supercells of 216 atoms for each PbX material,* and the formation energies of cationic and anionic vacancies and interstitials were evaluated for the stoichiometric solids. Results of these calculations are reported in Table 2.6, that shows that Pb vacancies are the most common defects of lead chalcogenides.

In their *ab initio* work on defects in PbTe, S. Ahmad et al. [164] were also able to show that the calculated density of states (DOS) of PbTe is modified by the presence of Pb vacancies, with an increase of states close to the top of the valence band.

The effect of the surface on the core properties of metallic nanoparticles was investigated in a recent work of Gupta et al. [16], who addressed the study at the defect properties of Al nanocrystals, using DFT calculations.

Bulk values of formation energies of vacancies and divacancies were initially estimated with supercells of 32 and 108 atoms using the LDA and PEB functionals and compared with the bulk experimental values (see Table 2.7) and with LDA values coming from a work of Carling et al. [166]. The fit is apparently excellent and confirms a value of 0.69 eV for the formation energy of a vacancy in the core of an Al nanocrystal.

The analysis was then extended to the effect of the surface, using layered slabs of defective (100), (110), and (111) surface supercells, for which the surface energies

---

TABLE 2.6

Calculated Formation Energies (eV) of Defects in Lead Chalcogenides

	V_{Pb}	V_X	Pb_i	X_i
PbS	1.78 *(1.36)*	2.08 *(1.77)*	2.41	2.91
PbSe	1.83	2.15	2.03	2.65
PbTe	1.90 *(2.19)*	2.13 *(2.14)*	2.20	2.76

In italics, literature values [163]

---

* Including PbO, about which details are not reported here.

TABLE 2.7
Calculated and Experimental Values of the Formation Energies of Vacancies and Di-Vacancies in Bulk Al Using LDA and PEB Functionals

T	LDA	PEB	Exp
$E_{f,V}$ (eV)	0.68	0.63	0.69 ± 0.03 (*)
	0.70 (**)		
$E_{f,dV}$ (eV)	1.45	1.33	

After Gupta et al. [16] (*) P. Tzanetakis [165], (**) K. Carling [166]

TABLE 2.8
Calculated Formation Energies of Vacancies on the Top and on the First Sublayer of (100), (110), and (111) of Al Surfaces

$E_{f,V}$ (eV)	LDA	PBE
100-00	0.44	0.35
100-01	0.75	0.68
110-00	0.15	0.11
110-01	0.57	0.52
111-00	0.60	0.55
111-01	0.72	0.66

After Gupta et al. [16]

and the formation energies of vacancies were calculated for the top layer (00) and the first subsurface (01) layer.

It can be seen in Table 2.8 [16] that the calculated formation energies of vacancies on the top (00) surface layers are lower than the calculated bulk values and strongly depend on the structure of the surface, but that the values relative to the first sublayer (01) are already close to the calculated and experimental core values.

Eventually it was also shown that starting from the third sublayer, the vacancy formation energies converge to the core values of 0.62 eV, slightly lower than the experimental and calculated values for bulk crystals.

If we assume that the accuracy of these results is ±0.01 eV, we could assume that the surface does slightly affect the properties of the core of the Al dot, by reducing by 0.05 eV the formation energy of the vacancy.

If this conclusion could be extended to all metallic and semiconductor NCs, we could assume that the properties of the core of an NP of nanoscopic size are probably slightly affected by the presence of a disordered surface.

2.6.5 STUDY OF SILICON SURFACE DEFECTS AT THE NANOSCALE

Silicon nanoparticles present not only point defects but also undercoordination defects and chemical defects at their surface, which play a critical role in their optical properties.

As an example, Ippolito et al. [167] studied the nature of undercoordinated Si atoms at the surface and into the core of silicon nanoparticles embedded in a-SiO$_2$ by MD simulations. They found that the number and the distribution of undercoordination defects depend on the size of the particle. For particles ≤1 nm the undercoordinated defects are spread in a region involving the particle and the SiO$_2$ matrix and are 50% of the total. Most of the undercoordinated Si atoms are three-fold coordinated, but in NCs smaller than 2 nm also two-fold coordinated atoms were found. Full coordination is observed only for distances larger than 5 A from the interface and four-coordinated crystalline cores are only observed for particles ≥10 nm.

DFT studies associated with scanning tunneling spectroscopy (STS) measurements, which permitted the production and simulation of the spatial maps of the electronic density of states (DOS) associated with oxygenated surface defects, were, instead, carried out by Kislitsyn et al. [168] on surface hydrogenated NC Si samples.

They found that Si–O–Si and Si–OH defects are probably responsible for the PL emissions of sufficiently small (less than 2.5 nm) samples. The photophysical properties of larger samples are, instead, almost insensitive to these defects.

REFERENCES

1. T. L. Hill (2001) A different approach to nanothermodynamics. *Nano Lett.* **1**, 273–275.
2. D. H. E. Gross (2001) *Microcanonical Thermodynamic Ensembles: Phase Transitions in "Small" Systems*, World Scientific, Singapore.
3. F. Ercolessi, W. Andreoni, E. Tosatti (1961) Melting of small particles of gold: Mechanism and size effects. *Phys. Rev. Lett.* **66**, 911.
4. A. P. Alivisatos (1996) Perspectives on the physical chemistry of semiconductor nanocrystals. *J. Phys. Chem.* **100**, 13226–13239.
5. G. Kaptay (2011) The Gibbs equation versus the Kelvin and the Gibbs-Thomson equations to describe nucleation and equilibrium of nano-materials. *J. Nanosci. Nanotechnol.* **12**, 1–9.
6. Ph. Buffat, J. P. Borel (1976) Size effect on the melting temperature of gold particles. *Phys. Rev. A* **13**, 2287–2298.
7. J. P. Borel (1981) Thermodynamical size effect and the structure of metallic clusters. *Surf. Sci.* **106**, 1–9.
8. P. Curie (1885) Sur la formation de cristaux et la constant de capillarite de leur different phases. *Bull. Soc. Fr. Min. Crist.* **8**, 145–150.
9. G. Wulff (1901) Zur Frage der Geschwindickeit der Wachstums and der Auflosung der Krystallflächen. *Z. Kristallogr.* **34**, 449–530.
10. M. Wautelet (1991) Estimation of the variation of the melting temperature with the size of small particles, on the basis of a surface-phonon instability model. *J. Phys. D: Appl. Phys.* **24**, 343–346.

11. J. S. Rowlison (1993) Thermodynamics of inhomogeneous systems. *Pure & Appl. Chem.* **65**, 873–882.
12. W. Martienssen, H. Warlimont (2005) *Springer Handbook of Condensed Matter and Materials Data*, Springer.
13. R. Koole, E. Groeneveld, D. Vanmaekelbergh, A. Meijerink, C. de M. Donegá (2014) Size effects on semiconductor nanoparticles. In: *Nanoparticles, Workhorses of Nanoscience*, M. Donegà Ed., Springer.
14. K. Sattler (2010) *Handbook of Nanophysics: Principles and Methods*. Vol. 7, Boca Raton, FL: CRC Press.
15. F. G. Shi (1994) Size dependent thermal vibrations and melting in nanocrystals. *J. Mater. Res.* **9**, 1307–1314.
16. S. S. Gupta, M. A. van Huis, M. Dijkstra, M. H. F. Sluiter (2016) Depth dependence of vacancy formation energy at (100), (110), and (111) Al surfaces: A first-principles study. *Phys. Rev. B* **93**, 085432.
17. W. P. Halperin (1986) Quantum size effects in metal particles. *Rev. Modern Phys.* **58**, 533–607.
18. Y. Volokitin, J. Sinzig, L. J. de Jongh, G. Schmidt, M. N. Vargaftik, I. I. Moiseev (1996) Quantum-size effects in the thermodynamic properties of metallic nanoparticles. *Nature* **384**, 621–623.
19. V. Novotny, P. P. M. Meinke (1973) Thermodynamic lattice and electronic properties of small particles. *Phys. Rev. B* **8**, 4186.
20. C. R. M. Wronski (1967) The size dependence of the melting point of small particles of tin. *British J. Appl. Phys.* **18**, 1731–1737.
21. A. N. Goldstein, C. M. Echer, A. P. Alivisatos (1992) Melting in semiconductor nanocrystals. *Science* **256**, 1425–1427.
22. M. Wautelet (1998) On the shape dependence of the melting of small particles. *Physics Lett. A* **246**, 341.
23. M. Wautelet, J. P. Dauchot, M. Hecq (2000) Phase diagrams of small particles of binary systems: A theoretical approach. *Nanotechnology* **11**, 6–9.
24. R. Vallée, M. Wautelet, J. P. Dauchot, M Hecq (2001) Size and segregation effects on the phase diagrams of nanoparticles of binary systems. *Nanotechnology* **12**, 68–74.
25. K. F. Peters, J. B. Cohen, Y-W. Chung (1998) Melting of Pb nanocrystals. *Phys. Rev. B* **57**, 13430.
26. Q. Jiang, H. X. Shi, M. Zhao (1999) Melting thermodynamics of organic nanocrystals. *J. Chem. Phys.* **111**, 2176–2180.
27. Z. Zhang, X. X. Lu, Q. Jiang (1999) Finite size effect on melting enthalpy and melting entropy of nanocrystals. *Physica B* **270**, 249–254.
28. M. Zhao, X.-H. Zhou, Q. Jiang (2001) Comparing of different models for melting point changes of metallic nanocrystals. *J. Mater. Res.* **16**, 3304–3308.
29. S. Xiao, W. Hu, J. Yang (2005) Melting behaviors of nanocrystalline Ag. *J. Phys. Chem. B* **109**, 20339–20342.
30. W. H. Qi (2006) Size- and coherence-dependent thermodynamic properties of metallic nanowires and nanofilms. *Mod. Phys. Lett. B* **20**, 1943–1951.
31. J. Sun, S. L. Simon (2007) The melting behavior of aluminum nanoparticles. *Thermochim. Acta* **463**, 32–40.
32. X. J. Liu, Z. F. Zhou, L. W. Yang, J. W. Li, G. F. Xie, S. Y. Fu, C. Q. Sun (2011) Correlation and size dependence of the lattice strain, binding energy, elastic modulus, and thermal stability for Au and Ag nanostructures. *J. Appl. Phys.* **109**, 074319.
33. S. Xiong, Y. Cheng, M. Wang, Y. Li (2011) Universal relation for size dependent thermodynamic properties of metallic nanoparticles. *Phys. Chem. Chem. Phys.* **13**, 10652–660.
34. R. Kumar, G. Sharma, M. Kumar (2014) Effect of size and shape on the vibrational and thermodynamic properties of nanomaterials. *J. Thermodyn.* art.ID 328051. doi:10.1155/2013/328051.

35. H. Li, H.-J. Xiao, T.-S. Zhu, H.-C. Xuan, M. Li (2015) Size consideration on shape factor and its determination role on the thermodynamic stability of quantum dots. *J. Phys. Chem. C* **119**, 12002–12007.
36. J. Zhu, Q. Fu, Y. Xue, Z. Cui (2016) Comparison of different models of melting transformation of nanoparticles. *J. Mater. Sci.* **51**, 4462–4469.
37. H. Reiss, I. B. Wilson (1948) The effect of surface on melting point. *J. Colloid Sci.* **3**, 551–361.
38. A. C- Lawson (2009) Physics of the Lindemann melting rule. *Philos. Mag.* **89**, 1757–1770.
39. Q. Jiang, Z. Zhang, J. C. Li (2000) Melting thermodynamics of nanocrystals embedded in a matrix. *Acta Mater.* **48**, 4791–4795.
40. N. F. Mott (1934) The resistance of liquid metals. *Proc. R. Soc. A* **146**, 465–472.
41. S. L. Lai, J. Y. Guo, V. Petrova, G. Ramanath, L. H. Allen (1996) Size-dependent melting properties of small tin particles: Nanocalorimetric measurements. *Phys. Rev. Lett.* **77**, 99.
42. S. Lai, J. R. A. Carlsson, L. H. Allen (1998) Melting point depression of Al clusters generated during the early stages of film growth: Nanocalorimetry measurements. *Appl. Phys. Lett.* **72**, 1098.
43. P. Pawlov (1909) The dependency of the melting point on the surface energy of a solid body. *Z. Phys. Chem.* **65**, 545–548.
44. E. Rie (1923) Influence of surface tension on melting and freezing. *Z. Phys. Chem.* **104**, 354–262.
45. G. Guenther, R. Theissmann, O. Guillon (2014) Size-dependent phase transformations in bismuth oxide nanoparticles. II. Melting and stability diagram. *J. Phys. Chem. C* **118**, 27020–27027.
46. M. N. Magomedov (2016) Size dependence of the shape of a silicon nanocrystal during melting. *Techn. Phys. Lett.* **42**, 761–764.
47. A. N. Goldstein (1996) The melting of silicon nanocrystals: Submicron thin-film structures derived from nanocrystal precursors. *Appl. Phys. A* **62**, 33–37.
48. M. Hirasawa, T. Orii, T. Seto (2006) Size-dependent crystallization of Si nanoparticles. *Appl. Phys. Lett.* **88**, 093119.
49. K. C. Fang, C. I. Weng (2005) An investigation into the melting of silicon nanoclusters using molecular dynamics simulation. *Nanotechnology* **16**, 250.
50. Q. Jiang, J. C. Li, B. Q. Chi (2002) Size-dependent cohesive energy of nanocrystals. *Chem. Phys. Lett.* **366**, 551–554.
51. W. H. Qi, M. P. Wang, G. Y. Xu (2003) The particle size dependence of cohesive energy of metallic nanoparticles. *Chem. Phys. Lett.* **372**, 632–634.
52. S. Xiong, W. Qi, Y. Cheng, B. Huang, M. Wang, Y. Li (2011) Modeling size effects on the surface free energy of metallic nanoparticles and nanocavities. *Phys. Chem. Chem. Phys.* **13**, 10648–10651.
53. S. Xiong, W. Qi, B. Huang, M. Wang, Y. Li S. Xiong, W. Qi, B. Huang, M. Wang, Y. Li (2010) Size and shape dependent Gibbs free energy and phase stability of titanium and zirconium nanoparticles *Mat. Chem. Phys.* **120**, 446–451.
54. G. Guisbiers, I. Buchaillot (2009) Modeling the melting enthalpy of nanomaterials *J. Phys. Chem C* **113**, 3566–3568.
55. G. Guisbiers, L. Buchaillot (2009) Universal size/shape-dependent law for characteristic temperatures. *Phys. Lett. A* **374**, 305–308.
56. G. Guisbiers (2010) Size-dependent materials properties: Toward a universal equation. *Nanoscale Res. Lett.* **5**, 1132–1136.
57. G. Guisbiers (2010) α shape, birth of one universal parameter? *Key Eng. Mater.* **444**, 69–80 Trans Tech Publications, Switzerland.

58. R. Kumar, M. Kumar (2012) Effect of size on the cohesive energy, melting temperature and Debye temperature of nanomaterials. *Indian J. Pure Appl. Phys.* **50**, 329–334.
59. I. H. Liang, B. Li (2006) Size-dependent thermal conductivity of nanoscale semiconducting systems. *Phys. Rev. B* **73**, 153303–153305.
60. H.-J. Xiao, T.-S. Zhu, H.-C. Xuan, M. Li (2015) Size consideration on shape factor and its determination role on the thermodynamic stability of quantum dots. *J. Phys. Chem. C* **119**, 12002–12007.
61. Z. Zhang, J. C. Li, Q. Jiang (2000) Modelling for size-dependent and dimension-dependent melting of nanocrystals. *J. Phys. D: Appl. Phys.* **33**, 2653.
62. C.-C. Chen, A. B. Herhold, C. S. Johnson, A. P. Alivisatos (1997) Size dependence of structural metastability in semiconductor nanocrystals. *Science* **276**, 398–401.
63. C. Q. Sun, H. L. Bai, S. Li, B. K. Tay, E. Y. Jiang (2004) Size-effect on the electronic structure and the thermal stability of a gold nanosolid. *Acta Mater.* **52**, 501–505.
64. C. Q. Sun (2004) Surface and nanosolid core-level shift: Impact of atomic coordination-number imperfection. *Phys. Rev. B* **69**, 045105.
65. C. Q. Sun, Y. Wang, B. K. Tay, S. Li, H. Huang, Y. J. Zhang (2002) Correlation between the melting point of a nanosolid and the cohesive energy of a surface atom. *J. Phys. Chem. B* **106**, 10701–10705.
66. C. Q. Sun (2014) *Relaxation of the Chemical Bond*, pp. 203–221. Springer.
67. C. Q. Sun, C. M. Li, H. L. Bai, E. Y. Jang (2005) Melting point oscillation of a solid over the whole range of sizes. *Nanotechnology* **16**, 1290–1293.
68. M. Cui, H. Lu, H. Jiang, Z. Cao, X. Meng (2017) Phase diagram of continuous binary nanoalloys: Size, shape, and segregation effects. *Sci Rep.* **7**, 41990.
69. T. L. Hill (1963) *Thermodynamics of Small Systems, Part I*, W. A. Benjamic Inc. Publishers, New York.
70. T. L. Hill (2001) Perspective: Nanothermodynamics. *Nano Lett.* **1**, 111–112.
71. D. H. E. Gross, E. V. Votyakov (2000) Phase transitions in "small" systems. *Eur. Phys. J. B* **15**, 115–126.
72. M. Wautelet, A. S. Shirinyan (2009) Thermodynamics: Nano vs. macro. *Pure Appl. Chem.* **81**, 1921–1930.
73. J. Pohl, C. Stahl, K. Albe (2012) Size-dependent phase diagrams of metallic alloys: A Monte Carlo simulation study on order–disorder transitions in Pt–Rh nanoparticles. *Beilstein J. Nanotechnol.* **3**, 1–11.
74. J. Steininger (1970) Thermodynamics and calculation of the liquidus-solidus gap in homogeneous, monotonic alloy systems. *J. Appl. Phys.* **41**, 2713–2724.
75. S. Xiao, W. Hu, W. Luo, Y. Wu, X. Li, H. Deng (2006) Size effect on alloying ability and phase stability of immiscible bimetallic nanoparticles. *Eur. Phys. J. B* **54**, 479–484.
76. W. H. Qi, B. Y. Huang, M. P. Wang (2009) Size and shape dependent formation enthalpy of binary alloy nanoparticles. *Phys. B Condens. Matter* **404**, 1761–1765.
77. S. Xiong, W. Qi, B. Huang, M. Wang (2011) Size-, shape-and composition-dependent alloying ability of bimetallic nanoparticles. *Chem. Phys. Chem.* **12**, 1317–1324.
78. H. N. G. Wadley, X. W. Zhou, R. A. Jonson, M. Neurok (2001) Mechanisms, models and methods of vapor deposition. *Progr. Mater. Sci.* **46**, 329.
79. L. H. Liang, D. Liu, Q. Jang (2003) Size-dependent continuous binary solution phase diagram. *Nanotechnology* **14**, 438–442.
80. H. Müller, Ch. Opitz, K. Strickert, L. Skala (1987) Abschatzung von Eigenschaften der Materie im hochdispersen Zustand - Praktische Anwendungen des analytischen Clustermodells (ACM). *Z. Phys. Chemie (Leipzig)* **268**, 623–646.
81. D. Tomanek, S. Mukherjee, V. Kumar, K. H. Bennemann (1982) Calculation of chemisorption and absorption induced surface segregation. *Surf. Sci.* **114**, 11–22.
82. H-M. Lu, F. Q. Han, X. K. Meng (2008) Size-dependent thermodynamic properties of metallic nanowires. *J. Phys. Chem. B* **112**, 9444–9448.

83. J. Park, J. Lee (2008) Phase diagram reassessment of Ag–Au system including size effect. *Calphad* **32**, 135–141.
84. F. Monji, M. A. Jabbareh (2017) Thermodynamic model for prediction of binary alloy nanoparticle phase diagram including size dependent surface tension effect. *Calphad* **58**, 1–5.
85. M. Ghasemi, Z. Zanolli, M. Stankovski, J. Johansson (2015) Size- and shape-dependent phase diagram of In–Sb nano-alloys. *Nanoscale* **7**, 17387–17396.
86. J. W. Hewage (2016) Core/shell formation and surface segregation of multi shell icosahedral silver-palladium bimetallic nanostructures: A dynamic and thermodynamic study. *Mater. Chem. Phys.* **74**, 187–194.
87. S. Divi, A. Chatterjee (2018) Generalized nano-thermodynamic model for capturing size-dependent surface segregation in multi-metal alloy nanoparticles. *RSC Adv.* **8**, 10409–10424.
88. H. Liao, A. Fisher, Z. J. Xu, (2015) Surface segregation in bimetallic nanoparticles: A critical issue in electrocatalyst engineering. *Small* **11**, 3221–3246.
89. Z. Liu, Y. Lei, G. Wang (2016) First-principles computation of surface segregation in L10 CoPt magnetic nanoparticles. *J. Phys.: Condens. Matter* **28**, 266002.
90. I. Yonenaga, M. Sakurai (2001) Bond lengths in $Ge_{1-x}Si_x$ crystalline alloys grown by the Czochralski method. *Phys. Rev. B* **64**, 113206.
91. A. R. Miedema (1978) Surface segregation in alloys of transition metals. *Z. Metallkd.* **69**, 455–461.
92. A. S. Shirinyan, M. Wautelet (2004) Phase separation in nanoparticles. *Nanotechnology* **15**, 1720–1731.
93. B. S. Meyerson (1992) UHV/CVD growth of Si and Si:Ge alloys: Chemistry, physics, and device applications. *Proc. IEEE* **80**, 1592–1608.
94. G. G. Jernigan, P. E. Thompson, C. L. Silvestre (1997) Quantitative measurements of Ge surface segregation during SiGe alloy growth. *Surf. Sci.* **380**, 417–426.
95. J. Nyéki, C. Girardeaux, G. Erdélyi, J. Bernardini (2003) Equilibrium surface segregation enthalpy of Ge in concentrated amorphous SiGe alloys. *Appl. Surf. Sci.* **212–213**, 244–248.
96. D. Chrastina, G. M. Vanacore, M. Bollani, P. Boye, S. Schöder, M. Burghammer, R. Sordan, G. Isella, M. Zani, A. Tagliaferri (2012) Patterning-induced strain relief in single lithographic SiGe nanostructures studied by nanobeam x-ray diffraction. *Nanotechnology* **23**, 155702.
97. I. Berbezier, M. Aouassa, A. Ronda, L. Favre, M. Bollani, R. Sordan, A. Delobbe, P. Sudraud (2013) Ordered arrays of Si and Ge nanocrystals via dewetting of pre-patterned thin films. *J. Appl. Phys.* **113**, 064908.
98. P. Chen, N. A. Katcho, J. P. Feser, Wu Li, M. Glaser, O. G. Schmidt, D. G. Cahill, N. Mingo, A. Rastelli (2013) Role of surface-segregation-driven intermixing on the thermal transport through planar Si=Ge superlattices. *Phys. Rev. Lett.* **111**, 115901.
99. R. J. Jaccodine (1963) Surface energy of germanium and silicon. *J. Electrochem. Soc.* **110**, 524–527.
100. G. Abudukelimu, G. Guisbiers, M. Wautelet (2006) Theoretical phase diagrams of nanowires *J. Mat. Res.* **21**, 2829–2834.
101. M. Balkanski, R. F. Wallis (2000) *Semiconductor Physics and Applications*, pp. 89–100, Oxford, UK: Oxford University Press.
102. J. Singh (2001) *Semiconductor Devices: Basic Principles*, pp. 71–82, J. Wiley &Sons.
103. E. F. Schubert (1990) Delta doping of III–V compound semiconductors: Fundamentals and device applications. *J. Vac. Sci. Technol. A* **8**, 2980.
104. J. D. Bryan, D. R. Gamelin. (2005) Doped semiconductor nanocrystals: Synthesis, characterization, physical properties, and applications. *Prog. Inorg. Chem.* **54**, 47–126.
105. R. B. Fair (1990) Point defect Charge-state effects on transient diffusion of dopants in Si. *J. Electrochem. Soc.* **137**, 667–671.

106. S. Ikhmayies (2014) Introduction to II–VI compounds. In: *Advances in the II-VI Compounds Suitable for Solar Cell Applications*, Publ. Research Signpost.

107. I. Teramoto (1972) Calculation of distribution equilibrium of amphoteric silicon in gallium arsenide. *J. Phys. Chem. Sol.* **33**, 2089–2099.

108. I. Teramoto (1974) Solid solubility of amphoteric silicon in gallium arsenide. *Jap. J. Appl. Phys.* **13**, 1817–1822.

109. D. Vazquez-Cortas, S. Shimomura, M. Lopez-Lopez,E. Cruz-Hernandez, S. Gallardo-Hernandez, Y. Kudriavtsev, V. H. Mendez-Garcia (2012) Electrical and optical properties of Si doped GaAs (631) layers studied as a function of the growth temperature. *J. Cryst. Growth* **347**, 77–81.

110. T. Ahlgren, J. Likonen, J. Slotte, J. Räisänen, M. Rajatora, J. Keinonen (1977) Concentration dependent and independent Si diffusion in ion-implanted GaAs. *Phys. Rev. B* **56**, 4597–4603.

111. Ph. Ebert (2002) Defects in III–V semiconductor surfaces. *Appl. Phys. A* **75**, 101–112.

112. D. Vazquez-Cortas, S. Shimomura, M. Lopez-Lopez, E. Cruz-Hernandez, S. Gallardo-Hernandez, Y. Kudriavtsev, V. H. Mendez-Garcia (2012) Electrical and optical properties of Si doped GaAs (631) layers studied as a function of the growth temperature. *J. Cryst. Growth* **347**, 77–81.

113. D. Li, J. Xu, P. Zhang, Y. Jiang, K. Chen (2018) Doping effect in Si nanocrystals. *J. Phys. D: Appl. Phys.* **51**, 233002 (23pp).

114. G. M. Dalpian, J. R. Chelikowsky (2006) Self-purification in semiconductor nanocrystals self-purification in semiconductor nanocrystals. *Phys. Rev. Lett.* **96**, 226802.

115. T. H. Yuan, X. D. Pi, D. Yang (2017) Nonthermal plasma synthesized boron-doped germanium nanocrystals. *IEEE J. Sel. Top. Quantum Electron.* **23**(5), 4800205.

116. E. Arduca, M. Perego (2017) Doping of silicon nanocrystals. *Mater. Sci. Semicond. Process.* 62, 156–170.

117. G. Cantele, E. Degoli, E. Luppi, R. Magri, D. Ninno, G. Iadonisi, S. Ossicini (2005) First-principles study of n- and p-doped silicon nanoclusters. *Phys. Rev. B* **72**, 113303.

118. S. Schulz, G. Czycholl (2005) Tight-Binding model for semiconductor nanostructures. *Phys. Rev. B* **72**, 165317.

119. N. T. T. Hang (2014) Size dependent melting of silicon nanoparticles. *Commun. Phys.* **24**, 207–215.

120. S. S. Dalgic, U. Domenikeli (2009) Melting properties of tin nanoparticles by molecular dynamics simulation. *J. Optoelectron. Adv. Mater.* **11**, 2126–2132.

121. M. A. Asoro, J. Damiano, P. J. Ferreira (2009) Size effects on the melting temperature silver nanoparticles: In-situ TEM observations. *Microsc. Microanal* 15(Suppl 2), 706–707.

122. K. K. Nanda, S. N. Sahu, S. N. Behera (2002) Liquid-drop model for the size-dependent melting of low-dimensional systems. *Phys. Rev. A* **66**, 013208.

123. J. A. Purton, C. S. Parker, N. L. Allan (2013) Monte Carlo simulation and free energies of mixed oxide nanoparticles. *Phys. Chem. Chem. Phys.* **15**, 6219–6225.

124. E. Golovneva, I. Golovnev, V. Fomin (2007) The calculation of thermodynamic properties of nanostructure by molecular dynamics method. *Phys. Mesomech.* **10**, 11–16. ISSN 1029-9599.

125. I. Golovnev, El. Golovneva, V. Fomin (2011) Molecular-dynamics calculation of nanostructures thermodynamics: Research of impurities influence on results. In: *Application of Thermodynamics to Biological and Materials Science*, M. Tadashi Ed., ISBN 978-953-307-980-6 InTec.

126. K. R. S. S. Subramanian, R. B. Venkat, J. Babu (2005) A molecular dynamics simulation study on melting of Pd-Pt nanoclusters. *Phys. Rev. B* **71**, 195415.

127. G. Guisbiers, G. Abudukelimu, D. Hourlier (2011) Size-dependent catalytic and melting properties of platinum-palladium nanoparticles. *Nanoscale Res Lett.* **6**, 396.

128. L. J. Lewis, P. Jensen, J. L. Barrat (1997) Melting, freezing and cohalescence of gold nanoclusters. *Phys. Rev. B* **56**, 2248.
129. O. Stepanyuk, D. Alekseev, A. Saletskii (2009) Calculation of the thermodynamic properties of copper by molecular dynamics simulation. *Moscow Univ. Phys. Bull.* **64**, 226–227, ISSN 0027–134.
130. E. K. Richman, J. E. Hutchison (2009) The nanomaterial characterization bottleneck. *ACS Nano* **3**, 2441–2446.
131. D. Turnbull (1950) Formation of crystal nuclei from liquid metals. *J. Appl. Phys.* **21**, 1022–1028.
132. L. Y. Chen, M-R. He, J. Shin, G. Richter, D. S. Gianola (2015) Measuring surface dislocation nucleation in defect-scarce nanostructures. *Nat. Mater.* **14**, 707–713.
133. G. Richter, K. Hillerich, D. S. Gianola, R. Mönig, O. Kraft, C. A. Volkert (2009) Ultrahigh strength single crystalline nanowhiskers grown by physical vapor deposition. *Nano Lett.* **9**, 3048–3052.
134. S. Pizzini (2017) *Point Defects in Group IV Semiconductors: Common Structural and Physico-Chemical Aspects*, Vol. 10. Materials Research Foundation. ISBN:978-1-945291-22-7.
135. S. Pizzini (2015) *Physical Chemistry of Semiconductor Materials and Processes*, pp. 93–98, J.Wiley & Sons, Chichester UK.
136. C. R. Mokry, P. J. Simpson, A. P. Knights (2009) Role of vacancy-type defects in the formation of silicon nanocrystals. *J. Appl. Phys.* **105**, 114301.
137. P. Knauth, H. L. Tuller (1999) Electrical and defect thermodynamic properties of nanocrystalline titanium dioxide. *J. Appl. Phys.* **85**, 897–902.
138. G. M. Dalpian, J R. Chelikowsky (2006) Self-purification in semiconductor nanocrystals. *Phys. Rev. Lett.* **96**, 226802.
139. H. Hossein-Babaei, H. R. S. Shirkoohi (2016) Defect-free nanocrystals are at thermodynamic equilibrium at room temperature. *arXiv preprint* arXiv:1608.05486.
140. M. Lannoo, J. Burgoin (1981) *Point Defects in Semiconductors I*, pp. 195–197, Springer Verlag.
141. G. Guisbiers (2011) Schottky defects in nanoparticles. *J. Phys. Chem. C* **115**, 2616–2621.
142. M. Müller, K. Albe (2007) Concentration of thermal vacancies in metallic nanoparticles. *Acta Mater.* **55**, 3237–3244.
143. A. B. Bortz, M. H. Kalos, I. L. Lebowitz (1975) A new algorithm for monte carlo simulation of ising spin systems. *J. Comput. Phys.* **17**, 10–18.
144. M. Salis, C. M. Carbonaro, M. Marceddu, P. C. Ricci (2013) Statistical thermodynamics of Schottky defects in metal nanoparticles. *Nanosci. Nanotechnol.* **3**, 27–33.
145. J. Makinson, J. S. Lee, S. H. Magner, R. J. DeAngelis (2000) X-Ray diffraction signatures of defects in nanocrystalline materials. JCPDS Centre for Diffraction Data. *Adv. X-Ray Anal.* **42**, 407–411.
146. J. Kuiala (2016) Point defects in nanocrystals and semiconductors studied with the annihilation spectroscopy. *Doctoral Dissertation 223–2016A*, Alto University, School of Science.
147. J. Kujala, J. Slotte, F. Tuomisto, D. Hiller, M. Zacharias (2016) Si nanocrystals and nanocrystal interfaces studied by positron annihilation. *J. Appl. Phys* **120**, 145302.
148. D. Wang, Z. Q. Chen, D. D. Wang, N. Qi, J. Gong, C. Y. Cao, Z. Tang (2010) Positron annihilation study of the interfacial defects in ZnO nanocrystals: Correlation with ferromagnetism. *J. Appl. Phys.* **107**, 023524.
149. S. Ögut, J. R. Chelikowky, S. G. Louie (1999) Optical properties of silicon nanocrystals: A first principles study. *MRS Online Proc. Lib. Arch.* doi:10.1557/PROC-579-81.
150. X. Zhao, C. M. Wei, L. Yang, M. Y. Chou (2004) Quantum confinement and electronic properties of silicon nanowires. *Phys. Rev. Lett* **92**, 236805.

151. S. Öğüt, J. R. Chelikowsky S. G. Louie (1997) Quantum confinement and optical gaps in si nanocrystals. *Phys. Rev. Lett.* **79**, 1770.
152. A. F. Wright (2006) Density-with DFT functional-theory calculations for the silicon vacancy. *Phys. Rev. B* **74**, 165116.
153. M. Tang, L. Colombo, J. Zhu, T. Diaz de la Rubia (1997) Intrinsic point defects in crystalline silicon: Tight-binding molecular dynamics studies of self-diffusion, interstitial-vacancy recombination, and formation volumes *Phys. Rev. B* **55**(14), 279.
154. S. A. Centoni, B. Sadigh, G. H. Gilmer, T. J. Lenosky, T. Diaz de la Rubia, C. B. Musgrave (2005) First-principle calculation on intrinsic defect volumes in silicon. *Phys. Rev B* **72**, 195206.
155. O. K. Al-Mushadani, R. J. Needs (2003) Free-energy calculations of intrinsic point defects in silicon. *Phys. Rev. B* **68**, 235205.
156. T. Sholihun, M. Saito, T. Ohno, T. Yamasaki (2015) Density-functional-theory-based calculations of formation energy and concentration of the silicon monovacancy. *Jpn. J. Appl. Phys.* **54**, 041301.
157. F. Corsetti, A. A. Mostofi (2011) System-size convergence of point defect properties: The case of the silicon vacancy. *Phys. Rev. B* **84**, 035209.
158. G. D. Watkins, J. W. Corbett (1964) Defects in irradiated silicon: Electron paramagnetic resonance and electron-nuclear double resonance of the Si-E center. *Phys. Rev.* **134**, A1359.
159. S. Dannefaer, P. Mascher, D. Kerr (1986) Monovacancy formation enthalpy in silicon. *Phys. Rev. Lett.* **56**, 2195.
160. Y. Shimizu, M. Uematsu, K. M. Itoh (2007) Experimental evidence of the vacancy-mediated silicon self-diffusion in single-crystalline silicon. *Phys. Rev. Lett.* **98**, 095901.
161. F. Corsetti (2012) On the properties of point defects in silicon nanostructures from ab initio calculations. PhD Thesis, Imperial College London, February 2012.
162. N. Fukata, A. Kasuya, M. Suezawa (2001) Formation energy of silicon vacancy determined by a new quenching method. *Jap. J. Appl. Phys.* **40**(Part 2, nr 8B), L854–L856.
163. F. Li, C.-M. Fang, M. Dijkstra, M. A. van Huis (2015) The role of point defects in PbS, PbSe and PbTe: A first principles study. *J. Phys. Cond. Matter* **27**, 355801.
164. S. Ahmad, S. D. Mahanti, K. Hoang, M. G. Kanatzidis (2006) Ab initio studies of the electronic structure of defects in PbTe. *Phys. Rev. B* **74**, 155205.
165. P. Tzanetakis, J. Hillairet, G. Revel (1976) The formation energy of vacancies in aluminium and magnesium. *Phys. Status Solidi B* **75**, 433–439.
166. K. Carling, G. Wahnström, T. R. Mattsson, A. E. Mattsson, N. Sandberg, G. Grimvall (2000) Vacancies in metals: From first-principles calculations to experimental data. *Phys. Rev. Lett.* **85**, 3862–3865.
167. M. Ippolito, S. Meloni, L. Colombo (2008) Interface structure and defects of silicon nanocrystals embedded into a-SiO_2. *Appl. Phys. Lett.* **93**, 153109.
168. V. Kocevski, J. M. Mills, S.-K. Chiu, C. F. Gervasi, B. N. Taber, A. E. Rosenfield, D. A. Kislitsyn, O. Eriksson, J. Rusz, A. M. Goforth, G. V. Nazin (2016) Mapping of defects in individual silicon nanocrystals using real-space spectroscopy. *J. Phys. Chem. Lett.* **7**, 1047–1054.

3 Preparation, Structural, and Physical Properties of Nanocrystalline Silicon, Germanium, and Diamond Nanocrystals, Films, and Nanowires*

3.1 NANOCRYSTALLINE SILICON: BACKGROUND CONCEPTS

Nanocrystalline silicon (nc-Si) represents an almost unique case of a nanomaterial of relevant scientific and technological interest, on which an extremely vast amount of literature is available concerning its preparation, morphology, structure, physical properties, and defectivity, that, however, implies a severe work of critical analysis to rationalize the main issues.

As is already well-known, nanocrystalline silicon could be prepared as a powder of individual nanocrystallites (NCs), or as nanocrystalline films and nanowires, mostly using plasma-assisted processes with silane (SiH_4) or fluorosilane (SiF_4) as the precursors.

It will be shown that nc-Si in nc-films [1] is a two-phase, single-component material, consisting of a distribution of silicon nanocrystals in an amorphous silicon matrix. Since this material is intrinsically passivated with hydrogen when prepared with plasma-assisted processes, it is conventionally also mentioned as nc-Si-H or micromorph μc-Si-H.

It will be seen that the morphological defects associated with the coexistence of two phases (amorphous + crystalline) and the hydrogen- or oxygen-passivation of nanocrystalline silicon surfaces do strongly affect its electronic properties [2, 3], depending on the processes adopted for their preparation.

It will, also, be seen that the plasma chemistry is very sensible to the nature of the precursors, and that the use of SiF_4 favors, for example, the onset of a large crystallinity [4] and shifts to lower temperatures the allowed temperature range for nc-Si deposition [5].

With this last process, under specific growth conditions, single-phase, nanocrystalline film could be also produced, as the stable product of the growth process.

---

* https://orcid.org/0000-0002-0542-3219.

A different technique, based on the thermal decomposition of SiO or of a sub-stoichiometric phase SiO$_x$* of SiO$_2$ is also currently used for the synthesis of Si–SiO$_2$ superlattices [6–18] or of silicon nanocrystals embedded in a SiO$_2$ matrix.

Silicon nanocrystals embedded in a SiO$_2$ matrix, after annealing at temperatures higher than 900°C, could be considered the prototype of hydrogen-free Si-NC, unlike their counterparts PECVD-prepared using SiH$_4$–H$_2$ mixtures.

Eventually, the synthesis of assemblies of individual nanowires could be typically carried out either by vapor phase deposition techniques on single-crystal surfaces, suitably seeded with metal dots, which work as nucleation sites, or by localized electrochemical etching [19–22].

Synthetic details, morphology, defects, and individual properties of these materials will be illustrated in the following sections.

3.2 COLLOIDAL SILICON NANOCRYSTALS AND SI QUANTUM DOTS

Colloidal silicon nanocrystals and silicon quantum dots (QD) belong to a vast family of colloidal nanocrystals (NC) suitably dispersed in a liquid supporting medium. They could be prepared using variants of empirical techniques well-known for millennia for the decoration of artifacts. These materials attracted renewed interest in the nineteenth century, when Faraday (1856), interested in understanding the particulate nature of matter, was able to carry out the synthesis of colored colloidal Au particles, and found that the color of these particles was due to their size, a property which now we know to be common to all colloidal particles of metals and semiconductors [23–26].

The success occurred, however, by chance after many unsuccessful attempts to use mechanical and electrochemical thinning methods. In fact, he succeeded only when he used phosphorous to reduce gold chloride in solution, and the resulting products were ruby or blue fluids, that dispersed the light passing through them, like conventional heterogeneous dispersions, but did not settle, like true solutions [27].

Faraday was then able to demonstrate, by evaporating the solutions, that the dried colloids† were particulates and that their size was at the origin of the color of the fluids; blue fluids had smaller particles than the purple ones. And he was also able to conclude that the color of the fluids was the consequence of the interaction of the light with the particles, not of a simple light dispersion.

Wet techniques could be applied for the preparation of a vast number of colloidal particles of metals reducible in hydrous media, or by precipitation/co-precipitation in non-aqueous phases [28], or by solvothermal synthesis in organic/inorganic solutions [29], as is the case of compound semiconductors (sulfides).

Dry techniques should be, instead, applied for the preparation of nanocrystals of elemental materials, like silicon or germanium, whose preparation from water

* We will show that SiO$_x$ is not a homogeneous phase.
† Using today's notation.

solutions is thermodynamically impossible, although wet techniques are used for processes addressed at the isotropic or anisotropic etching of bulk silicon.

Individual silicon nanocrystals are known to behave as single quantum dots and, thus, as (efficient) photoluminescence sources [30–32] with light emission in the visible range, once confinement effects become active due to their nanometric or sub-nanometric size, to the presence of nanometric or sub-nanometric surface defects, or to specific impurity contamination, including hydrogen or oxide passivation [33].

Nanocrystalline silicon powders could be, also, prepared from porous silicon submitted to various surface chemistries [34] to obtain their colloidal solutions. Porous silicon itself, with its peculiar light emission arising from pore sizes compatible with quantum confinement [35], is prepared by the anisotropic anodic etching of silicon in HF solutions; see Section 3.4.4, where details on its preparation and properties can be found.

We anticipate, however, here that the formation of the peculiar morphology of porous silicon, with its complex architecture consisting of interconnected silicon nano-dendrites, implies the assistance of holes in correspondence with the un-passivated tip of the pores, where anodic etching occurs

$$Si + 2HF + 4F^- + 4h \rightleftharpoons H_2SiF_6 \qquad (3.1)$$

while the pore walls are oxygen-passivated [35].

The direct, thermal decomposition of silane (SiH_4) in a flow reactor leads, instead, to the synthesis of nanocrystalline silicon powders, from which the separation of individual nanocrystals is, however, almost impossible due also to the extreme ease of oxidation of silicon nanoparticles. The process consists of the thermal pyrolysis of silane to silicon

$$SiH_4 \rightleftharpoons Si + 2H_2 \qquad (3.2)$$

and the properties of the product depend entirely on the kinetics of the gas phase process, where homogeneous and heterogeneous nucleation processes do occur simultaneously, before the crystallization.

If the process is, instead, carried out in a plasma phase of silane and argon activated by a radiofrequency (RF) source and confined in the inner space of the two electrodes shown in Figure 3.1, at a pressure of 4 mbar [36], a powder consisting of individual silicon nanocrystals is created inside the plasma phase [36–41], whose size and rate of formation depend on the residence time and on the gas mixture composition.

The particles collected on the silica tube walls, or at the bottom grid of the reactor, are luminescent particles which present a core ordered structure and a defective surface (see Figure 3.2) from Transmission Electron Microscopy (TEM) measurements.

Relatively little is known about the process of nucleation and growth of silicon nanoparticles in these kinds of non-thermal reactors, although photoluminescence (PL) and Raman spectroscopies bring a rich contribution to the understanding of their properties.

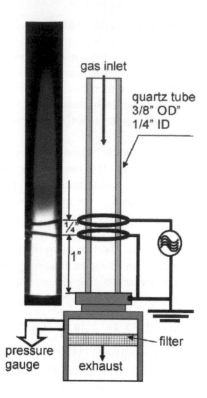

FIGURE 3.1 Schematic diagram of a continuous-flow nonthermal plasma reactor. (*Reprinted with permission from* L. Mangolini, E. Thimsen, U. Kortshagen (2005) High-yield plasma synthesis of luminescent silicon nanocrystals *Nano Letters* **5** (4), 655–659. *Copyright (2005) American Chemical Society Dec 03, 2019.*)

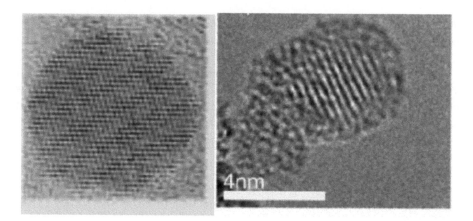

FIGURE 3.2 TEM images of silicon nanoparticles prepared in a plasma reactor. (*Reprinted with permission from* L. Mangolini, E. Thimsen, U. Kortshagen (2005) High-yield plasma synthesis of luminescent silicon nanocrystals Nano letters **5** (4), 655–659. *Copyright (2005) American Chemical Society Dec 03, 2019.*)

It is, instead, known that these reactors are capable of producing a range of nanopowders from fully amorphous to crystalline ones, depending on the input-RF power. It is supposed that nanocrystals are created in the plasma through electron impact dissociation of SiH_4 and subsequent clustering of the fragments [42–43].

This issue will be discussed in the next section, since dust production is also common in plasma reactors used for the synthesis of nc-Si films.

It is also known from aerosol sampling procedures that amorphous particles are nucleated first and that crystallization does occur during their transit in the plasma phase, which typically extends a few cm before the first ring electrode to a few cm after the second electrode, see Figure 3.3, with a calculated plasma residence time of a few msec.

The TEM diffraction patterns reported on top of each micrograph displayed at the bottom of Figure 3.3 show that the morphology of the nanopowders streaming across the plasma phase evolves from an amorphous (a) to crystalline structure, well-evident from the Raman spectra reported in Figure 3.4. There, one can observe the presence of a broad band within 400 and 550 cm^{-1} typical of amorphous Si (a-Si) in the sample (a), the co-presence of the amorphous Si band and of crystalline silicon band peaked at 520 cm^{-1} in the sample (b), and the main presence of the band of crystalline silicon in the sample (c). The shoulder at 400–500 cm^{-1} in the Raman spectrum of the sample (c) shows, however, the co-presence of a-Si also in this last sample.

The main conclusion drawn from these experiments is that a-Si and c-Si coexist in silicon nanodots prepared with non-thermal plasma reactors, due to a kinetic control of the aSi → cSi transformation.

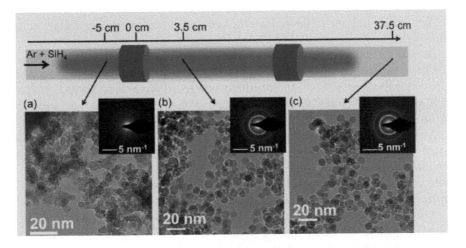

FIGURE 3.3 *Top* Schematic drawing of the plasma region of the plasma reactor; *bottom* TEM images of the morphology and size of the powder extracted in the different regions. Inserts display the Electron diffraction patterns (*After* T. Lopez, L. Mangolini (2014) On the nucleation and crystallization of nanoparticles in continuous-flow nonthermal plasma reactors *J. Vacuum Sci. & Technology B* **32**, 061802. *Reproduced under a Creative Commons Attribution (CC BY) license.*)

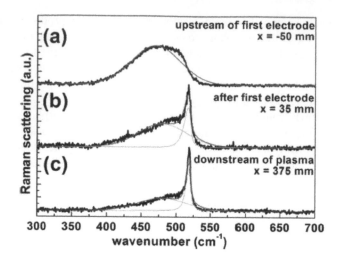

FIGURE 3.4 Raman spectra of the Si nanocrystal samples streaming across the plasma phase. (*After* T. Lopez, L. Mangolini (2014) On the nucleation and crystallization of nanoparticles in continuous-flow non-thermal plasma reactors *J. Vacuum Sci. & Technology B* **32**, 061802. *Reproduced under Creative Commons Attribution (CC BY) license.*)

3.3 NANOCRYSTALLINE SILICON FILMS

3.3.1 INTRODUCTION

Nanocrystalline silicon (nc-Si) films were considered, for more than two decades, to be promising candidates for low-cost, flexible, thin-film solar cells [44–50], under the hypothesis that their conversion efficiency would be eventually found comparable with that of multicrystalline silicon solar cells.

Despite the challenge of silicon being an indirect gap semiconductor, the fabrication of light-emitting diodes (LEDs) was also considered a potentially interesting application of nc-Si [6, 51–52], under the hypothesis that efficient quantum confinement effects could be demonstrated achievable with Si nanocrystals, with a substantial shift to the blue of the light emitted.

Although nc-Si, like a-Si, does suffer from light-induced degradation (LID) associated with the Wronsky–Staebler effect [53], it was shown that LID could be prevented in nc-Si when the impinging photons' energy is maintained lower than the energy gap of a-Si, proving that LID in nc-Si is associated only with defects present in the traces of a-Si always present in nc-Si films [54].

Also, the application of nc-Si in its *black silicon* configuration on the surface of solar cells gained increasing interest from the photovoltaic industry [55, 56], as soon as the decreasing thickness of solar cell bodies made the conventional antireflection texturization no longer applicable.

Black silicon (BS) is nothing other than an assembly of silicon nanowires or pillars (see Figure 3.5) with an average length of a few micrometers, produced by a variant of the metal-assisted wet etching technique, which will be dealt with in

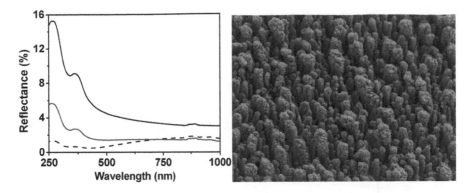

FIGURE 3.5 *(Left panel)* Reflection spectra of an array of silicon nanorods of different lengths (2.1 μm (solid line), 3.4 μm (thin line), and 7.1 μm dotted line) in the UV-VIS-NIR region: *(Right panel)* Typical plan view of an array of nanowires. *(Reproduced by permission of the Royal Society of Chemistry after* Y. Li, J. Zhang, S. Zhu, H. Dong, Z. Wang, Z. Sun, J. Guo, and B. Yang (2009) Bioinspired silicon hollow-tip arrays for high performance broadband anti-reflective and water-repellent coatings *J. Mater. Chem.* **19**, 1806–1810 Copyright Order detail ID: 71932719; ISSN: 1364–5501 27/06/2019.)

Section 3.4.4, but its application to solar cell technology required more than one decade of dedicated application before success. The advantage of BS over conventional antireflection texturization is the uniform black color of the cells and the excellent antireflection properties of the layer (see their reflection spectra in Figure 3.5), which improve with the increase of the height of the pillars.

Although nc-Si thin films for solar cell and optoelectronic applications have been mostly manufactured with variants of the plasma-enhanced CVD route, using silane (SiH_4) as the precursor in (SiH_4–H_2–Ar) mixtures [47, 48] or SiF_4 in SiF_4–H_2–Ar mixtures [5], operating at low temperatures (max 300°C)* and offering the advantage of very clean preparation conditions, alternative preparation methods have been, also, envisaged.

An example, the magnesio-thermic reduction of quartz [57] or of lime glass [58], occurring according to the reaction

$$SiO_2 + 2Mg \rightleftharpoons Si + 2MgO \qquad (3.3)$$

($\Delta G = -375.01$ kJmol^{-1} at 520°C) [58] has been considered potentially useful to obtain nc-Si thin films at relatively low temperatures (500–750°C), maintaining the original micro- or nanostructures of the parent SiO_2 phases [59]. This process, however, occurs in competition with that leading to magnesium silicide (Mg_2Si) in the same temperature range [59, 60]

$$SiO_2 + 4Mg \rightarrow Mg_2Si + 2MgO \qquad (3.4)$$

---

* Point defect formation is favoured by high temperatures.

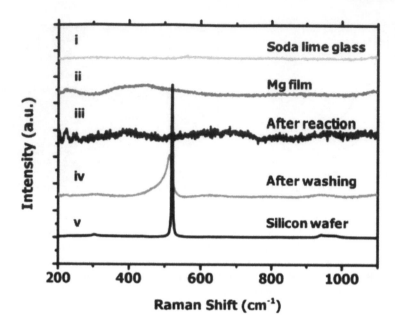

FIGURE 3.6 Raman spectra of the original lime glass (i), of the Mg film (ii), and of the reaction product before (iii) and after acid leaching (iv). (*After* D. P. Wong, H-T. Lien, Y-T. Chen, K.-H. Chen, and L-C. Chen (2012) Patterned growth of nanocrystalline silicon thin films through magnesiothermic reduction of soda lime glass *Green Chem.*14 896–900. *Reproduced with permission of the Royal Society of Chemistry, c*onfirmation Number: 11827292 Order Date: Jun 27, 2019.)

but the process under Equation 3.3 is thermodynamically favored, since the Reaction 3.5

$$Mg_2Si + SiO_2 \rightleftharpoons 2MgO + 2Si \tag{3.5}$$

which occurs simultaneously, should enable at least the partial decomposition of magnesium silicide (Mg_2Si) ($\Delta G = -196$ kJmol^{-1} at 520°C), provided kinetic control can be applied [59].

The solution successfully adopted by Wong et al. [58] was to deposit onto a glass substrate a layer of Mg thin enough to control the reaction rate. It is, however, obvious, that a Mg contamination of the surface of this material should be expected.

With this process, after acid leaching off the by-product MgO, 500 nm thick, free-standing nc-Si films, showing a granular mesostructure, with an average grain size around 60 nm*, were eventually obtained. From the Raman spectrum of one of these films after washing to remove traces of the starting material, see Figure 3.6, one could observe the presence of a sharp peak at 520 cm^{-1}, typical of bulk crystalline Si (see the spectrum at the bottom of the figure).

---

* That does not exclude that these grains would be aggregates of nanocrystallites.

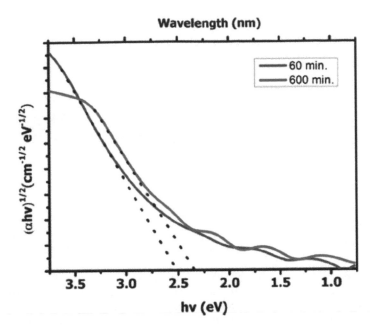

FIGURE 3.7 Tauc's plot of the optical absorption coefficient of two nc-Si films prepared *via* the magnesio-thermic route. (*Reproduced with permission of the Royal Society of Chemistry after* D. P. Wong, H-T. Lien, Y-T. Chen, K.-H. Chen, and L-C. Chen (2012) Patterned growth of nanocrystalline silicon thin films through magnesiothermic reduction of soda lime glass *Green Chem.14 896–900, License number11827298, license date Jun 27, 2019.*)

It is, however, apparent that the Raman peak shows the typical broadening of the Raman spectra of nc-Si samples obtained with Hot Wire Chemical Vapour Deposition (HWCVD) or Plasma Enhanced CVD (PECVD) processes, suggesting the co-presence of a-Si.

From optical absorption measurements on these films, a Tauc's gap of 2.3–2–5 eV, depending slightly on the duration of the thermal treatment, could be deduced (see Figure 3.7), interpreted by the authors as the energy gap of the material. As will be seen in Chapter 5, Section 5.1, an energy gap of 2.3 eV should be observed in the case of a distribution of nanocrystals with a mean size of 2.3 nm, which should also present phonon confinement effects with an important red-shift of the Raman peak. Since a red-shift is not observed in the Raman spectra of this material, as it clearly seen in Figure 3.6, where the peak energy fits well that of bulk crystalline silicon, the level at 2.3–2.5 eV should be due to surface defects, possibly associated with minute surface MgO contamination, which would, in fact, lead to a luminescence emission at ≈2.1 eV, according to the results of Soma and Uchino [61].

nc-Si arrays for light-emitting diodes (LED) applications have been, instead, manufactured by the Zacharias [6, 62, 63] and Priolo [64–67] groups with variants of the thermal decomposition of a sub-stoichiometric (SiO_x) silica suitably prepared, as an example, by reactive sublimation of SiO in oxygen atmosphere followed by annealing at 1100°C in inert gas atmosphere to promote the thermal decomposition of SiO_x to Si and SiO_2 and the crystallization of amorphous silicon nanoclusters.

FIGURE 3.8 TEM image of a Si/SiO$_2$ superlattice after crystallization annealing. (*Reproduced with permission of American Physical Society after* M. Zacharias and P. Streitenberger (2000) Crystallization of amorphous superlattices in the limit of ultrathin films with oxide interfaces *Phys. Rev. B* **62**, 8391 *copyright of American Physical Society (2000) License Number: RNP/19/JUN/016021 License date: Jun 25, 2019 DOI: 10.1103/ PhysRevB.62.*)

In both cases the final device structure is a superlattice consisting of alternate layers of nc-Si, with an average size of 3–5 nm embedded in an SiO$_2$ matrix, and of SiO$_2$ (see Figure 3.8). Details on the chemistry of SiO$_x$ preparation and annealing and on the structural properties of these Si-NC arrays are reported in Section 3.5.

A laser ablation technique under increasing oxygen pressures (10^{-4} to 0.5 mbar) has been also successfully applied by Dey and Khare [68] for the preparation of thin SiO$_x$ films, which look like a distribution of nanometric aggregates (clusters) of individual Si nanocrystals in a matrix consisting of a-Si and SiO$_2$. With the increase of the oxygen pressure, the calculated size of the Si nanocrystals within the clusters decreases from 20 to 2 nm (see Figure 3.9), with minute differences when applying the Richter, Wang, and Ley (RWLM) model or the bond polarizability (BPM) model.*

The solar cell promises of nc-Si failed (together with those of thin film solar cells), with the almost full collapse of industrial activities in 2017, after seven years of decline [69].

Intrinsic physical drawbacks, unfavorable costs, poor efficiency competition with high-efficiency silicon solar cells, and wrong investment policy were the main reasons, together with the success of Chinese manufacturers in the mass production of low-cost, high-quality silicon for solar cells.

Instead, nc-Si light emitting diodes still deserve attention [70–73] "since the luminescence intensity level of nc-Si is comparable with that of direct gap quantum dots."†

---

* See details in the Dey and Khare paper [68].
† Original sentence in F.Priolo et al. [73].

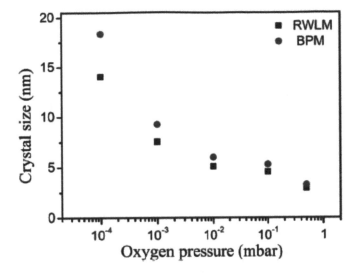

FIGURE 3.9 Crystal size distribution of nanometric Si particles in micrometric Si/SiO$_2$ clusters as a function of the oxygen pressure. (*After* O.P. Dey, A. Khare (2017) Fabrication of photoluminescent nc-Si:SiO$_2$ thin films prepared by PLD *Phys.Chem.Chem.Phys.* **19** 21436–21445. *Reproduced with permission of the Royal Society of Chemistry Order detail ISNN 1463–9084.*)

New industrial applications of nc-Si are, also, now on the way, as, for example, the application of nc-Si anodes in Li-ions batteries, which forecasts radically new preparation techniques, with the use of a molten salt medium for its synthesis from silicon tetrachloride (SiCl$_4$) [74] or directly by electroreduction of SiO$_2$ in molten LiCl–Li$_2$O [75].

3.3.2 GROWTH AND STRUCTURE OF NANOCRYSTALLINE SILICON FILMS PREPARED VIA HOT-WIRE AND PECVD PROCESSES

Most of the knowledge available on defects in nc-Si is the indirect outcome of years-long studies aimed at optimizing the quality, and then the efficiency, of thin film nc-Si solar cells, with the reduction or suppression of structural and topological drawbacks as well as of point-defects induced by the growth process.

Two main routes were followed for the synthesis of nc-Si thin films, the plasma-enhanced (PE)-CVD deposition, including the low-energy, plasma-enhanced CVD (LEPECVD) process, and the hot-wire (HW)-CVD method, using silane (SiH$_4$) or tetrafluoro silane (SiF$_4$) or both, as precursors, suitably diluted in hydrogen, argon (or helium).

As will be discussed in detail in the next section

- While in the plasma phase processes the precursors are dissociated with the formation of radicals, these last being the active species in the growth process, the growing film is subjected to argon ion bombardment, with the potential formation of defects.

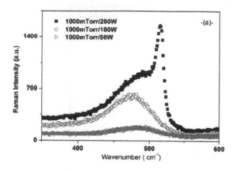 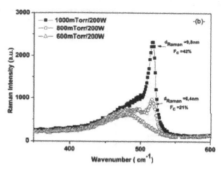

FIGURE 3.10 Effect of varying the RF power (left) and pressure (right) on the amorphous–crystalline transition from Raman spectroscopy measurements. (*After* R. Amrani, F. Pichot, L. Chahed, Y. Cuminal (2012) Amorphous-Nanocrystalline Transition in Silicon Thin Films Obtained by Argon Diluted Silane PECVD in *Crystal Structure Theory and Applications,* **1** 57–61 http://dx.doi.org/10.4236/csta.2012.13011 *Published Online December 2012 (www. SciRP.org/journal/csta).)*

- In the HWCVD process the precursors are catalytically dissociated at an high-temperature filament, and the deposition process does occur in the absence of ionic bombardment over the growing film surface, limiting the formation of defects in the films and, thus, the deterioration of device performances [76].

Conversely, while the effective energy transfer between impinging ions and the substrate during processing can induce ad-atoms motion on the growing layer, improving the structural quality of the material, the absence of ion bombardment during processing can lead to the presence of interconnected nanovoids (cracks) through the entire film thickness [76].

In both processes, the deposition temperature ranges from room temperature to $\approx 300°C$, and the deposited material contains variable amounts of hydrogen.* In the case of samples prepared by LEPECVD, the average hydrogen content measured with IR spectroscopy at 640 cm^{-1} ranges around 2.3% [77]. Like in case of single crystal or polycrystalline silicon, hydrogen is supposed to passivate the surface and bulk defects of nc-Si.

Almost independent of the process chemistry, the morphology of the nc-Si film manifests a progressive transition from amorphous to crystalline as a function of the power input and of the reactive gas pressure, as could be deduced from the example reported in Figure 3.10, that displays the evolution of the Raman spectra of an nc-Si film deposited with a PECVD process as a function of the process conditions, considering that a fully amorphous material presents only a broad Raman band centered at 480 cm^{-1}, whereas the presence of crystalline silicon is manifested by a shallow Raman band peaking at 520 cm^{-1}.

---

* Possibly arising from the decomposition of SiH$_4$ when the process is carried out using noble gases as diluents.

It can be seen that an amorphous–crystalline transition does occur with the increase of the RF power and the decrease of the pressure, and that the final structure of nc-Si is a heterogeneous mixture of c-Si and a-Si [78].

According to various authors active in the field [76, 79–84], the HWCVD approach presents several advantages over the conventional PECVD method, for the following reasons:

- A plasma-free process leads to improved stability against light-induced degradation (LID) of µc-Si:H HWCVD-deposited films in comparison with PECVD-deposited ones [53–54].
- The LID increases with the amorphous silicon content in the deposited µc-Si films, independently of the growth method [82]. A volume defect density around $5 \cdot 10^{16}$ cm^{-3} was measured with photoconductance measurements in fully amorphous a-Si:H films, grown with HWCVD, corresponding to the onset of a sub-gap level around E_c- 0.8 eV [84] which manifests in PL measurements as an optical emission band.
- The HWCVD process favors high deposition rates, due to the efficient catalytic cracking of the feed molecules at the hot filament surfaces [80] and, therefore, warrants a more efficient use of feedstock gases [83]. HWCVD-grown films show lower residual film stress than that measured on films deposited by PECVD method [85].

Eventually:

- Both a-Si:H and µc/nc-Si:H films can be prepared at low substrate temperatures without losing the material quality [86–87]. From the LID behavior of nc-Si it is possible to preliminarily conclude that, independently of the growth process used, nc-Si films *mostly suffer of defects present in the amorphous phase*, whose saturated amount decreases with the increase of the crystallinity of the material. We will see in Chapter 5, Section 5.3.3, details on the effect of a-Si on the optical properties of nc-Si films. Very little, instead, is brought to evidence about defects present in the crystalline phase, and of defects localized at the amorphous/crystalline interfaces.

A phenomenon called *crystalline volume evolution* is typical of nc-Si, which consists of the increase of the crystalline silicon volume (and a corresponding decrease of the a-Si content) with the increase of the film thickness, associated with a progressive degradation of the photovoltaic efficiency [46]. The amorphous-to-crystalline transition and then the crystalline fraction of the sample are shown to depend on the hydrogen dilution ratio. Therefore, a suitable profiling of the hydrogen dilution ratio during the film growth was found by Yan et al. [46] to be an effective way to maintain constant the crystallinity of the film.

The decrease of the photovoltaic efficiency with the decrease of the a-Si content indicates that a-Si plays an active function, possibly minimizing the recombination of minority carriers if the nc-Si surfaces are passivated with a layer of a-Si.

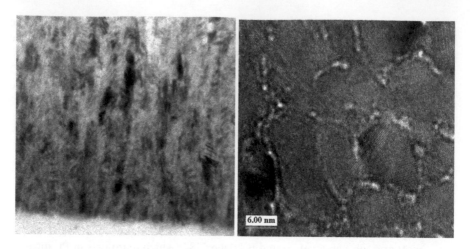

FIGURE 3.11 *(Left panel)* TEM image of the columnar structure of an nc-Si film grown with the LEPECVD process. *(Right panel)* TEM plan view of a section of an nc-Si grown with LEPECVD. *(Final progress report: Nanophoto Project no. 013944 (2005).)*

The morphology of nc-Si films grown with the LEPECVD* process should present favorable characteristics in this respect, since they display a columnar structure Figure 3.11 (left panel), with crystallites entirely embedded in amorphous silicon. The TEM image of a section of the same film, see Figure 3.11 (right panel), shows that the crystalline fraction of the film consists of arrays (clusters) of nanocrystallites having a mean size of 3–10 nm, and that a thin layer of amorphous material lies at the interface of each cluster. Details relevant to these nc-films will be discussed in Section 3.3.3.3.

Although a columnar structure is typical of nc-Si films grown with PECVD processes [88–90] a recent work based on a process making use of a mixture of SiH_4 and He, with He as the only diluent of the plasma [91], proves, instead, that nc-Si samples consisting of a three-dimensional array of pseudo-spherical dots, with a size around 10–50 nm, could be grown at 200°C, demonstrating the key role of a neutral diluent gas, instead of hydrogen, in the morphology of the deposited film.

A further example demonstrating, instead, the role of precursor gases is given by the use of silicon tetrafluoride (SiF_4) in admixture with silane and hydrogen, attempted by Cicala et al. [92], to explore the potentialities of fluorine in favoring the amorphous/crystalline transition. The plasma-deposited nc-Si-H,F films at 450°C are, in fact, eminently crystalline, with silicon grains having an XRD radius of 100 A. They are heavily doped with hydrogen (4.5%), which is incorporated in the crystal with the Si–H and SiH_2 configurations, as results from IR spectroscopy measurements.

The use of a mixture of SiF_4 and SiH_4 in a PECVD growth with H_2 as cover gas presents the additional advantage [5] of allowing the deposition of nc-Si films also at

---

* See details in the next section.

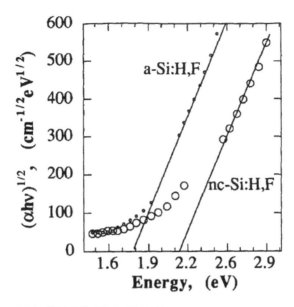

FIGURE 3.12 Tauc's gaps of nanocrystalline and amorphous silicon film grown by SiF$_4$,SiH$_4$, H$_2$ plasma. (*From* G. Cicala, G. Bruno, P. Capezzuto, L. Sciavulli, V. Capozzi, G. Perna (1997) Deposition of photoluminescent nanocrystalline silicon films by SiF$_4$,SiH$_4$ H$_2$ plasmas, Conference Paper in MRS Online Proceeding Library Archive · November 1997:DOI: 10.1557/PROC-452–809. *Reproduced with permission of Cambridge University press License number 4724240042876, License date Dec 08, 2019.*)

very low temperatures (95°C). Unlike the single-phase films [92] deposited at high temperatures, nc-Si films grown in the temperature range 95–250°C are a mixture of crystalline and amorphous Si. In good agreement with the Cicala's suggestion, it is shown that the crystallinity $\chi_c = \dfrac{m_c^{\text{Si}}}{m_c^{\text{Si}} + m_a^{\text{Si}}}$, where m_c^{Si} is the mass of crystalline silicon and m_a^{Si} is the mass of amorphous silicon, increases with the increase of temperature and with the increase of the H$_2$/SiF$_4$ ratio, as indicated by the increase of the Raman peak intensity at 520 cm^{-1}, characteristic of crystalline Si.

Though the average grain size is of the order of 10 nm, the Tauc's gap of this material is shifted well above the quantum confinement regime (see Figure 3.12), with energy gap values of 2.2 eV.

We will see in Section 5.3.1 that the optoelectronic properties of these fluorinated nc-Si films strongly differ from those grown from SiH$_4$ and H$_2$.

3.3.3 Growth and Structure of Nanocrystalline Silicon Films Prepared with LEPECVD

3.3.3.1 Basic Chemistry of the LEPECVD Process

This and the following sections deal with the physics and chemistry of nanocrystalline silicon films grown from SiH$_4$/H$_2$ mixtures using the low-energy plasma-enhanced

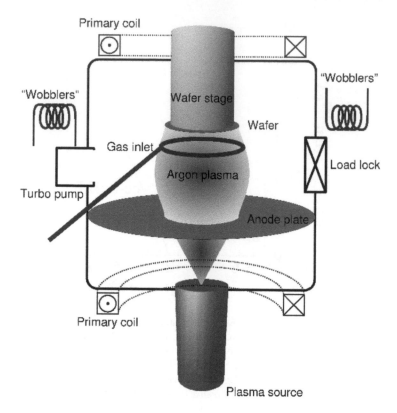

FIGURE 3.13 Schematic drawing of the LEPECVD furnace used for the deposition of nc-Si films. (*Reproduced from Final progress report: Nanophoto Project no. 013944 (2005).*)

CVD (LEPECVD) technique,* already used with excellent results for SiGe epitaxy [93], on which a vast amount of theoretical and experimental information is available [90, 94–112], but not yet discussed under an unitary view point.

A schematic drawing of the LEPECVD chamber used to grow nc-Si films is shown in Figure 3.13. The system is preliminary evacuated with a turbo-pump at a base pressure of 10^{-9} mbar, to minimize residual contaminations of the chamber, and works at a pressure of about 10^{-2} mbar, sufficiently low to avoid the formation of dust (see preceding section).

Gaseous precursors SiH_4 (for Si deposition), B_2H_6, and PH_3 (diluted in Ar) for doping, along with H_2 and Ar can be introduced into the system *via* mass-flow controllers under computerized control.

The precursors are decomposed into reactive radicals by the Ar plasma, leading to quite efficient growth processes, with about 5–10% of the chemical precursors incorporated into the growing film,[†] regardless of the substrate temperature.

---

* In the frame of an European Research Project, The Nanophoto Project no. 013944 (2005) sponsored by the European Commission within the SIXTH FRAMEWORK PROGRAMME: PRIORITY 3-NPM Research Area 3.4.2.1 Development of fundamental knowledge: 3.4.2.2.1–2 Modelling and design of multifunctional materials. (Most of the pictures reported in this section are drawn from the final Project Report.)

† With lower production costs.

LEPECVD shares with conventional PECVD processes the capability to carry out a preliminary surface cleaning with hydrogen plasma,* but has the additional advantage of suppressing the ion damage of the surface and the consequent defects formation during the nc-Si deposition in a SiH_4/H_2 plasma, due to the control of the energy of the impinging ions.

Defect formation in epitaxial SiGe layers due to high-energy ions was, in fact, already observed by Ramm et al. [113] and Rosenblad et al. [114], who remarked that stacking faults (SF) formation occurs at a depth of 90 nm in the growing films, for ion energies larger than 15 eV. Wang et al. [76] confirmed that defect formation occurs as well in the plasma deposition of nc-Si from SiF_4/H_2 mixtures, without control of the ion energy.

Unlike classical CVD and HWCVD processes, that rely on the catalytic decomposition of the precursors ($SiHCl_3$ or SiH_4) on a hot substrate, in the case of plasma-enhanced CVD processes, conventionally called *athermal*,† the precursors are decomposed into the plasma phase under electron-impact conditions, with the formation of radicals, and the properties of the deposited film depend on the chemical reactions occurring at the surface of the film and on surface diffusion processes, at the substrate temperature.

Therefore, the growth kinetics is substantially different in the two cases.

The low -temperature, reaction-limited CVD-growth rate‡ is, in fact, given by the following equation

$$\frac{dx}{dt} = C K_g p_{SiH_4} (1 - \theta_H) \tag{3.6}$$

where C is a constant, K_g is the temperature-dependent reaction constant, θ_H is the hydrogen coverage, and p_{SiH_4} is the partial pressure of silane.

Instead, the reaction rate of the analogous process, carried out under plasma enhancement of the ionization (I) and dissociation (DE) of the precursor molecules, is given by the following equations

$$\frac{dx}{dt}\bigg|_{I,DE} = K_{I,DE} N_a \tag{3.7}$$

$$K_{I,DE} = \sqrt{\frac{2}{m_e}} \int_0^{E_{max}} \sqrt{\varepsilon}\, \sigma_{I,DE}(\varepsilon) f(\varepsilon)\, d\varepsilon \tag{3.8}$$

where N_a is the density of molecular/atomic species in the plasma phase, $K_{I,DE}$ is the reaction constant, $\sigma_{I,DE}$ is an energy-dependent cross section, ε is the electron energy, m_e is the electron mass, and

$$N_e = \int_0^{E_{max}} f(\varepsilon)\, d\varepsilon \tag{3.9}$$

is the electron density in the plasma phase [112].

---

* Which would also induce surface passivation and reconstruction [115].
† Since they occur without external heating.
‡ The precursor decomposition is the overall slower step.

As expected from Equation 3.7, and as one can see in Figure 3.14 and in Table 3.1, the growth rates of LPCVD (low-pressure CVD) processes decrease with the decrease of the deposition temperatures, down to values lower than 0.01 nms⁻¹. Furthermore, according to Chung [115] the PECVD deposition rates are favored by reactive precursors, as is the case polysilanes, but very low levels of water vapor are also required to avoid the formation of hillocks and other defects typical of epitaxial films. The main advantage of LEPECVD deposition (see Table 3.1) is that in the same temperature range of low-temperature CVD processes, deposition rates up to 5–10 nms⁻¹ are systematically achievable [113, 116].

It could be, therefore, supposed that also the deposition of nc-Si could be favored by using a LEPECVD process, as is discussed in the next section.

3.3.3.2 Detailed Physico-Chemical Aspects of the nc-Si Growth with the LEPECVD Process

The experimental evidence of the satisfactory operation of the LEPECVD process for the growth of nc-Si films is enriched by a vast amount of experimental and theoretical results concerning the physico-chemical aspect of the deposition process.

While experimental results deal with the deposition growth rate as a function of the process parameters, as well with the structural and spectroscopic properties of the nc-Si, the main result of theoretical studies consists of

- A detailed fluido-dynamic model of the plasma reactor and a chemical gas-phase kinetic scheme consisting of electronic, ionic, and radical reactions [97, 107], capable of forecasting the concentration of the radicalic

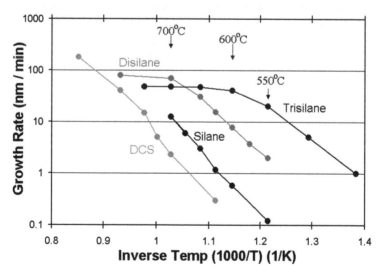

FIGURE 3.14 Dependence of the growth rates of CVD epitaxial silicon deposition from the process temperature and the nature of precursors. (*After* K.H.Chung (2010) Silicon-based epitaxy by chemical vapor deposition using novel precursor neopentasilane *PhD Thesis*, Princeton University, Dept. Electrical Eng. *Reproduced with kind permission of Keith Chung, May 30, 2018.*)

TABLE 3.1
Comparison between Reaction Rates of LPCVD and LEPECVD Processes for the Epitaxial Deposition of Silicon

Material	Process	Gas Mixture	T °C	Growth Rate (nm s⁻¹)	Base Pressure	Working Pressure (mbar)	Residual Water Contamination	References
Si	LPCVD	SiH_4/H_2	600	0.016	UHV	1.33	ppb	115
Si-Ge (strained)	LPCVD	SiH_4-GeH_4/H_2	500–750	0.016 0.2	UHV	1.33	ppb	115
Si	LEPECVD	SiH_4/H_2	600	5–10 (*)	UHV (10–9 mbar)	10^{-2}		114
Si-Ge	LEPECVD	GeH_4/H_2	550	5–10 (*)	UHV	10^{-2}		114

species in the plasma and at the sample surface as a function of the inlet gas concentration.

- Atomistic models capable of forecasting the rates of the reactions occurring at the surface of the growing films, also in the presence of strain effects at the nanocrystalline/amorphous interfaces, associated with different growth substrates.

Simulations and mass-spectrometric measurements show that SiH_3^* is the radicalic species dominating the chemistry of nc-Si deposition at low SiH_4 fluxes (see Figure 3.15), while the SiH^* species dominate at the largest H_2 inputs.

The predictive value of the model is demonstrated by the good fit of the calculated values of the key ionic species with their mass-spectrometric abundance measured in the growth furnace.

The knowledge of surface radicalic fluxes and of the mean surface composition during nc-Si growth is needed for surface chemistry simulations. The calculated surface fluxes as a function of the silane dilution are reported in Figure 3.16, which shows the large excess of hydrogen radicals at the sample surface at each silane dilution and leads to the assumption that the surface is fully hydrogenated, at least in the absence of Ar^+ bombardment.

The results of further studies addressed to the calculation of the H-desorption rate as a function of the Ar^+ ions energy and flux rate, of the surface orientation and of the local strain [98, 99] reported in Figure 3.17, show that in the typical energy range used in the growth process (10–15 eV), the probability that an Ar^+ ion could remove a surface hydrogen atom is rather low, regardless of the surface orientation or of the strain condition at the surface. These results were obtained by exploiting the empirical potential developed by Izumi et al. [118], which provides by far the best description of the Si–H interactions, when compared with other potentials available in the literature.

This allows, preliminarily, the conclusion that the observed high growth rates in LEPECVD are not due to a strong "cleaning action" produced by Ar^+ ions at the temperatures of interest (below 300°C), and that the growth process occurs on a highly hydrogenated surface.

On these highly hydrogenated surfaces, silyl radicals $SiH_{x(x=1,2,3)}^*$ are adsorbed, but the SiH_3^* radicals preferentially chemisorb, by sticking on surface H-sites, with the formation of silane and dangling bonds (DB)

$$H_{surf} + SiH_3^* \rightarrow SiH_4^g + DB \tag{3.10}$$

at energies close to the thermal energy (0.025 eV) as shown in Figure 3.18.

In turn, DBs are reactive sites, which do allow the chemisorption of SiH_3^* from the plasma phase

$$SiH_3^* + DB \rightarrow SiH_3^{surf} \tag{3.11}$$

which occurs with a sticking probability assumed to be unitary [97, 109]. SiH_3^{surf} is not an epitaxial species, but the incorporation of Si in an epitaxial position is assisted

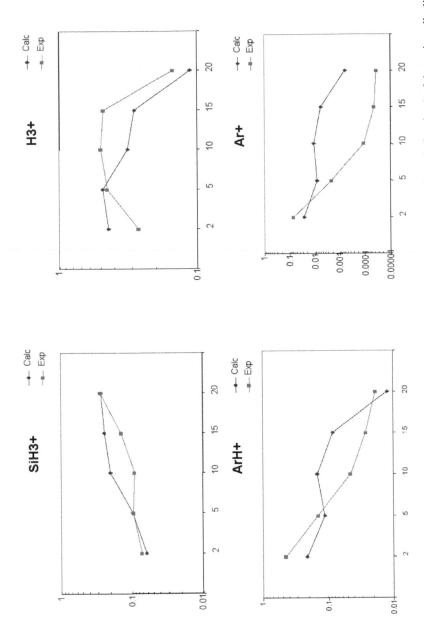

FIGURE 3.15 Comparison between experimental (mass-spectrometric) and calculated concentration (in mole fractions) of the main radicalic species present in the plasma phase as a function of the SiH_4 flow rate [given in standard cubic cm per minute]. (*Final progress report: Nanophoto Project no. 013944 (2005).*)

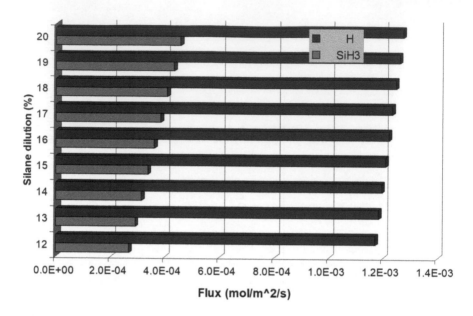

FIGURE 3.16 Calculated fluxes (in moles/m² s⁻¹) of relevant radicals SiH_3^* and H^* at the growth surface as a function of the silane dilution. (*Final progress report: Nanophoto Project no. 013944 (2005)*.)

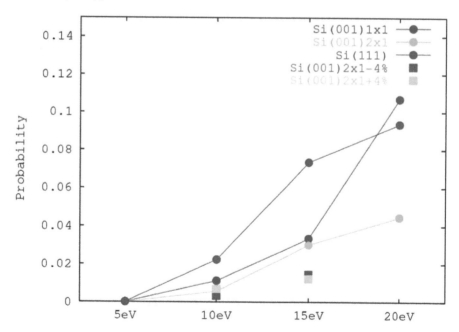

FIGURE 3.17 Effect of ion energy on the desorption yield of surface H atoms by Ar⁺ ion impact on several silicon surfaces, including two 4% biaxially strained ones (tensile strain is positive, compressive is negative). (*Final progress report: Nanophoto Project no. 013944 (2005)*.)

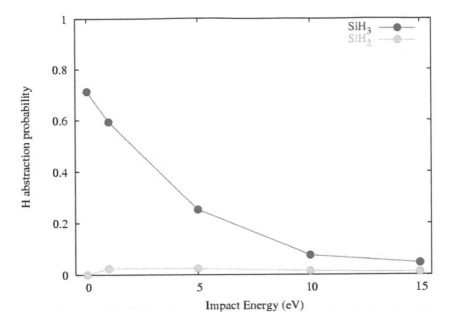

FIGURE 3.18 Chemisorption probability of $SiH_3^{\bullet}$and $SiH_2^{\bullet}$ silyl radicals with different impact energies on a Si(001)2x1:H surface.* (*Final progress report: Nanophoto Project no. 013944 (2005).*)

by hydrogen radicals in the plasma phase, which etch out hydrogen from chemisorbed SiH_3^{surf} and promote a first transformation of the chemisorbed SiH_3^{surf} species in a truly epitaxial SiH_2^{epi} species [108]

$$SiH_3^{surf} + H^{\bullet} \rightarrow SiH_2^{epi} + H_2 \qquad (3.12)$$

followed by the dissociation of the SiH_2^{epi} species

$$SiH_2^{epi} \rightarrow Si_{Si} + H_2 \qquad (3.13)$$

and incorporation of Si atoms in epitaxial positions, surmounting a kinetic barrier of only 1 eV, accessible at the experimental temperatures of nc-Si growth [98–99, 117].

This mechanism is responsible for the growth of crystalline silicon aggregates, while the non-epitaxially adsorbed silyl species are involved in the growth of amorphous silicon islands, with a process which would involve their thermal dissociation

$$SiH_x^{ads} \rightarrow Si_a + \frac{x}{2}H_2 \qquad (3.14)$$

---

* These results were obtained by exploiting the empirical potential developed by Izumi *et al.* [118[It was found out that this potential provided by far the best description of the Si-H interactions, when compared with other potentials available in the literature

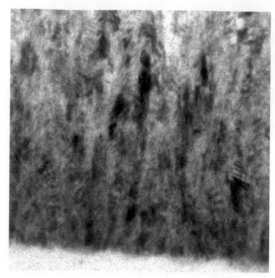

FIGURE 3.19 Comparison between the modeled plain view images of a nc-Si film *(left)*. *(After* P.L. Novikov, A. Le Donne, S. Cereda, Leo Miglio, S. Pizzini, S. Binetti, M. Rondanini, C. Cavallotti, D. Chrastina, T. Moiseev, H. von Kaenel, G. Isella, and F. Montalenti (2009) Crystallinity and microstructure in PECVD-grown Si films: a simple atomic-scale model validated by experiments, *Appl. Phys.Lett.* **94** 051904. *Reproduced with permission of AIP, License Number 4723740599871, license date Dec 07, 2019)* and the TEM image of a vertical section of a nc-Si film *(right)*. *(Final progress report:- Nanophoto Project no. 013944 (2005).)*

Since the calculated SiH$_3^*$ sticking yield is three times larger on amorphous silicon aggregates, we would expect an expansion of the amorphous regions at the expense of the crystalline ones.

A further set of Car–Parrinello simulations showed, instead, that atomic H can etch out silicon atoms adsorbed at the surface of an amorphous Si phase, restoring local crystallinity [95]. On that basis, the simulation of nc-Si films growth using an atomic-scale Kinetic Monte Carlo (KMC) model,* that considers H, SiH$_3$, and SiH$_2$ the active species in the process dynamics [108, 109], leads to the result reported in Figure 3.19, where the calculated image of a section of an nc-Si film is compared with the TEM image of a vertical section of a nc-Si film and in Figure 3.20, where the calculated plain view image of an nc-Si film is compared with the TEM image of a section of a nanocrystalline Si film grown under the same process conditions [96, 119]. White dots in these images consist of a-Si.

These results show an excellent qualitative agreement between theory and experiment, confirm that nc-Si grows inside an amorphous silicon matrix, and demonstrate that the basic physics of LEPECVD-grown nc-Si films were understood.

The KMC method was also used to simulate the evolution the crystallinity $\chi_c = \dfrac{m_c^{Si}}{m_c^{Si} + m_a^{Si}}$ of the film as a function of the silane dilution $d = \dfrac{\Phi_{SiH_4}}{\Phi_{SiH_4}\Phi_{H_2}}$, where

* KMC is a kinetic Monte-Carlo simulation carried out by imposing a diamond lattice and using a model cell with 128 × 128 sites per layer.

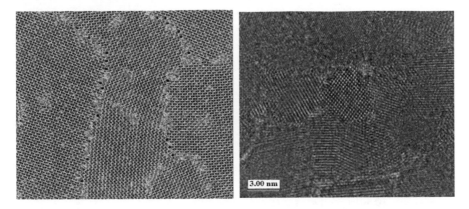

FIGURE 3.20 Comparison of the modeled plain view image of a nc-Si film grown with a silane flux of 20 sccm, $d=30\%$, at 280°C with the TEM image of a section of a nanocrystalline Si film grown with the same conditions. (*Final progress report: Nanophoto Project no. 013944 (2005).*)

the crystallinity χ_c is the average mass of the crystalline content in an nc-Si sample and is a key figure of merit of the material, since it accounts for the light absorption and electrical conductivity contributions of the crystalline and amorphous components of the nc-material. In turn, χ_c is experimentally determined by the deconvolution of the Raman spectra and by X-ray diffraction (XRD) measurements.

The results of the dependence of the crystallinity on the dilution ratio, calculated for a process carried out at a typical temperature of growth (280°C), are reported in Figure 3.21 together with the experimental values of the Raman crystallinity of films grown at the same temperature.

This figure shows that theory and experiments do agree well for dilution ratio values d below 30%, while large experimental deviations occur for d values higher than 30%, where amorphous and crystalline phases in variable amounts do coexist.

Not only the dilution $d = \dfrac{\Phi_{SiH_4}}{\Phi_{SiH_4}\Phi_{H_2}}$, but also the absolute values of the silane flux Φ_{SiH_4}* are shown to influence the crystallinity and the structural homogeneity of the samples. Table 3.2 shows, in fact, that low silane fluxes Φ_{SiH_4} lead to higher crystallinity values together with better volume homogeneity, even in the range of d between 30% and 50%.

It was also found, see Figure 3.22, that the material is completely amorphous in correspondence of the substrate but turns out to have a more crystalline structure near the film surface [101] at SiH_4 fluxes of 20 sccm, and presents almost constant Raman features throughout the entire thickness of the sample, if the silane flux is held constant at 12 sccm.

It should be also observed that the Raman peak position at the surface of the samples systematically takes values around 510 cm^{-1}, which is evidence of strong phonon confinement [120, 121].

---

* Given in sccm = standard cubic cm per minute.

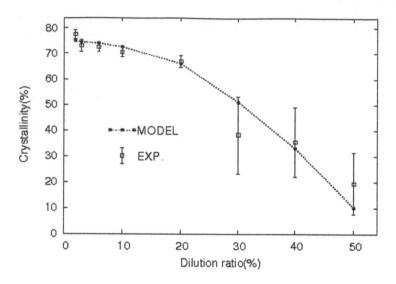

FIGURE 3.21 Dependence of the film crystallinity on the dilution ratio at a growth temperature T = 280°C and at a silane flux of 20 sccm. Comparison between model and experiments. (*After* P.L. Novikov, A. Le Donne, S. Cereda, Leo Miglio, S. Pizzini, S. Binetti, M. Rondanini, C. Cavallotti, D. Chrastina, T. Moiseev, H. von Kaenel, G. Isella, and F. Montalenti (2009) Crystallinity and microstructure in PECVD-grown Si films: a simple atomic-scale model validated by experiments, *Appl. Phys. Lett.* **94** 051904. *Reproduced with permission of AIP, License Number 4723740599871, license date Dec 07, 2019.*)

A similar, excellent, consistency of the simulation with the experimental results occurs for the measured and calculated growth rates, as shown in Figure 3.23, for two dilution ratios $d = \dfrac{\Phi_{SiH_4}}{\Phi_{SiH_4}\Phi_{H_2}}$ of the inlet gases, as a function of the silane flux.

Eventually, also the structural evolution of nc-Si nuclei embedded in an amorphous Si phase during a thermal annealing process was successfully modeled by

---

TABLE 3.2

Evolution of the Crystallinity and Bulk Homogeneity as a Function of the Silane Flux for Two Values of the Dilution Factor *d*

d	Φ (SiH$_4$) [sccm]	Crystallinity χ_c	Bulk Homogeneity
30%	20	28%	Bad
30%	16	27%	Good
30%	12	41%	Good
50%	20	36%	Bad
50%	16	27%	Acceptable
50%	12	39%	Good

Substrate ITO/glass. Deposition temperature 250°C

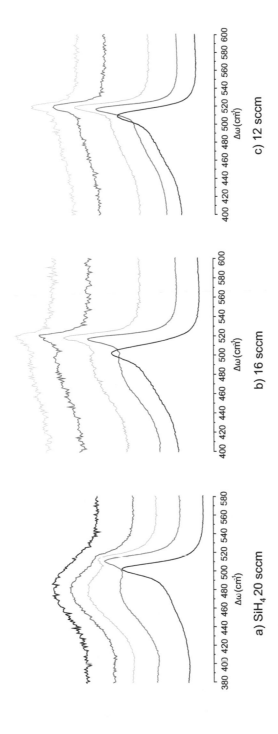

FIGURE 3.22 Effect of silane flux on the Raman depth profiles measured on films grown at 250°C. The spectra on the top are always measured in correspondence of the substrate interfaces *(Final progress report: Nanophoto Project no. 013944 (2005)).*

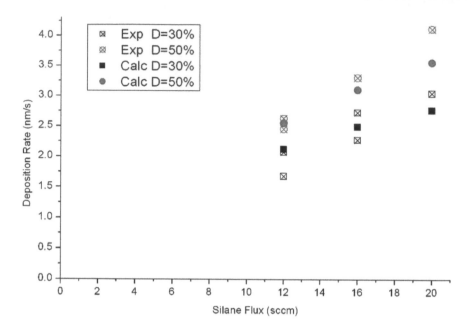

FIGURE 3.23 Comparison between calculated and measured average film growth rates as a function of the silane flux, for two different dilution ratios *d*. (*Final progress report: Nanophoto Project no. 013944 (2005)*.)

molecular dynamics simulations by Mattoni and Colombo [106, 119], as an example of a solid/solid phase transformation, whose kinetic behavior could be described using the Kolmogorov–Johnson–Mehl–Avrami (KJMA) theory [122, 123]. The underline hypothesis is that, initially, all the amorphous grains randomly nucleate a crystalline nucleus, such that after a nucleation time τ_N the crystallization is controlled only by the rate constant of the crystal growth process.

Figure 3.24 displays the results of the KJMA theory applied to a 3.5 ns annealing at 1200 K of a fixed number of Si grains distributed in an amorphous Si matrix.

During the thermal annealing, the nanocrystal size increases with time, associated with the decrease of the amorphous silicon mass, and a-Si begins to behave as a capping layer of Si-nanocrystals.

Similar results were recently obtained by Zhang et al. [124] who, using a kinetic Monte Carlo Method, studied the topological evolution of PECVD-grown silicon films driven by the presence of atomic hydrogen, which promotes the thermally enhanced evolution of the films from amorphous to crystalline.

It is supposed that hydrogen breaks surface Si–Si bonds and forms an intermediate Si–H–Si bridging configuration, that allows hydrogen to migrate almost freely along the surface. It is also supposed that Si–H–Si species do allow the incorporation of the silyl radicals, even in the absence of dangling bonds at the surface, thus promoting the film growth.

By increasing the temperature of the growth process, eventually, the hydrogen-promoted H-desorption from the hydride bonds of the amorphous phase is enhanced, leading to Si–Si bond reconstruction with the formation of a crystalline Si network.

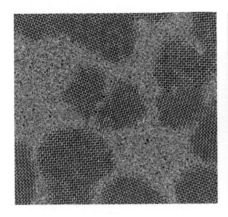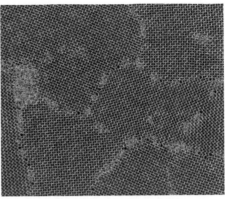

FIGURE 3.24 Monte Carlo simulation of the crystallization of a cluster of nanograins in an amorphous matrix, induced by a 3.5 ns-long thermal annealing at 1200 K. White regions consist of a-Si. (*Final Report: Nanophoto Project no. 013944 (2005).*)

Figure 3.25 displays the calculated evolution of the crystallinity of a-Si film with the temperature which leads to the formation of crystalline silicon islands.

3.3.3.3 Microstructural Characterization of nc-Si Samples

XRD, TEM, high-resolution (HR)TEM, and atomic force microscopy (AFM) measurements did allow a complete structural characterization of the nc-Si films, LEPECVD-grown on oxidized silicon substrates.

The morphology of the films deposited on silicon substrates, in a region of high crystallinity and good homogeneity, is well-represented by the images which were displayed in Figure 3.11, already preliminary discussed in Section 3.3.2.

The TEM image on the left of the figure is a longitudinal section of the sample, that shows a columnar structure orthogonal to the substrate surface, consisting of elongated grains having a section of a few nm (more than 6 nm), which well corresponds the average crystal size resulting from XRD measurements. The TEM image on the right is a plain view of a section of the sample, which shows that each grain is separated from the neighboring grains by a white tissue, consisting of amorphous silicon.

The HRTEM images of sections of the same sample displayed in Figure 3.26 show, additionally, that each grain is a cluster of misoriented and defected nanocrystallites with an average size of 3 nm. Apparently, some these nanocrystallites are separated from the neighboring ones by a-Si, and a few of them also show the presence of stacking faults, as is shown in the right-hand-side image.

This kind of mosaic structure is typical of hierarchical materials found in nature [125], that generally consist of two phase components with different chemical properties, and the fact that it develops spontaneously in a synthetic material opens new perspectives for understanding the physics behind it.

Grain misorientation does also occur along the columns, as could be observed in the HRTEM image displayed in Figure 3.27, that shows, in fact, that their preferential

116   Defects in Nanocrystals

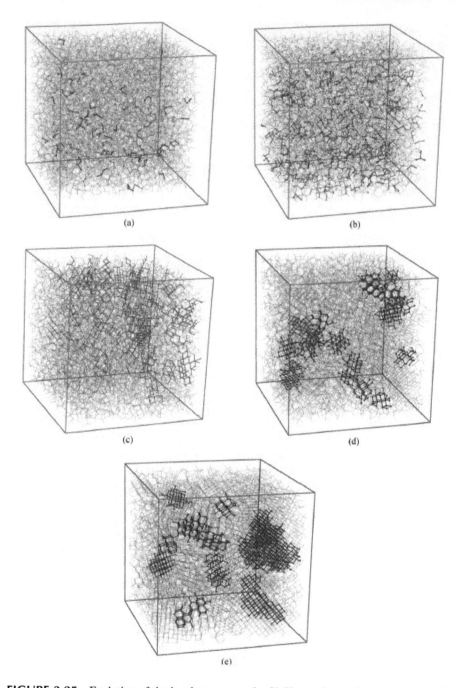

FIGURE 3.25 Evolution of the local structure of a-Si films at increasing temperatures (a) 300 (b) 400 (c) 500 (d) 550 (e) 650 K. (*After* Y. Zhang, H. Wang, and S. Jiang (2018) Evolution of structural topology of forming nanocrystalline silicon film by atomic-scale-mechanism-driven model based on realistic network *AIP Advances* **8** 095321 *Licensed under a Creative Common Attribution (CCBY) license.*)

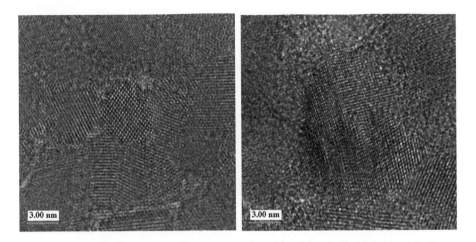

FIGURE 3.26   HRTEM images of the sample of Figure 3.11 The left image shows a cluster of misoriented grains with an average size of 3 nm, while the right image shows the presence of stacking faults inside a nanocrystallite. (*Final report: Nanophoto Project no. 013944 (2005).*)

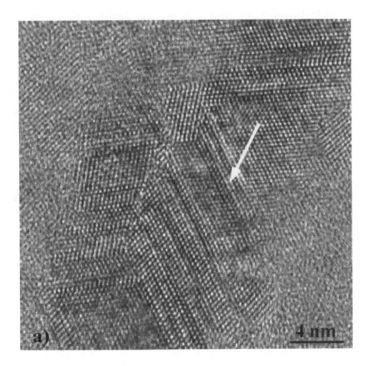

FIGURE 3.27   HRTEM micrograph of a section of a nc-Si film, showing details of the columnar structure far from the growth interphase. Substrate CZ-Si; the white arrow indicates the growth direction. (*Final report: Nanophoto Project, no. 013944 (2005).*)

TABLE 3.3

Effect of the Dilution Ratio d% on the Grain Size (GS) and Crystallinity along Three Different Growth Directions L

d%	L$_{(111)}$ GS(nm)	L$_{(220)}$ GS(nm)	L$_{(311)}$ GS(nm)	L$_{mean}$ GS(nm)	χ%
10	14.5	18,6	9	14	70
20	8.1	24.5	15.5	16	65
30	(5.6)	24	20.5	16.5	40±10
40	9.1	14.5	–	12	40±10
50	5.8	–	–	5.8	20±

orientation is along (111), but that also (200) and (311) oriented grains are distributed within the columns, as can be seen in Table 3.3.

Table 3.3 shows also that the grain size decreases with the increase of the dilution factor d%, getting a limiting value of 5.8 nm with d = 50% [102].

The silane flux has also a substantial role in the microstructure and in-depth homogeneity of the LEPECVD-grown samples, high silane fluxes favoring low crystallinity features and large in-depth inhomogeneities [101].

As an example, Figure 3.28 displays the Raman depth profiles of two nc-Si samples grown under a silane flux of 20 and 12 sccm, respectively. It is apparent that the sample grown with the largest silane flux is crystalline only close to the film surface and presents a large in-depth homogeneity, while the other is crystalline along the entire thickness.

It is possible also to observe that the sample grown with the largest silane flux presents an important red-shift (9 cm^{-1}) of the Raman peak, which is almost negligible (2 cm^{-1}) in the case of the other sample. Since a Raman red-shift of 2 cm^{-1} is

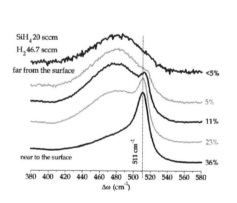
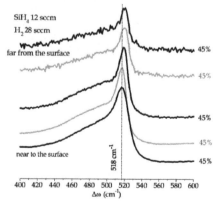

FIGURE 3.28 Raman depth profiles of two nc-Si samples grown under a silane flux of 20 (a) and 12 (b) sccm. (*After* A. LeDonne, S. Binetti, G. Isella, S. Pizzini (2008) Structural homogeneity of nc-Si films grown by Low Energy PECVD *Electrochem. Solid State Letters* **11** P5–P7 *Free permission of ECS to authors.*)

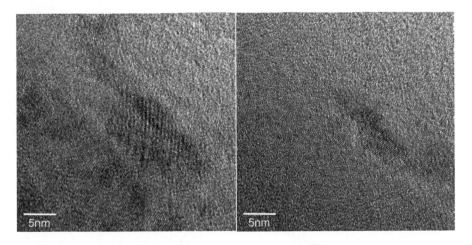

FIGURE 3.29 TEM images of silicon nanocrystals present in nc-Si films grown with a silane flux of 12 sccm (left) and 20 sccm. (*Final report: Nanophoto Project no. 013944 (2005).*)

associated with a nanocrystal size of 10 nm, while a red-shift of 9 cm^{-1} is relative to an average nanocrystal size of 3 nm [126]* the nanocrystals present close to the substrate interface of the sample (a) are smaller in size and could present quantum confinement conditions.

This conclusion is confirmed by the HRTEM measurements carried out on two nc-Si films grown with a silane flux of 20 and 12 sccm (see Figure 3.29), which show that the nanocrystal size of the sample grown with the largest silane flux is about 5 nm in length, while that of the other sample is more than 10 nm.

One can, thus, conclude that the intrinsic structural defectivity of nc-Si films consists of the presence of an a-Si matrix which embeds crystalline silicon clusters, consisting of misoriented and defected nanocrystallites, some of which could be, as well, surface-decorated with a-Si. Eventually, it is demonstrated that crystallinity and the nanocrystallite size and orientation strongly depend on the local chemistry of the deposition process.

Laser annealing has been demonstrated as a powerful tool to tune the crystallinity of a starting material consisting of a-Si LEPECVD grown at 70°C from pure silane [110], which presents a Raman spectrum typical of a-Si (see Figure 3.30, spectrum (a)).

The effect of laser annealing at low fluence (2 kWcm^{-2}) is the development of a Raman peak close to the Raman frequency of the c-Si emerging from the broad band of a-Si (spectrum (b), Figure 3.30) and the presence of the c-Si peak alone at a fluence of 12kWcm^{-2} (spectrum (c) in Figure 3.30).

In both cases the Raman peak shows an important red-shift from its standard position at 520 cm^{-1}, consistent with the presence of nanocrystals with a size of 10 nm at 30% of the laser power and 2 nm at 100% of laser power (12 kWcm^{-2}).

---

* According to Doerk et al. [127] silicon nanowires with a diameter of 50 nm do not present significant red-shifts of their Raman peaks.

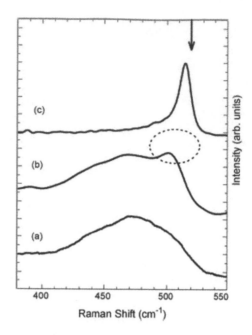

FIGURE 3.30 Raman spectra of laser irradiated a-Si samples as a function of the laser fluence (a) as grown sample (b) 2 kW cm^{-2} (c) 12 kWcm^{-2}. (*Final rReport: Nanosilicon Project no. 013944 (2005)*.)

3.4   SILICON NANOWIRES: GROWTH PROCESSES AND STRUCTURAL PROPERTIES

Unlike nanocrystals, which are 0D materials, nanowires are 1D materials and their properties are experimentally demonstrated to depend on their length/diameter ratio.

Two main processes are used for the growth of silicon nanowires (Si NWs) [19, 20, 127–132]. The first, systematically adopted in the past, is a metal-assisted, high-temperature, gas phase process, using tetrachlorosilane (SiCl$_4$) or silane (SiH$_4$) as the precursors. In this process, the metal catalyzes the decomposition of the precursors and forms a Si-supersaturated liquid Si–Me alloy at the surface of a suitable substrate.

Consequent to the segregation of the excess Si, one obtains a random or ordered distribution of nanometric wires or whiskers, depending on the temperature and on the surface distribution of the metal.

The second is a metal-assisted electrochemical process, by means of which ordered arrays of nanowires could be produced at room temperatures.

In literature metal-assisted processes for the growth of elemental or compound semiconductor nanowires are mentioned as metal-catalyzed processes, assuming that the primary role of the metal is to catalyze the decomposition of the precursors. Actually, the role of the metal is also to behave as the solvent of the elemental or compound semiconductor components in Vapour-Liquid-Solid (VLS) processes, where the solid wire segregates from a liquid phase, or to deliver an active surface to absorb the semiconductor components in the case of a Vapour-Solid-Solid (VSS) process, occurring below the eutectic temperature of a binary or pseudo-binary phase [131], when the wire segregates from a solid solvent phase.

It is also important to note that in the literature the effect of size of the metal particle on the melting temperature of a generic Me-Si eutectic is generally neglected, assuming that all processes which experimentally do occur below a bulk eutectic temperature are VSS processes. This is, of course, an improper assumption, on the base of the concepts and experimental evidences detailed reported in Chapter 2.

A few other techniques are also used, among them the laser ablation process presents different thermochemical features with respect to the other processes, that need to be discussed in detail.

In the following of this section we will discuss the physico-chemical aspects of the different growth processes which have a relevant role in the NWs size and defectivity. The structural, morphological, and topological features of silicon nanowires will be, instead, deeply discussed at the end of this section.

3.4.1 AU-ASSISTED VLS-CVD PROCESSES

Silicon whiskers and silicon nanowires could be grown with a VLS mechanism using tetrachloro-silane ($SiCl_4$)* [133, 134] or silane (SiH_4) as the precursors, and nano-size metal particles deposited on the surface of a single crystal silicon substrate [135, 136]. The Au-assistance has been mostly employed, but Al, Zn, and Pt have also been used.

The metal could be deposited at the surface of a silicon substrate as a thin metallic layer (0.6 nm of Au according to Westwater [137]) or as an ordered distribution of preformed metallic clusters. In both cases, at temperatures above the eutectics (363°C for Au, 577°C for Al, and 419°C for Zn[†]), randomly distributed or ordered arrays of nano-droplets of a liquid Me–Si alloy spontaneously form on the silicon surface, which will act as seeds for the silicon growth when the gaseous precursors are introduced in the reaction chamber.

The liquid Me–silicon alloys catalyze the decomposition of silane to elemental silicon and hydrogen on its surface,[‡] and behave as silicon sinks with the formation of a super-saturated metal–silicon liquid alloy. Thermodynamic equilibrium is eventually achieved with the surface segregation of the excess Si and the epitaxial, unidirectional, or anisotropic growth of a whisker or a wire, at temperatures above the eutectic, although hypoeutectic temperatures have been also operatively used [137]. In this last case the growth occurs with a VSS process, with a supersaturated Au/Si solid solution which works as the seed of the Si NWs.

The thermodynamics of Si nanowire growth could be deduced from the phase diagram of the (bulk) Au–Si system, see Figure 3.31, from which one can observe [138] that the lower growth temperature under VLS conditions is that of the Au–Si eutectic. Furthermore, if the NW growth is assumed to occur from an Au–Si solution

---

* Due to the associated lower operation temperatures, silane is the preferred gaseous precursor.
[†] These temperatures are bulk values; we will see that the eutectic temperatures for nanosized systems are different.
[‡] In the presence of H_2 as reactant in the case of $SiCl_4$.

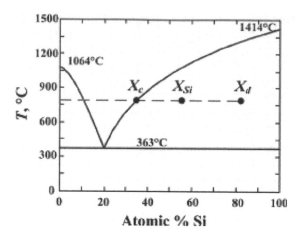

FIGURE 3.31 Phase diagram of the Au–Si system. (*After* A. Sarikov (2011) Metal induced crystallization mechanism of the metal catalyzed growth of silicon wire-like crystals *Appl. Phys. Lett. 99 143102. Reproduced with permission of AIP Publishing, License Number 4724250803562 License date Dec 08, 2019.*)

supersaturated in Si, of concentration X_d, it occurs under a gradient of chemical potential $\Delta\mu = RT \ln \dfrac{X_d}{X_c}$ at the melting temperature of an alloy of equilibrium concentration X_c.

Process temperatures as high as 950°C are employed for the growth of silicon whiskers with $SiCl_4$ [133, 134] or within 320 and 600°C for the growth of Si NWs in the case of silane.

It is experimentally known that the growth temperature has a key role on the morphology, size, and defectivity of the wires. In fact, if the growth temperature is held within 450–460°C, the crystal structure of the wires is cubic, while they keep the hexagonal structure of wurtzite at temperatures between 500 and 600°C [139].

The nature of the process occurring at growth temperatures between 320 and 950°C could not, however, be discussed without considering that the eutectic temperature of the Au--Si system depends on the size and on the shape, as shown by Hourlier et al. [140–141] who showed that in the case of nanocrystals with a radius of 5 nm, the calculated eutectic temperature (250°C) is lower than that of bulk (362°C), while in the case of a nanowire it is higher (450°C) than the bulk eutectic temperature, see Figure 3.32.* Similar results were obtained by Kim et al. [142], who calculated a lowering of the eutectic temperature for the nanoscale Au–Si system by ≈ 240°C.

Therefore, if we take as good the results of Hourlier [140–141], NWs grown at temperatures below 450°C are grown at hypoeutectic temperatures, and this could

---

* A similar behavior is observed for the Au–Ge system.

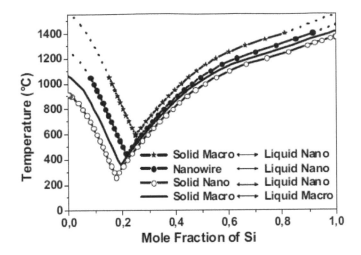

FIGURE 3.32   Evolution of the eutectic temperature of the Au–Si system from bulk to nano conditions (r = 5 nm). (*From* D. Hourlier, P. Lefebre-Legry, P. Perrot (2009) Preparation of silicon-based nanowires and the thermochemistry of the process JEEP 00002 EDP Science *Open Access Journal.*)

be the reason why NWs grown at temperatures around 450–460°C on silicon substrates present a cubic structure, since the growth occurs with an epitaxial solid to solid transformation.

From the arguments discussed in Chapter 2, concerning the relationship between nanocrystal size and melting temperature, and from the diagram of Figure 3.33, which displays the melting temperatures of Si NWs as a function of their diameter [143], it could be, also, argued that the NWs diameter should decrease with the decrease of the process temperature.

Not only does this assumption seem intuitively to be correct, but the few experimental data found in the literature reported in Table 3.4 show, also, that the expected trend is qualitatively confirmed.

The silane gas pressure, the substrate orientation, and the growth direction have, instead, a critical role on the nanowire growth kinetics, and on their size and defectivity.

As an example, Westwater et al. [137] show that using a Si (111) substrate, the wire orientation is <111> and that a kink-free regime occurs at 600 and 520°C at low silane partial pressures (up to 0.05 torr). At 320°C (below the eutectic temperature) only randomly oriented whiskers with a diameter of 15 nm could be, instead, grown.

As a further example, Li et al. [144] investigated the conditions of thermodynamic stability of semiconductor nanowires and showed that above a critical length h^* the NW is thermodynamically unstable and that the value of the critical length decreases with the increase of the NW diameter as can be seen in Figure 3.34 for Si and ZnSe NWs.

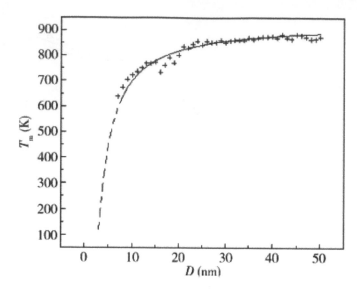

FIGURE 3.33 Size-dependent melting temperatures of silicon nanowires. (*After* Z J. Yanfeng, Z. Yamin Influence of gold particles on melting temperature of VLS grown silicon nanowires (2010) *J. Semiconductors* **31** 1 012002–1/5. *Reproduced under IOP Creative Commons CC BY 3.0 license.*)

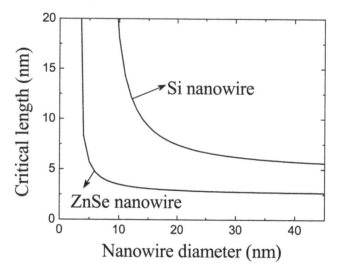

FIGURE 3.34 Calculated diameter dependence of the critical length h^* of Si and ZnSe nanowires. (*After* X. Li, J. Ni, R. Zhang (2017) A thermodynamic model of diameter- and temperature-dependent semiconductor nanowire growth *Scientific Reports* **7** 15029 *Open access article distributed under the terms of the Creative Commons CC BY license.*)

Li et al. [144] also show that the energy per unit length of a NW depends on the wire orientation and that the dependence is critical when the NW diameter is less than 25 nm. In these conditions, see Figure 3.35, a change of growth direction could arise to allow favorable thermodynamic stability conditions.

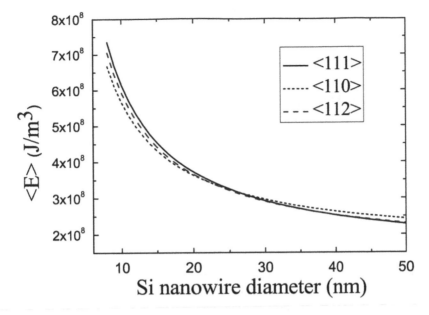

FIGURE 3.35 Size and orientation dependence of the thermodynamic stability of silicon nanowires. (*After* X. Li, J. Ni, R. Zhang (2017) A Thermodynamic Model of Diameter- and Temperature-dependent Semiconductor Nanowire Growth *Scientific Reports* **7**, 15029 *Open access article distributed under the terms of the Creative Commons CC BY license.*)

Eventually, according to Xu et al. [145], tapering* of NWs could also occur as consequence of the reduction of the metal particle size during the growth and of lateral overgrowth caused by the direct (uncatalyzed) deposition of Si species onto the nanowire sidewalls.

In the CVD-VLS process the diameter of the nanowire depends on the size of the metallic (gold) nanocluster used to seed the wire. With a well-controlled CVD growth process, using silane as the precursor and hydrogen as the carrier gas, seeding the process with Au nanoclusters having a size of a few nanometers, nanowires with a constant diameter down to 3 nm could, in fact, be grown [127–129].

The diameter of the nanowire could be, however, larger than that of the original Au seed, consequent to the formation, at the beginning of the process, of a solution of Si in the gold seed, with a consequent volume increase.

It has been, also, demonstrated that the predominant growth direction is the <111> for NWs with a diameter larger than 20 nm, while the <100> direction becomes dominant for diameters smaller than 10 nm.

In every case, Si NWs are grown under (quasi-)equilibrium conditions, and we expect that not only the bulk but also the surface of the wire should present physical properties compatible with equilibrium conditions.

We will, actually, see in Chapter 5 that the optical properties of Si NWs grown with the VLS-CVD process are consistent with a growth carried out in thermodynamic equilibrium conditions, unlike those grown with the MACE process.

---

* Tapering is the process of lateral overgrowth.

TABLE 3.4

Temperature Dependence of the Diameter of Si NWs ((*) Kink-Free Regime at Low SiH$_4$ Partial Pressures (0.01–0.05 torr) (**) B- or P-Doped (***) Kinks at Every Partial Pressure

T°C	NW Diameter	Process	References
950	100	CVD (SiCl$_4$)	134
600	120 (*)	CVD (SiH$_4$)	137
600	30 (**)	CVD (SiH$_4$)	148
520	120–20 (*)	CVD (SiH$_4$)	137
440	20–40 (***)	CVD (SiH$_4$)	137
320	15 (wiskers)	CVD (SiH$_4$)	137

The growth kinetics was investigated among others by Jeong et al. [146] for the case of Pt- and Au-catalyzed growth in SiCl$_4$/H$_2$ atmosphere at 900–1100°C, and the results suggest that the rate-determining step is the incorporation of Si atoms in the crystalline lattice of the wire at the liquid/solid interface. Furthermore, the nature of the metal catalysts is shown to affect the rate of this step, and that its activation energy is larger (130 kJ/mol) for Au in comparison with Pt (80 KJ/mol).

Considering that the activation energy for the CVD deposition of Si from SiH$_4$ below 650°C is 213.38 kJ/mol [147] and that the activation energy for the incorporation of Si atoms from the catalyst in the crystalline lattice of the wire should not depend on the nature of the silicon precursor (SiH$_4$ instead of SiCl$_4$) and holding 130 kJ/mol for Au catalyst, it is clearly apparent that the catalyzed growth of Si-NWs is much faster than the lateral, uncatalyzed growth at the NWs walls, thus favouring conditions of constant diameter wires.

Although Au presents several advantages with respect to Zn, Al, and Pt, among which the lowest eutectic temperature,* and thus, the lowest growth temperature from a liquid alloy, this would not avoid a possible Au contamination of the Si NW, making the use of Au incompatible with microelectronic or optoelectronic applications. This conclusion is supported by a work of Sato et al. [148], who showed with DLTS measurements that in both B- and P-doped Si NWs, grown at 600°C using silane as the precursor, a donor level (0/+) at $E_v + 0.36$ eV is observed, associated with the presence of a gold–hydrogen complex [149]. The concentration of this complex in B-doped NWs is $5 \cdot 10^{15}$ cm^{-3}, two orders of magnitude higher than the gold solubility in bulk silicon at 600°C ($3 \cdot 10^{13}$ cm^{-3}). No conclusive explanation of this excess solubility of gold in B-doped Si NWs is given, but a possible role of B–H complexes in this enhancement is suggested. Since the complex is observed in a portion of the NW far from the Au-tip, the conclusion is that Au diffuses along the NW during its growth, possibly at the NW surface, in good agreement with Ross et al. [150], who demonstrate that the Au diffuses at the surface and induces surface faceting.

---

* The eutectic temperatures of the Zn–Si, Al–Si, and Pt–Si systems are 419, 577, and 980°C, respectively.

As already mentioned, the crystallographic structure of CVD-grown Si NWs is not necessarily that of diamond. Fabbri et al. [139], in fact, show that the undoped material is a 90–10% mixture of wurtzite- and diamond-cubic Si NWs, and that this ratio is modified by B-doping, which also favors the presence of non-negligible amounts of defective (flawed) NWs. These last, in turn, consist of a crystalline silicon core embedded in a shell of a-Si. A discussion about the puzzling structural properties of silicon NWS and on recent theoretical modeling of non-cubic Si NWs is included in Section 3.4.6

3.4.2   Al-Assisted VSS Processes

Despite several attractive properties,* including its P-doping properties, aluminum found less success than gold as a metal catalyst for NWs growth.

This is mainly due to the fact that Al is easily oxidable, and that the formation of aluminum oxide can adversely influence the NW growth, according to Hainey and Redwing [151], who show also that the process could occur with the parasitic formation of silicon whiskers, due to the Al-reduction of the silica sidewalls of the process reactor.

A successful Al-catalyzed process relies, therefore, on the capability to mitigate[†] the aluminum oxide formation, using a hydrogen atmosphere or a UHV growth.

This could be done using a VLS process, like in the case of Si NWs growth with Au as the catalyst, employing SiH_4 as the precursor and carrying the process above 577°C, the eutectic temperature of the Al–Si system, but also this process has been proven unsatisfactory and difficult [152].

Alternatively, a VSS process could be exploited, thanks to the known fast diffusion of silicon in solid Al [151–153], at hypoeutectic temperatures.

In this case, the active phase is a solid solution of Si in Al (the α-Al phase), which is prepared by depositing a thin layer of Al onto a single crystal silicon substrate, which works also as the Si source. After the low temperature deposition, the wafer is annealed at temperatures above the eutectic and the Al layer melts and breaks in a number of nanometric droplets of a Si–Al alloy, with a size of ≈ 35 nm.

Then the wafer is cooled at the growth temperature (490 or 430°C), to start a localized epitaxial process in correspondence of the α-Al dots. According to the results of Levitas and Samani [154] and Lai et al. [155], at these temperatures the alloy is still solid since the size-effect on melting temperature becomes substantial only for sizes below 10 nm, as can be seen in Figure 3.36.

The morphology of CVD-grown Al-catalyzed wires is strongly growth-temperature dependent; see Figure 3.37 [151]. One can observe that below 500°C the wires are vertically aligned, but tapered, tapering being the typical defect of Si NWs VSS-grown with Al as the catalyst [152]. Apparently, the best results are obtained at 490°C, while at 520°C and above only nanopyramids with a triangular section could be grown, with a temperature limit of 600°C.

---

* Low-temperature eutectic, great earth abundance, and lower cost.
† Due to the high Gibbs free energy of formation of Al_2O_3, only a kinetic control of its formation could be a satisfactory solution for mitigation, since thermodynamic suppression is impossible in practical conditions.

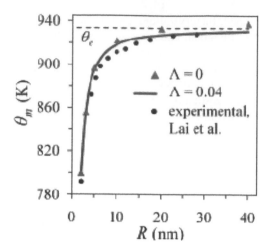

FIGURE 3.36 Theoretical and experimental size dependence of the melting temperature θ_m of Al nanoparticles. (*After* V.I. Levitas and K. Samani (2011) Coherent solid/liquid interface with stress relaxation in a phase-field approach to the melting/solidification transition *Phys. Rev.B* **84** 140103; Λ is a kinetic parameter. *Reproduced with permission of APS, License number* RNP/19/DEC/020994 *License date: Dec 07, 2019.*)

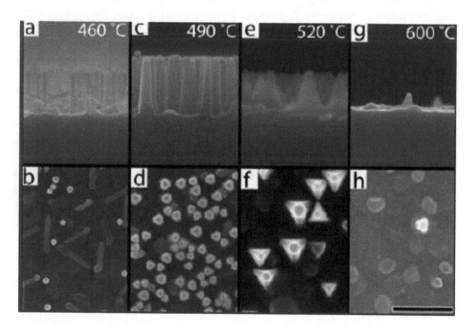

FIGURE 3.37 Growth-temperature dependence of the morphology of Al-catalyzed Si nanowires. (*From* M.F. Hainey, J.M. Redwing (2016) Aluminum-catalyzed silicon nanowires: Growth methods, properties and applications *Appl.Phys.Rev.* **3** 040806. *Reproduced with permission of AIP, License Number 4723701287015: License date Dec 07, 2019.*)

As expected, the NWs are strongly contaminated with Al, up to 10^{20} cm^{-3}, depending on the temperature, well above the saturation concentration of Al in Si (5–6·10^{18} cm^{-3} at 500°C). Since Al is a p-type dopant but also a deep-level impurity [156], devices fabricated with Al-catalyzed Si NWs demonstrated significantly reduced performances even when compared with Au-catalyzed Si NWs [151].

3.4.3 Metal Catalyst-Free Growth of Silicon and Germanium Nanowires: Oxide- or Sulfide-Catalyzed Growth

Silicon nanowires could be also CVD-grown without a metal catalyst. As an example, the anisotropic, non-epitaxial growth of silicon (and germanium) whiskers can be obtained using as the seeds silicon (or germanium) nanoparticles segregated by thermal decomposition of a layer of non-stoichiometric SiO$_x$ (GeO$_x$), grown onto a single crystal silicon (or germanium) substrate by reaction with ultrapure water at 100°C [157].

The CVD growth of Si NWs is typically carried out at 490°C in a silane/hydrogen atmosphere, after a short stage at 520°C, with growth conditions similar to those typical for Au-catalyzed NWs.*

CVD growth was also carried out at higher temperatures (620–660°C) in the presence of HCl which helps in preserving conditions of constant diameter growth in this temperature range.

The final product is a dense mat of nanowires.

A TEM image of a single Si NW grown at 620°C displayed in Figure 3.38 [157] shows that its surface is free of an amorphous Si overcoating and that the wire diameter is close to 10 nm.

Two main questions are left open concerning the nucleation and growth mechanism of catalyst-free grown Si NWs.

The first concerns the nucleation of the nanocrystalline Si seeds and the further the NW growth, that is suggested to occur with a vapor-solid (V-S) mechanism.

The second concerns the mechanism involved in the retardation of the lateral growth associated with the uncatalyzed Si deposition on the NW sidewalls, that could be understood as due to the surface oxidation of the wire by minute oxygen contamination of the growth atmosphere.

Presumably, the CVD growth takes place under self-catalyzed conditions in correspondence of the nc-Si seeds, that define also the NW diameter like in the case of Au or Ag tips, and where the decomposition of the silane diffusing through the porous SiO$_x$ layer does occur. This qualitative picture does not however explain why the nucleation and the anisotropic growth of a wire is thermodynamically or kinetically favored with respect to the 3D growth of the nucleus, with the formation of a nanoparticle.

It is also unexplained why the preferred growth direction is the <112> or the <110>, whereas the <111> and <110> are predominant in the metal-assisted Si NWs

---

* Kim et al. [157] suppose that the seeding process does occur as a consequence of the reaction of SiH$_4$ with the reactive SiO$_x$ layer. We believe that the seeding occurs during the initial step of the process and follows Reaction 3.1.

FIGURE 3.38 TEM image of a single crystal silicon nanowire catalyst free grown at 620°C. (*After* B.-S. Kim, T.-W. Koo, J.-H. Lee, D. S. Kim, Y. C. Jung, S. W. Hwang, B.L. Choi, E. K. Lee, J- M. Kim, and D. Whang (2009)Catalyst-free Growth of Single-Crystal Silicon and Germanium Nanowires *Nano Lett.* **9** (2), 864–869 scale bar 10 nm. *Reproduced with permission of ACS 09 Dec 07, 2019.*)

growth, depending on wire diameter. We remind, however that the calculated thermodynamic stability of these configurations is very similar; see Figure 3.35.

A different method has been used by Ishiyama et al. [158] for the (formally) catalyst-free growth of Si NWs, based on the use of silicon sulfides (SiS and SiS_2) at temperatures higher than the melting point of SiS_2 (1090°C).

The process is carried out by sealing in a quartz capsule evacuated at 20 Pa the silicon substrate with sulfur powder, and the epitaxial growth does occur in the presence of molten SiS_2 which behaves as the catalyst and of SiS vapors which behave as precursors, although no chemical model is given by the authors, in view of the poor knowledge available about the Si–S system.

Actually, the solid monosulfide is unstable, but vapors of SiS could be obtained by heating a mixture of Si and SiS_2

$$Si + SiS_2 \rightleftharpoons 2SiS \tag{3.15}$$

In turn, SiS_2 is the stable silicon–sulfur compound that is obtained by reacting sulfur with silicon [159]

$$Si + 2S \rightleftharpoons SiS_2 \tag{3.16}$$

FIGURE 3.39 SEM images of arrays of <111> oriented Si NWs grown with the sulfide process. (*After* T. Ishiyama, S. Nakagawa, T. Wakamatsu (2016) Growth of epitaxial silicon nanowires on a Si substrate by a metal-catalyst-free process *Scientific Reports* **6**:30608 | DOI: 10.1038/srep30608 *Open access paper.*)

On that basis, it could be supposed that heating the silicon substrate with sulfur powder leads to the formation of a surface layer of SiS_2 which behaves as the source of SiS vapors and as a liquid phase seed (at temperatures higher that its melting point), where the tip of the wire nucleates when the liquid is saturated with silicon.

Looking at the SEM images displayed in Figure 3.39 of an array of Si NWs grown with this process, one can observe that the silicon surface is very rough, possibly due to SiS_2 etching, and it could be, therefore, supposed that the liquid phase seeds are randomly distributed onto the silicon substrate surface, which may present a fractal distribution of covered and uncovered regions.

The further growth of the wire arises thanks to SiS vapors, which are the gaseous precursors of silicon, like SiH_4 in CVD growth, which ensure silicon saturation conditions in the liquid SiS_2 seed with the reverse of Equation 3.15.

The wires present a core-shell structure (see Figure 3.40 left panel), with a core consisting of silicon and a shell consisting of silicon oxide (see Figure 3.40 right panel), and their diameter in the middle section of the wire averages 500 nm. The oxide shell arises from the residual oxygen content in the growth capsule.

Apparently, the wire diameter is well inside a mesoscopic size and the wires are heavily contaminated with oxygen, but potentialities exist for qualitative and quantitative improvements, although the surface roughening by chemical etching could be an obstacle.

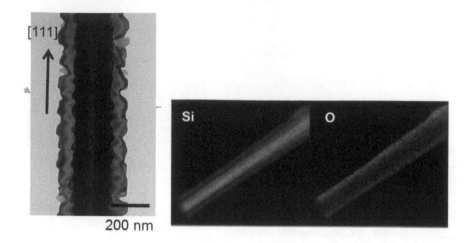

FIGURE 3.40 (*Left side*) TEM image of a section of a Si nanowire grown with a metal-catalyst-free process. (*Right side*) Scanning TEM surface maps of silicon and oxygen contamination of the wire. (*After* T. Ishiyama, S. Nakagawa, T. Wakamatsu (2016) Growth of epitaxial silicon nanowires on a Si substrate by a metal-catalyst-free process *Scientific Reports* 6:30608 | DOI: 10.1038/srep30608 *Open access paper.*)

3.4.4   SILICON NANOWIRES OBTAINED WITH A LOW-TEMPERATURE METAL-ASSISTED CHEMICAL ETCHING (MACE) PROCESS

Metal-assisted chemical etching (MACE) processes are used to selectively etch a silicon substrate and obtain arrays of vertically aligned nanowires arranged with a random fractal geometry [160], following an electroless variant of the route originally designed for the synthesis of porous silicon by anodic etching in HF, recently improved by using a MACE process [161].

The chemistry of the process, which is a room-temperature, electroless galvanic process, consists in the anodic dissolution of Si in an electrolytic solution of HF and H_2O_2, activated by a distribution of noble metal (Ag or Au) grains at the surface of a single crystal silicon wafer

$$Si + 2HF + 4F^- + 4h \xrightarrow{\text{Au}} H_2SiF_6 \tag{3.17}$$

coupled to the following reactions

$$2H_2O_2 + 4H^+ \xrightarrow{\text{Au}} 4H_2O + 4h \tag{3.18}$$

$$4H^+ + 4e \xrightarrow{\text{Au}} 2H_2 \tag{3.19}$$

of which the first determines the NW growth rate, delivering the holes of Equation 3.17.

The process proceeds with the formation of an array of pores which underline a meso- or nanostructured array of filaments (in the case of porous silicon) or of constant diameter nanowires, if the etching is isotropic.

According to Gösele and Lehman [35] the basic condition for the electrochemical formation of constant diameter pores is the passivation of the surface of the pores by H_2O_2, such that the electrochemical dissolution process could occur only at the pore tips, where silicon is in contact with the catalyst metal. Since holes are involved in the anodic dissolution of silicon in a fluorinated electrolyte, a "passivated" condition of a silicon electrode does only occur if the surface is depleted of holes.

Alternative views found in the literature suggest that the diffusion model explains quantitatively the morphology of porous silicon [162, 163] and that the activation of the process is due either to the injection of holes from silicon (reaction 3.17) [164] or to the delivery of holes from H_2O_2 (reaction 3.18) [161].

If the process is considered from a purely electrochemical viewpoint, one should first account for the fact that the reactions under Equations 3.17 and 3.18 do occur in short circuit conditions localized at the metal/silicon interface, and thermodynamically driven by the Gibbs free energy $\Delta G = -1935.09$ kJ mol^{-1} of the process resulting from the electroneutral coupling of reactions 3.17 and 3.18

$$Si + 6HF + 2H_2O_2 \rightleftarrows H_2SiF_6 + 4H_2O \qquad (3.20)$$

Further, considering that the electrical conductivity of gold ($\sigma = 4.10 \cdot 10^7$ S/cm^2) is $\approx 10^{10}$ larger that the conductivity of silicon ($\sigma = 1.53 \cdot 10^{-3}$ S/cm^2), the reaction rate at the metal/silicon interface would be 10^{10} higher than at the naked walls of the NWs, even in conditions of negligible reaction overvoltages.

Therefore, holes are not necessarily involved in the silicon oxidation process and hole depletion is not required to limit the reaction rate at the silicon walls.

While pristine Si NWs were grown with a two-stage method, starting with the electroless deposition of Ag at the surface of a silicon substrate from a solution of AgNO$_3$, followed by the electroless NWs growth using a solution of hydrogen peroxide and HF, Pal et al. [165]* and Irrera et al. [164, 166–169] replaced the electrochemical deposition of Ag or Au with the vacuum evaporation of an ultrathin (2–10 nm) layer of metal (Ag or Au) on the <100> oriented surface of single crystal silicon, previously oxidized and then HF-etched to remove the oxide. <100> oriented surfaces are preferred due to the lower density of surface back-bonds needed to be broken in the dissolution process, in comparison to other surfaces.

The Au (or Ag) films have a characteristic nanoscale morphology, with a fractal distribution of covered and uncovered regions, of which a typical example is given in Figure 3.41.

This kind of texture is due to a self-organization of the gold (silver) distribution in the ultrathin Au(Ag) layer, that is close "to the percolation threshold, with a gold filling fraction of 54,6% on a silicon (111) surface" [168].†

Metal-covered regions behave as the starting point of the chemical etch process, which proceeds for a depth of several micrometers, leading to the vertically oriented array of ultrathin (5–9 nm) NWs shown in Figure 3.42. At the end of the etch process the metal can be removed by a final chemical etch with KI.

Since the entire process is carried out at room temperature, metal contamination can be considered virtually‡ absent in MACE-grown Si NWs. Another advantage of MACE due to the low-temperature operation is that the NWs maintain the structural and chemical properties of the silicon substrate.

---

* Who report also about previous papers on the topic.
† Original sentence in the B. Fazio et al. paper [168].
‡ Actually, even negligible, contamination invariably occurs driven by a concentration gradient, since gold is a fast diffuser.

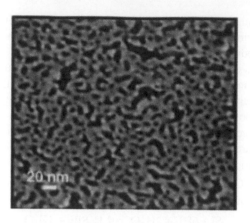

FIGURE 3.41   SEM image of the nanoscale morphology of a ultrathin gold deposit on a silicon sample: dark regions are metal-free regions. (*After B. Fazio et al.* (2016) Strongly enhanced light trapping in a two-dimensional silicon nanowire random fractal array *Light Science & Applications* **5**, e16062. *Reproduced under Creative Commons Attribution 4.0 International License.*)

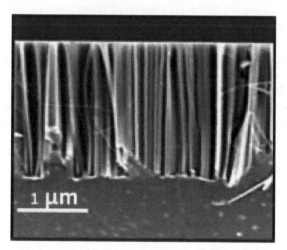

FIGURE 3.42   SEM image of an array of Si NWs produced with the MACE process. (*After B. Fazio et al.* (2016) Strongly enhanced light trapping in a two-dimensional silicon nanowire random fractal array *Light Science & Applications* **5**, e16062. *Reproduced under Creative Commons Attribution 4.0 International License.*)

Eventually, TEM images of MACE-grown NWs show that they present a core-shell structure, with a core consisting of crystalline silicon and an amorphous shell consisting of SiO_2 [167], formed by reaction with atmospheric oxygen.

Si NWs grown with the MACE process present an exceptionally strong Raman peak at 520 cm^{-1} and a very strong luminescence signal at 690 nm [168], with which we will deal in Chapter 5.

The mechanism of self-assembly of Si NWs prepared with the MACE process and the nano- architectonics involved has been recently discussed by Carapezzi and

Cavallini [169]. It has been found that these systems are fractal in nature, but that a multi-fractal analysis should be applied to describe their complex morphology.

3.4.5 NANOWIRES GROWN WITH THE LASER ABLATION METHOD

Laser ablation processes were also used for the synthesis of silicon nanowires [170–172], using as targets sintered mixtures of silicon and of the metal catalyst, which could be Au, Fe, or Ca.

The thermodynamics behind this process are the phase diagrams of the Si–Me systems, which do present a wide interval of stability of a liquid Si–Me alloy.

Morales and Lieber [170] used for the Fe- catalyzed growth of Si NWs a target consisting of a sintered mixture of silicon and iron with a $Si_{0.9}Fe_{0.1}$ composition, which is laser-ablated at 1200°C in a tubular furnace, fluxed by an inert gas. The wire formation process is supposed to occur entirely in the liquid phase, with the preliminary volatilization of nanometric droplets of a liquid Me–Si alloy

$$Si_{0.9}Fe_{0.1}(s) \xrightarrow{\ h\nu\ } Si_{0.9}Fe_{0.1}(l) \qquad (3.21)$$

and the successive segregation of Si under form of individual nanowires from the Si-supersaturated liquid Si-Fe alloy (see Figure 3.31 the analogous case of Au-Si alloys) following the mechanism (a) in Figure 3.43, when the liquid droplets dispersed

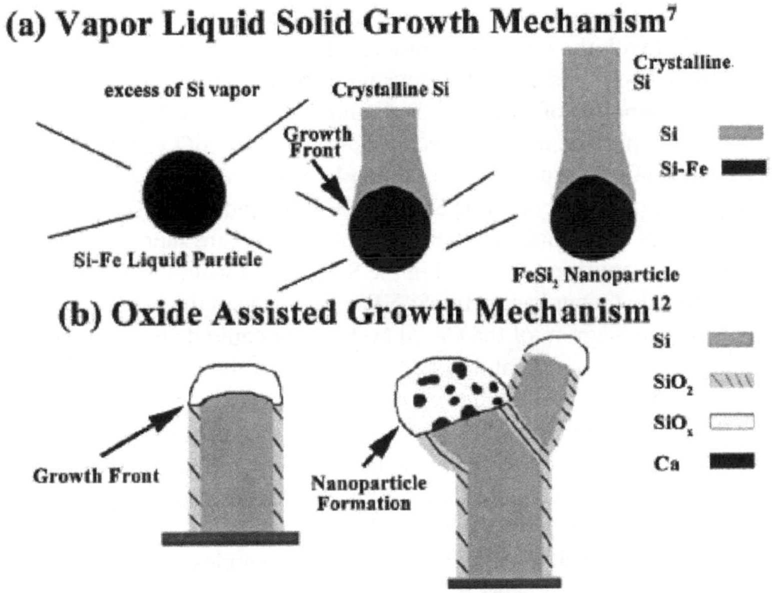

FIGURE 3.43   Growth mechanisms of Si NWs carried out with the laser ablation method (a) Fe-assisted growth (b) oxide-assisted growth. (*Reproduced with permission of AIP publishing* from R.J. Barsotti, Jr., J. E. Fischer, C.H. Lee, J. Mahmood, C.K.W. Adu, P.C. Eklund (2002) Imaging, structural, and chemical analysis of silicon nanowires *Appl. Phys.* Lett. **81** 2866 *License number 4700851406545 License date Nov 02, 2019.*)

in the gas phase stick on the walls of the furnace. Since it has been experimentally demonstrated that the growth process does occurs only at a temperature not lower than 1150°C , it is supposed that the alloy remains in the liquid state at temperatures below that of the eutectic $\beta - SiFe_2 - Si$ at 1207°C. in view of the nanometric size of the Fe_xSi_{1-x} droplets.

The TEM image of an individual wire shows that it consists of a uniform diameter core, surrounded by an amorphous SiO_2 layer. The diameter of the core of this nanowire was 7.8 nm and the thickness of the sheet about 5 nm. The average diameter of the Si NWs grown with this process was around 10 nm, with a length from 1 to 10 μm.

Similar results were obtained by Shi et al. [172] and Barsotti et al. [173], who used, as well, a $FeSi_2$ target and a furnace temperature of 1100°C, lower than that used by Morales, with the explicit hypothesis that "catalysis by nanoscale Fe particles would still be possible because the melting point of $\beta - FeSi_2$ would be considerably reduced"* from its equilibrium temperature.

The material prepared this way consists of a random distribution of wires, some of which are nanochains of silicon nanowires connected by many nanoparticles. The nanowires present a core-shell structure with a silicon core having an average diameter of 6.7 ± 2.9 nm and a SiO_x shell of few nm in thickness, oxygen being the main contaminant of the gas atmosphere. The average length of the wires is around 10 μm. Some of the wires were shown to be contaminated by Ca, which was not intentionally added, but whose presence was supposed to be due to Ca contamination from the balls used to mill the Fe–Si mixture.

For this reason, Barsotti et al. [173] suppose that a Ca-assisted process could also occur in parallel with the Fe-assisted process, following the mechanism depicted in Figure 3.43b. The process starts to occur with the formation of SiO vapors, originated by the sublimation of silicon in the liquid Si-Fe alloy and its successive oxidation by oxygen impurities in the argon used as cover gas. SiO vapors condensate on the quartz walls of the furnace as a solid non-homogeneous phase [174–175] that is, in fact, a mixture of nanocrystalline Si and SiO_2 [176] and behaves as the seed for the growth of Si wires with a SiO_2 shell. Since vapors of SiO are supposed to be contaminated by Ca, Ca impurities dissolved in the SiO droplet would lead to the formation of a Ca–Si alloy that would assist the Si NW growth, as in the case of Fe–Si alloys.

From the phase diagram of the Ca–Si system one could deduce that it allows the operation at a process temperature of 1100°C, also in the absence of eutectic temperature depression induced by the size of the particles.

According to Bhattacharya et al. [174] who studied quantum confinement effects in Si nanowires prepared by laser ablation, some energy upshifts observed are difficult to quantify, due to the contribution of both the core and the oxide layer. More details about these results will be discussed in Chapter 5.

Different structural behavior is presented by two samples of Si NWs prepared, as well, with a laser ablation process by Yu et al. [177], using as the target a hot-pressed (at 150°C) mixture of Si with 3% of Fe and of Ni and Co nano-powders as catalysts (type A samples) or a hot-pressed target without Ni–Co catalyst (type B samples), at a process temperature of 1200°C.

The reaction product is a distribution of Si NWs, with an average diameter around 13 nm for type A samples and a diameter between 15 and 60 nm for the B type

---

* Original sentence in ref. 172.

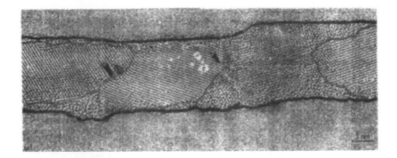

FIGURE 3.44 TEM micrograph of a polycrystalline Si NW grown with a laser ablation process. (*Courtesy of Prof. S-L. Zhang, Dept. Physics, Beijing University.*)

TABLE 3.5
Structure of High-Pressure Polymorphs of Silicon

Designation	Structure	Pressure Region (GPa)	Notes
I	Cubic (diamond)	$0 \rightarrow 12.5$	
II	Body-centered tetragonal (β-Sn)	8.8–16	
III	Body-centered cubic (BC8)	>18	Metastable at P ambient
IV	Wurtzite (hexagonal) diamond	>18	Metastable at P ambient
V	Primitive hexagonal	$\approx 14 \rightarrow 40$	
VII	Hexagonal close packed	40–78.3	
VIII	FCC cubic	≥ 78.3	

Source: Properties of Silicon, *EMIS Data Review Series*, Vol. 4, INSPEC: London, UK (1988)

samples. XR diffraction measurements on these wires show that they consist of crystalline Si without additional impurity phases.* TEM measurements on type A nanowires, see Figure 3.44, show that they are polycrystalline in nature, with nanometric grains exhibiting different orientations [178].

Eventually, the Raman spectra of A and B samples show prominent Raman features at 504 and 511 cm^{-1}, respectively [179], evidencing a substantial red-shift from that of crystalline silicon and the set-up of phonon confinement features inside the grains of the wire, not along the wire [179].

3.4.6 STRUCTURAL PROPERTIES OF SILICON NANOWIRES

Silicon is a typical polymorphic material, see Table 3.5, with several high-pressure phases and two low-pressure metastable phases, one of which plays a role in the of morphology of silicon nanowires.

As already mentioned in Section 3.4.1 [148], TEM measurements carried out on CVD-grown, Au-catalyzed Si NWs show that undoped Si NWs are a 90–10% mixture of wurtzite and diamond cubic Si NWs, and that this ratio is modified by

* The presence of some % of impurity phases could not be excluded from XRD measurements.

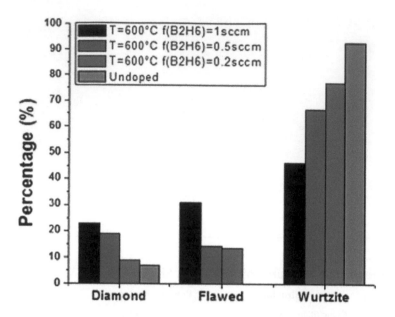

FIGURE 3.45   Statistical distribution of Si polytypes in CVD-grown Si NWs. (*After* F. Fabbri, E. Rotunno, L. Lazzarini, N. Fukata, G. Salviati (2014) Visible and infra-red light emission in boron-doped wurtzite silicon nanowires *Scientific Reports* 4, 3603, (Nature Publishing Group) *Open Access Journal.*)

B-doping, which also favors the presence of non-negligible amounts of defective (flawed) NWs, featuring a crystalline silicon core embedded in a shell of a-Si [148, 180]. P-doping, instead, favors the growth of diamond cubic NWS. A schematic diagram of the statistical distribution of silicon polytypes in a mat of Si NWs is reported in Figure 3.45, which demonstrates the critical role of B-doping.

We have already shown that it is experimentally known that the growth temperature has a key role in the structure, size, and defectivity of the wires. In fact, if the growth temperature is held within 450–460°C, the crystal structure of the wire is cubic, while it turns out to the hexagonal structure of wurtzite at temperatures between 500 and 600°C [139].

The different structure of the phases is obviously reflected by the micro-Raman spectra of the two polytypes, peaking at 520 cm^{-1} in the case of cubic silicon and at ≈495 cm^{-1} in the case of hexagonal silicon.* These results well agree with those of Foncuberta et al. [181] who in addition showed that undoped Si NWs, CVD-grown in SiH$_4$ between 500 and 650°C, present complex structural relationships with their diameter. In fact, all the NWs with a diameter larger than 10 nm and the 56% of the NWS with a diameter smaller than 10 nm exhibit the hexagonal wurtzite structure of the Si polymorph Si IV,† while the 44% of Si NWS having a diameter less than 10 nm are grown with the diamond structure of the Si polymorph Si I.

---

* Also the optoelectronic properties are influenced by the structure, as will be discussed in detail in Chapter 5.

† Actually, they showed that two slightly different Si IV structures are present.

They also show that the cubic NWs present a sharp Raman peak at 520 cm^{-1}, while those with the wurtzite structure present a sharp Raman peak at 503 cm^{-1}. All samples do also appear free of structural defects.

According to Doerk et al. [182] silicon nanowires with a diameter of 50 nm do not present significant red shifts of their Raman peaks, and crystallize with the diamond structure.

It is also known that Si NWs with the structure of the metallic, β-Sn isomorph (Si II) could be only prepared by submitting cubic Si NWS to a high-pressure (P > 9.9 GPa) treatment [183–184], while wurtzite silicon NWS could be prepared by laser ablation of crystalline silicon samples [185].

Apparently, the fact that the hexagonal Si IV polymorph is metastable at ambient pressure qualitatively explains why Si NWs with this structure could be found in a mat of Si NWs, but the reason why Si NWs could be grown with the hexagonal structure should be understood as a kind of thermodynamic stabilization of a non-cubic NW core due to the increasing role of surface and edges when the diameter decreases. On that basis, Zhang et al. [186] were able to demonstrate the stability of NWs with the wurtzite structure with MD calculations.

The effect of a thermodynamic stabilization of an Si NW with the hexagonal structure could be also suggested on the basis of the results of Yang and Jiang [187], who show that the melting temperature depression of silicon nanocrystals with a radius of 6 nm ($T_m \approx 1450$ K) is consistent with the application of a pressure of ≈ 6 GPa. The spontaneous set-up of such a pressure arising from surface stress would be close to that needed to stabilize the wurtzite structure during the growth of Si NWs.

It could be, therefore, concluded that a pronounced size effect is involved in the stabilization of the wurtzite structure of Si NWs.

It is eventually important to note that the use of Au as the catalyst in the CVD growth not only induces the presence of an Au-H deep center [148] but influences also the morphology of Si NWS,* as shown by Lee et al. [188], who confirm the Ross et al. [150] hypothesis that Au favors the faceting of the wires, but also induces the presence of a high density of stacking faults leading to onset of a crystalline core with the diamond structure and of highly defective Au-decorated sidewalls.

3.4.7 GROWTH-DEPENDENT TOPOLOGICAL DIFFERENCES AND PHONON CONFINEMENT IN SILICON NWs

Significant topological differences are presented by metal-assisted and MACE-grown Si NWs. In fact, MACE-grown wires take the shape originated by the etch cutting procedure, since reconstruction could not occur in the low-temperature environment typical of their growth. In addition, their orientation is that of the substrate from which are etched.

The experimental evidence indicates, instead, that silicon nanowires prepared with the metal-assisted, high-temperature processes "are rodlike structures constructed around a bulk single-crystalline core which grow along well defined crystalline

---

* With diameters within 100 and 200 nm.

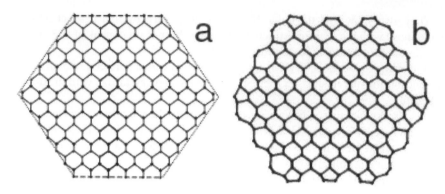

FIGURE 3.46 Hexagonal <110> silicon wire with unreconstructed (a) and reconstructed facets. (*After* Y. Zhao, B.I. Yacobson (2003) What is the ground-state Structure of the Thinnest Si Nanowires? *Phys.Rev.Lett.* **91** 035501. *Reproduced with permission of APS,* License Number: RNP/19/DEC/020992 License date: Dec 07, 2019.)

directions"* [22]. The experimental evidence indicates, also, that the orientation of metal-assisted grown Si NWs depends on the wire diameter, and theoretical work demonstrates that the experimentally determined orientation fulfils thermodynamic equilibrium conditions [189], as discussed in Section 3.4.1.

High growth temperatures, eventually, induce surface reconstruction effects, that lead to characteristic shapes of the cross sections, with different faceting arrangements. The energy balance associated with reconstruction could be calculated making use of the Wulff criterion, that relates the equilibrium shape with the surface energy γ_s of the facets involved, the extension of their surface being measured by the number s of unit cells. The Wulff criterion, in the case of macroscopic solids, leads to the definition of the Wulff energy $F = \sum_s s\gamma_s$.

According to Zhao and Yacobson [190] the Wulff energy, in the case of the thinnest nanowires, could be calculated by adding to the Wulff energy $\sum s\gamma_s$ a term E_e, that accounts for the energy of matching adjacent facets, and a term^{s}of bulk energy ΔE_b that accounts for the changes induced by strain and extended defects (e.g. stacking faults) consequent from faceting, on the bulk energy

$$F = \sum_s s\gamma_s + E_e + \Delta E_b \tag{3.22}$$

The reconstruction of a hexagonal <110> wire with six facets is displayed, as an example, in Figure 3.46, which shows the network of edges formed on the original facets.

The shape-dependent excess surface energies $\Gamma_o = \dfrac{F - s\gamma_s}{s}$, calculated by energy minimization, of square, diamond, hexagonal, and pentagonal wires are reported in Figure 3.47, that shows that small pentagonal wires present the lowest excess surface energy.

---

* Original sentence in Rurali's manuscript.

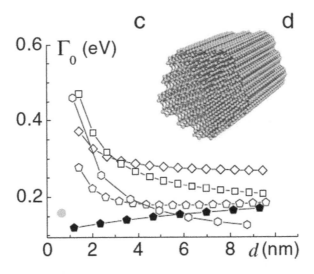

FIGURE 3.47 Size-dependent excess surface energy for wires having square, hexagonal, diamond, and pentagonal shapes. On top of the figure (d), the structure of a low energy pentagonal wire is displayed. (*After* Y. Zhao, B.I. Yacobson (2003) What is the ground-state structure of the thinnest Si nanowires? *Phys.Rev.Lett.* **91** 035501. *Reproduced with permission of APS, License Number: RNP/19/DEC/020992 License date: Dec 07, 2019.*)

Raman spectroscopy is used as a direct probe of phonon confinement in Si NWs, just in the same way as in the cases of dots and of nanocrystalline silicon.

Unlike dots and nanocrystals, and independently of their growth processes, nanowires are, however, not confined along the length, with potential influences on the experimental correlation among Raman peak position and wire diameter.

In fact, several results found in the literature [189, 191] report peak positions within 500 and 510 cm^{-1} for wires having a diameter of 10–15 nm, while the shift associated with phonon confinement calculated by Piscanec et al. [192, 193] using the Richter [194], Campbell and Fauchet [195] (RCF) model for NWs having a diameter from 25 to 2 nm should amount only to a few cm^{-1}, following the trend shown in Figure 3.48.*

Furthermore, Zhang et al. [196] show that the Raman peak shift toward lower frequencies depends also on the excitation wavelength, inconsistent with the fundamental Raman scattering features.

Eventually, Piscanec et al. [192] show also that the peak frequency of the Raman spectrum of a mat of silicon nanowires depends on the exciting laser beam power, leading to Raman shifts of more than 10 cm^{-1} toward lower frequencies and to substantial peak broadening with the increase of the power, as is apparent in Figure 3.49, where a peak frequency of 510 cm^{-1} is obtained by using a laser power of 2.5 mW, while the peak frequency of the nanowire fits with that of c-Si for a laser power of 0.02mW.†

---

* It should be noted that the curve is calculated taking the Raman peak of crystalline silicon at 523 cm^{-1}; therefore it displays a trend, not the real values.
† Note again that the peak frequency of c-Si is given at 523 cm^{-1} against the literature value 520.09 cm^{-1}.

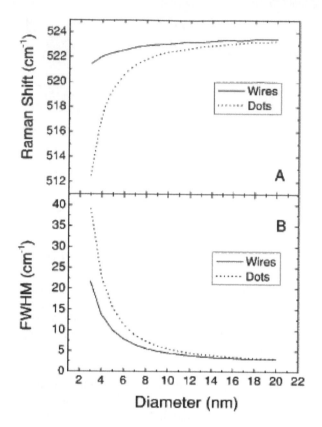

FIGURE 3.48   Calculated trend of the Raman shift (A) and of the full width at half maximum (FWHM) (B)with the decrease of diameter of nanowires and dots. (*After* S. Piscanec, M. Cantoro, A, C. Ferrari, J.A. Zapien, Y. Lifshiz, S.T. Lee, S. Hoffmann, J. Robertson (2003) Raman spectroscopy of silicon nanowires *Phys.Rev. B* **68** 241312. *Reproduced with permission of APS License Number:* RNP/19/DEC/020978.)

The experimental shift well corresponds to the calculated one by assuming a substantial laser-induced heating (up to 900 K) of the sample, possibly due a poor contact of the NW with the substrate and to the lower thermal conductivity of the wire as compared to bulk silicon.

A lower thermal conductivity could be associated, in the case of NWs with diameters from 15 to 35 nm, for which phonon localization is weak, with anharmonic scattering events, as shown by Antidormi et al. [197] and Cartoixà et al. [198] for porous silicon and for porous and non-porous nanowires.

Cartoixà et al. [198] carried out molecular dynamics calculations on silicon NWs, with a diameter from 15 to 35 nm, which shows that the thermal conductivity should decrease with the decrease of the diameter and with the increase of the length.

The hypothesis of a lower thermal conductivity is, also, compatible with phonon localization effects, which lead, in fact, to reduced thermal conductivity of a semiconductor material, as shown by Giri et al. [199] for amorphous heterostructures and by D. Banerjee, et al. [200] for silicon nanowires.

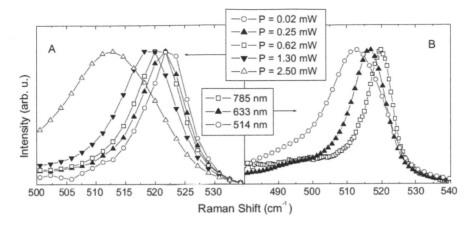

FIGURE 3.49 Laser power (A) and excitation frequency (B) dependence of the Raman peak frequency of silicon NWs. (*After* S. Piscanec, M. Cantoro, A, C. Ferrari, J.A. Zapien, Y. Lifshiz, S.T. Lee, S. Hoffmann, J. Robertson (2003) Raman spectroscopy of silicon nanowires *Phys.Rev. B* **68** 241312. *Reproduced with permission of APS License Number:* RNP/19/DEC/020978.)

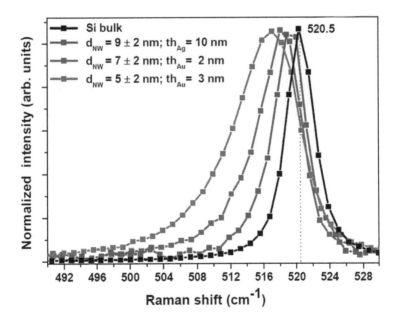

FIGURE 3.50 Raman spectra of three silicon nanowires grown onto an Ag layer 2, 3, and 10 nm thick. For comparison also the Raman spectrum of bulk silicon is reported. (*With courtesy of Dr. Barbara Fazio.*)

Banerjee et al. [200] demonstrated, as the result of high-resolution micro-Raman spectroscopy measurements on MACE-grown wires, with a diameter between 40 and 50 nm, that phonons are localized only at tip of the nanowire.

Thus, Raman measurements on Si NWs should be necessarily carried out at low laser power in order to avoid spurious effects and to detect the true relationships between wire diameter and phonon localization effects.

Working under low laser power (4 μW) conditions, Irrera et al. [167] carried out Raman spectroscopy measurements on silicon nanowires grown with the MACE technique, whose diameter is experimentally shown to depend on the thickness of the Au or Ag film and decreases with the decrease of the metal film thickness.

As expected, the Raman peak is red-shifted from the position of c-Si, with a maximum shift at 517 cm^{-1} for NWs with a diameter of 3 nm (see Figure 3.50) [164], in excellent agreement with the results of Piscanec [192] for wires of the same diameter L.

3.5 THIN AND ULTRA-THIN FILMS OF SILICON NANOCRYSTALS EMBEDDED IN A DIELECTRIC MATRIX

2D silicon-based quantum-confined structures could be prepared by thermal annealing of SiO films or of films of silicon-rich metastable oxides, nitrides, and carbides, though we will limit here our interest to SiO and to silicon-rich oxide (SRO), *formally* consisting of a sub-stoichiometric, amorphous SiO$_x$ phase.

We have seen in Section 3.3.1 that the thermal annealing of SRO leads to the formation of silicon nanocrystals and SiO$_2$

$$SiO_x \rightleftharpoons \left(1 - \frac{x}{2}Si\right) + \frac{x}{2}SiO_2 \qquad (3.23)$$

and is currently used for the synthesis of thin films of silicon nanocrystals embedded in a dielectric SiO$_2$ matrix [18, 67, 71, 201–202].

Similarly, the thermal annealing of SiO$_x$ – SiO$_2$ multilayers, deposited by reactive sublimation or RF sputtering of SiO, is used [6, 62, 63] for the preparation of Si – SiO$_2$ superlattices, with silicon nanocrystals embedded in an amorphous SiO$_2$ matrix.

Despite the excellent choice of SiO and SRO as the precursors of heterogeneous mixtures of silicon nanocrystals and amorphous silica, their very nature as multiphase materials has been generally neglected, and the formation of silicon nanocrystals and amorphous silica during thermal annealing is considered the result of a true decomposition process.

Synthetic SiO powder, in fact, is a heterogeneous mixture of (cubic) SiO, Si, and SiO$_2$, and amorphous silicon nanoclusters are *ab initio* present in the SiO$_x$ deposited film, due to the fact that RF sputtered SiO vapors spontaneously disproportionate to Si and SiO$_2$ during their condensation [203].

Also, the as-deposited SRO is a multiphase material, as can be seen from the XPS spectrum of Figure 3.51 [204], and consists of a heterogeneous mixture of sub-nanometric amorphous silicon clusters and amorphous silicon oxide.

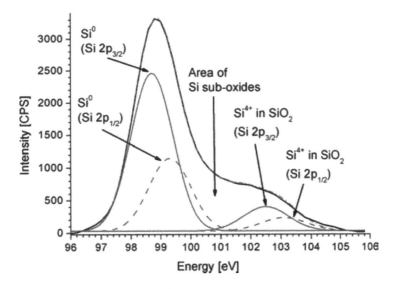

FIGURE 3.51 XPS spectrum of SRO and its deconvolution . (*After* J.A. Salazar, R.L. Estopierr, E.Q. Gonzales, A.M. Sanchez, J.P. Chavez, I.E.Z. Huerta, M.A. Mijares (2016) Silicon-rich oxide obtained by Low-Pressure CVD to develop silicon light sources INTEC *Open literature.*)

Therefore, in both cases, thermal annealing of SiO_x would not cause a phase decomposition, just because it is not a homogeneous phase, but only drives the growth of amorphous silicon clusters *via* an Ostwald ripening process and their further crystallization in the amorphous silica matrix in which they are embedded.

Zacharias and Streitenberg [62] studied the crystallization scheme of amorphous Si nanoclusters in amorphous SiO_x multilayers* of an average thickness of a few nm, seen in the TEM micrograph of Figure 3.8. It was preliminarily observed that crystallization does occur above 800°C, while the growth of silicon nanoclusters alone does occur at lower temperatures.

Following a trend common to all nanomaterials, the crystallization temperature was shown to depend on size. It was, in fact, observed, see Figure 3.52, that the crystallization temperature T_c exponentially increases with the decrease of the layer thickness $d^\dagger$, a behavior well-fitted by the equation

$$T_c = T_b + (T_m - T_b)\exp{-\frac{d}{C}} \tag{3.23}$$

where T_b is the crystallization temperature of a thick, freestanding layer of SiO_x, T_m is the melting temperature of crystalline silicon, and C is a constant, which holds 2.56 nm.

---

* Deposited by reactive sublimation of an SiO powder in oxygen atmosphere on a silicon substrate maintained at 100°C.

† The same behaviour is exhibited by GeO_x / GeO_2 superlattices.

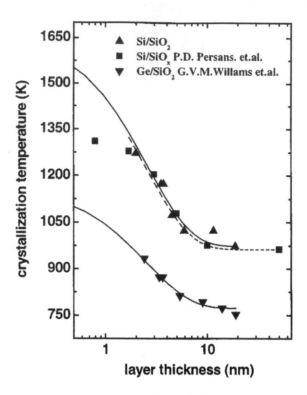

FIGURE 3.52 Dependence of the crystallization temperature as a function of the layer thickness of SiO$_x$ and GeO$_x$ materials. Full lines display the theoretical behavior. (*After* M. Zacharias and P. Streitenberger (2000) Crystallization of amorphous superlattices in the limit of ultrathin films with oxide interfaces *Phys. Rev. B* **62** 8391. *Reproduced with permission of APS, License Number RNP/19/JUN/016024 License date: Jun 25, 2019 DOI: 10.1103/ PhysRevB.62.8391.*)

Apparently, crystallization does not occur at layer thicknesses below 1.6 nm, when the annealing temperature tends to the melting temperature of crystalline silicon, limiting as well the nanocrystal size at 1.6 nm.

The crystallization process has been modeled by accounting for the strain field at the SiO$_x$ / SiO$_2$ interfaces, with an excellent fit of the experimental results displayed in Figure 3.52.

Daldosso et al. [67, 201] and Iacona et al. [71] studied, instead, the thermal annealing scheme of SiO$_x$ films, deposited by plasma-enhanced CVD on silicon or quartz substrates at 300°C, using SiH$_4$ or SiH$_4$/H$_2$ mixtures and N$_2$O as the oxidant, and further annealed in flowing nitrogen atmosphere up to 1250°C .

The silicon content in the as-deposited layers was 46%, as measured by Rutherford backscattering spectroscopy. The film thickness ranged around 80 nm [71] and between 200 and 1000 nm [201], in both cases orders of magnitude larger than those of Zacharias.

The temperature-driven structural evolution of the 80 nm thick SiO$_x$ films was followed [71] by energy-filtered (EF)TEM measurements, that allowed excellent

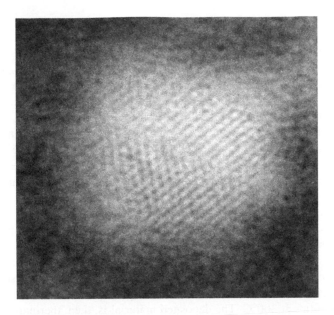

FIGURE 3.53 Energy filtered TEM image of a silicon nanocrystal nucleated from an amorphous Si nanocluster after annealing at 1250°C. (*Reproduced with permission of APS after* N. Daldosso et al. (2003) Role of interface region on the optoelectronic properties of silicon nanocrystals embedded in SiO$_2$ *Phys.Rev B* **68** 085327 *License Number: RNP/19/ JUN/016006 License date: Jun 25, 2019 DOI: 10.1103/PhysRevB.68.0hy.*)

images of the films to be obtained at different annealing temperatures. Clustering is very evident at annealing temperatures of 900°C and higher, where well-defined nanocluster structures do appear. Their size is shown to increase with the annealing temperature, rising from 1 nm (± 0.2 nm) at 1000°C to 2 nm (± 0.4 nm) at 1250°C.

Since also the as-deposited films present a non-uniform, nanogranular background, some clustered material is present already in the as-grown film, in good agreement with our previous comments on the morphology of the as-grown films.

From further EFTEM measurements it became also evident that the thermal annealing of SiO$_x$ in the temperature interval 600–1100°C induces only the growth of amorphous silicon nanoclusters. Instead, at temperatures of 1100°C and higher (up to 1250°C), the formation of crystalline Si nuclei does occur. The energy-filtered TEM image of a silicon nanocrystal grown after annealing at 1250°C displayed in Figure 3.53* [201] shows a well-developed crystalline lattice in the nanocrystal core and a nanometric-thick amorphous silicon surface layer [71].

Further information about the chemical evolution of SiO$_x$ films deposited from SiH$_4$/H$_2$/N$_2$O mixtures, induced by thermal annealing, was obtained from FTIR and Raman spectroscopy measurements.

The FTIR spectra of SiO$_x$ annealed at temperatures below 900°C show the pres ence of both H-related modes and O-related modes, as an indication that the material is, in fact, hydrogenated SiO$_x$: H.

---

* The same image is displayed in ref. 71.

The evolution of the FTIR spectra at annealing temperatures above 900°C show, instead, that the H-related modes almost totally disappear, while the frequency of the $Si - O - Si$ stretching modes gradually shift, reaching a value of 1090 cm^{-1} at 1250°C. Since this value is a little larger than the stretching mode of stoichiometric silica, and also the width of the stretching absorption peak is larger than that of good quality silica, it has been supposed that a non-negligible Si excess remains in the amorphous silica matrix embedding the nanocrystals. Therefore, its chemical composition is not that of stoichiometric amorphous silica, as it is normally assumed.

The Raman spectrum of a material annealed at 1000°C and above is characterized by a broad gaussian distribution centered at 480 cm^{-1}, that is the contribution of amorphous silicon, and by a peak at 515 cm^{-1} typical of quantum-confined Si nanocrystals with an average diameter of 3 nm.

It is, therefore, apparent, that traces of a-Si are present at least at temperatures up to 1000°C and that the crystallization of a-Si clusters is not complete at these temperatures, in good agreement with the morphology of the Si-NC displayed in Figure 3.53.

Similar results were obtained by Ma et al. [18, 202], who made use of gaseous $N_2O-SiH_4-H_2$ mixtures to PECVD-deposit on quartz substrates under-stoichiometric $SiO_x : H$ films at 300°C. The deposited material is, then, thermally annealed in flowing nitrogen, and its chemical and structural evolution is followed with FTIR and Raman spectroscopies [202]. It could be observed that above 900°C hydrogen is completely driven out, and that SiO_x is converted to a mixture of Si-NCs and of SiO_2 at 1170°C. The evolution of the $Si - O - Si$ stretching band with the annealing temperature is used to follow the evolution of the stoichiometry of the SiO_x material above 900°C, which turns out to stoichiometric SiO_2 (1084 cm^{-1}) at 1170°C.

The thermally annealed films have an average thickness between 1.12–1.62 μm, lying in the range of very thick films, while the average diameter of the nanocrystals, estimated by phonon confinement of Raman scattering, ranged between 4 and 6 nm.

Apparently, in all the cases examined, the morphological evolution of the SiO_x films at temperatures below the crystallization range remains an unknown.

3.6 GERMANIUM NANOWIRES: GROWTH AND STRUCTURAL ISSUES

Although Si NWs deserved, and still deserve, major attention for their immediate potential of optoelectronic applications, research on Ge NWs has attracted substantial interest in the last few years due some specific advantages offered by Ge, such as its high carrier mobility, its low effective mass, and its exciton Bohr radius which is larger than that of Si, easing the set-up of QC effects [205].

An additional advantage of Ge is the chance to achieve direct band gap transition by elastic strain engineering, favored by its being a pseudo-direct band gap semiconductor [206] due to an energy difference of only 136 meV between the direct- and indirect-valleys minima. Theoretical calculations show that the crossover for direct gap transition is obtained at a uniaxial stress of 4%, but at least so far, the practical application of this concept with the application of a large mechanical strain to Ge NWs has been attempted with partial success, since the effect is not permanent and vanishes with the end of strain application. Also the use of permanent stressors, created by means of core-shell structures, has been attempted, though with moderate success. Since a level of strain of 4% is already present in Ge NWs with a diameter less than 200 nm [207], it could not be excluded that success might arise by devoting more attention to *ad hoc* dedicated Ge NWs growth processes.

Most of the work carried out so far for the growth of Ge NWs was done using the metal-assisted CVD technique [208–214] or a metal-assisted MBE process [215], although also the MACE technique has been recently used [216–217].

The metal-assisted CVD growth is carried out with modalities close to those employed for Si NWs, using germanium chloride ($GeCl_4$), iodide (GeI_4), or, preferably, germanium hydrides (GeH_4 or Ge_2H_6) as the precursors and H_2–Ar mixtures as carrier gases, at temperatures ranging between 250 and 450°C. Au is generally employed as the metal catalyst, deposited as a nanometer-thick layer on a Si- or Ge-single crystal substrate, or on a Ge buffer layer over a Si substrate.*

By heating the substrate above the (bulk) eutectic temperature of the Au–Ge system (361°C) the formation of liquid Au–Ge alloy droplets of equilibrium composition does occur as consequence of local Ge dissolution. Once germane is introduced in the growth chamber, the liquid Au–Ge alloy catalyzes the decomposition of germane to elemental Ge and hydrogen, and behaves as a Ge sink with the formation of a super-saturated Au–Ge liquid alloy. Thermodynamic equilibrium is eventually achieved with the surface segregation of the excess Ge and the epitaxial, unidirectional, or anisotropic growth of a whisker or of a wire with a conventional VLS process. If the growth is carried out at hypoeutectic temperatures a VSS process does occur, but these process conditions are questionable, since Kotakamba et al. [211] were able to show that the eutectic temperature drops to 240°C for catalyst size of 14–34 nm.

Epitaxial growth, like in the case of Si NWs, could be obtained within 250–450°C using single crystal substrates, while higher temperatures (720°C) are used to obtain endotaxial growth, i.e. growth parallel to the substrate surface. Vertically aligned wires grow preferably along the (111) orientation on Si and Ge substrates, but tapering is the most common defect of CVD-grown NWs.

Vertically aligned, tapering-free NWs, 10–30 nm in diameter, were grown using bio-templated Au nanoparticles [214]. It was observed that the wire diameter is larger than that of the original Au nanoparticles due size expansion arising from the Au–Ge alloy formation.

Tapering-free Ge NWs have been also grown by Periwal et al. [209] using a mixture of GeH_4 and PH_3, a Si substrate at 400°C and colloidal Au particles of 50 nm in diameter as catalysts. It was shown that PH_3 is adsorbed at the wire surface with the formation of surface P-adatoms and P–P bonds, which prevent, however, sidewall Ge deposition and tapering only at low growth temperatures (275°C).

The presence of HCl in the GeH_4–PH_3 mixture does improve the morphology of the wires, as could be seen in Figure 3.54, were it is apparent that tapering is absent with a HCl flux of 20 sccm, while wire roughening is present at intermediate HCl fluxes and Ge etching at the highest HCl fluxes.

---

* Ni, Cu, and Ag are, however, also used.

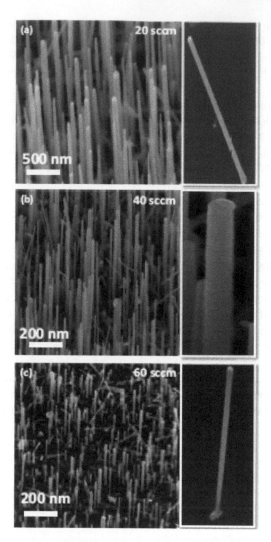

FIGURE 3.54 SEM image of Ge NWs arrays grown under increasing HCl flow rates. (*After* P. Periwal, T. Baron, P. Gentile, F. Bassani (2014) Growth strategies to control tapering in Ge nanowires *APL Materials* **2** 46105(1–8) *Open Access Journal.*)

Ge NWs growth was also carried out at 300 to 450°C* in a liquid phase of ciclo-hexane heated and pressurized above its critical point by Hanrath and Korgel [213], using as catalysts gold nanoparticles 2–7 nm in size suitably protected with an alka-nethiol layer. With this technique a mesh of nanowires is obtained with a non-constant wire diameter, due to the aggregation of the catalyst during the initial period of reaction.

---

* Above 500°C a structural degradation is observed.

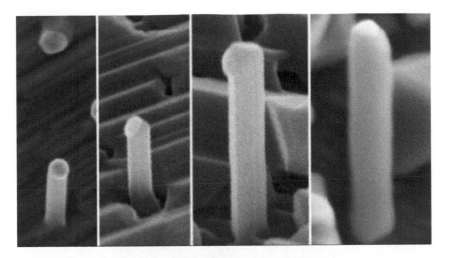

FIGURE 3.55 Growth temperature dependence of the morphology of MBE grown Ge NWs. From left T = 220°C, 380°C, 470°C. (*After* Z. Zhu, Y. Song, Z. Zhang, H. Sun, Yi. Han, Y. Li, L. Zhang, Z. Xue, Z. Di, S. Wang (2017) The vapor-solid-solid growth of Ge nanowires on Ge (110) by Molecular Beam Epitaxy (arXiv.org/pdf/1706.0165).)

Molecular beam epitaxy (MBE) in ultra-high vacuum conditions was eventually used by Zhu et al. [215] at temperatures ranging between 220°C and 470°C on Ge (110) substrates, using Au droplets as catalysts, showing that the wire diameter decreased with the decrease of the growth temperature without loss of uniformity as can be seen in Figure 3.55. The final conclusion of the authors is that the best conditions to obtain assemblies of vertically oriented nanowires is to operate at hypoeutectic temperatures (220°C) with solid Au droplets and a VSS process. This temperature, however, is very close to the eutectic temperature at the nano-size and the assumption of hypoeutectic conditions is questionable.

Formally hypoeutectic temperatures were also used by Kotambaka et al. [211] using as catalyst Au droplets with a diameter ranging from 14 to 34 nm. These authors were able to demonstrate that the eutectic temperature for droplets of this size is 106°C lower than the bulk eutectic temperature, by examining a series of SEM images acquired during the wire growth. It is possible to see that when the Au catalyst is liquid (frame A in Figure 3.56) the wire shape is cylindrical, while faceting and tapering occurs when the catalyst is solid (frame B of Figure 3.56).

Eventually, Li et al. [210] carried out Ge NW deposition on H-terminated* or oxidized silicon substrates, confirming Kodambaka's [211] arguments about a substantial decrease of the Au–Ge eutectic temperature at the nano-size. They showed as well that the morphology of the wires depends on the H- or O-termination of the silicon substrate as well as on the presence of the catalyst on the tip of the wire, as is shown in Figures 3.57 and 3.58.

---

* After HF etching.

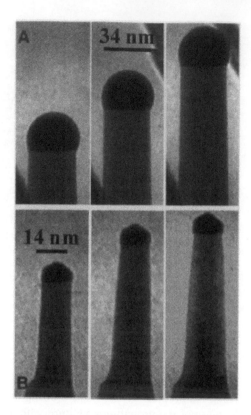

FIGURE 3.56 SEM images of time evolution of two different Ge NWs grown at T = 340°C with a liquid catalyst (A) and a solid catalyst (B). (*Reprinted with permission of the American Association for the Advancement of Science, from* S. Kodambaka, J. Tersoff, M.C. Reuter, F.M. Ross (2007) Germanium nanowire growth below the eutectic temperature *Science* **316,** (5825) 729–732 *license nr. 4567141051713, license date Dec 02, 2019.*)

Figure 3.57a displays the SEM micrograph of a thin, un-tapered, constant diameter, 20 μm long wire, grown on a H-terminated silicon substrate, while Figure 3.57b displays the side view micrograph of 200 nm thick, tapered, and rough surface wires grown on a oxidized silicon substrate.

The effect of the O-termination of the substrate on the morphology of the NWs is apparent, and it could be supposed that the growth does occur with an oxide-assisted process.

The effect of the presence of the catalyst on top of the wires is shown in Figure 3.58, which displays an array of nanowires 3 um long and 30 nm in diameter grown at 300°C on a SiO$_2$ substrate. It is possible to remark that NWs with the Au catalyst on the tip present a smooth surface and an almost constant diameter along the entire length. Different is the morphology of the NWs without the catalyst on the tip, which exhibit a rough surface and an increasing diameter along the length. While in the first case the growth occurs only in correspondence of the Au–Ge

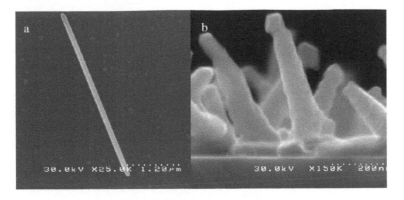

FIGURE 3.57 SEM images of Ge nanowires grown at 350°C on H-terminated (a) and oxidized Si substrate (b). (*From* C. Li, H. Mizuta, S. Oda (2011) Growth and characterisation of Ge nanowires by chemical vapour deposition *in Nanowires – Implementations and Applications* Abbass Hashim Edt. Intechopen, pp.487–508. *Reproduced under Creative Commons Attribution-NonCommercial-ShareAlike-3.0 License.*)

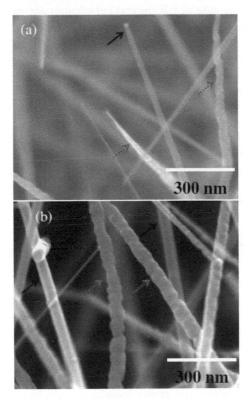

FIGURE 3.50 SEM images of Ge nanowires grown at 300°C on a SiO_2 substrate (a) top part (b) middle part of the nanowires. Black arrows show NWs with the catalyst on the tip, while blue arrows show strongly corrugated NWs. (*From* C. Li, H. Mizuta, S. Oda (2011) Growth and characterisation of Ge nanowires by chemical vapour deposition *Nanowires: Implementations and Applications* Abbass Hashim Edt. Intechopen, pp.487–508. *Reproduced under Creative Commons Attribution-NonCommercial-ShareAlike-3.0 License.*)

alloy on top of the wire, in the second case a preferential VSS radial growth does occur.

Like in the case of Si NWs, and despite the lower growth temperatures, the metal catalysts should contaminate the wire, inducing the presence of deep recombination centers, which could make critical their optoelectronic application.

MACE growth of Ge nanowires has been also attempted [216–217], with unsatisfactory results, due to the poorer electrochemical properties of Ge,* as compared with Si. Therefore, no valuable experience exists on structural properties and defects of Ge NWs grown with this method.

3.7 DIAMOND NANOWIRES: GROWTH AND STRUCTURAL ISSUES

Due to the excellent biocompatibility and superior physical and optical properties of diamond, diamond nanowires (DNWs) are ideal candidates as voltammetric electrodes for sensors applications, and as basic components of UV light emitters and quantum information devices, as suggested by Szunerits et al. [218–219], Peng et al. [220], and by Hausmann et al. [221].

However, unlike the case of silicon and germanium nanowires, the synthesis of diamond nanowires presents reproducibility problems that are not yet overcome, as shown in a recent work of Shellaiah and Sun [222]. At present, two main processes are applied to the growth of DNWs, the plasma-assisted reactive ion etching (RIE) and the chemical vapor deposition on suitable templates, though template-free processes are also attempted.

RIE processes with oxygen- or fluorine-plasma are carried out covering the surface of the diamond substrate with metallic- or non-metallic masks, which define the regions where the diamond substrate should be etched off, and then also the diameter of the nanowires. Al- masks are generally used, but non-metallic masks are preferred. These last could be prepared by wetting the diamond surface with a suspension of oxide or diamond nanoparticles with a well-defined range of size distribution, typically within 8–10 nm.

Figure 3.59 displays the schematic illustration of a process carried out with a mask prepared with diamond seeds, which leads to an assembly of pseudoconical wires. Here it is well-apparent the difference with electrochemical MACE processes, which do allow the fabrication of vertically aligned constant diameter wires, due to the fact that the metal dots, where the electrochemical process occurs, maintain constant from the beginning to the end of the process the diameter of the hole, and thus the vertical alignment of the wires. The morphology of DNWs prepared with this process can be observed in the SEM image displayed in Figure 3.60, from which one can easily observe their conical shape, which is, however, not a drawback for the use of DNWs as base components of electrochemical sensors [219].

The Raman spectra of diamond films grown on silicon substrates (BDD) and of diamond NWs (BDD NWs) RIE grown from diamond films [223] show that at the low and moderate doping the spectra of the films and of the NWs present the narrow

---

* With respect to metal-assisted chemical etching processes.

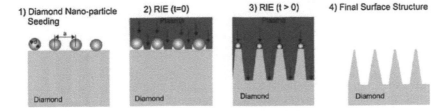

FIGURE 3.59 Growth of vertically aligned diamond nanowires using an RIE process and a diamond nanoparticle mask consisting of a distribution of diamond nanoparticles. (*Reproduced from* M. Shellaiah, K.W. Sun (2019) Diamond Nanowire Synthesis, Properties and Applications http://dx.doi.org/10.5772/intechopen.78794 *Open access publication.*)

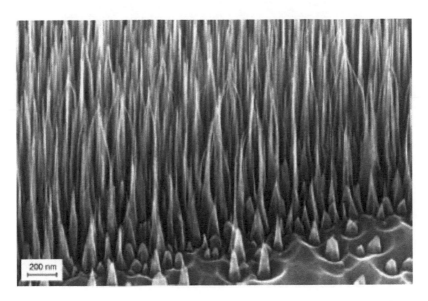

FIGURE 3.60 SEM image of vertically aligned diamond nanowires prepared using an RIE process and diamond seeds in the mask. (*Reproduced from* S. Szunerits, Y. Coffinier and Rabah Boukherroub (2015) Diamond nanowires: A novel platform for electrochemistry and matrix-free mass spectrometry *Sensors,* **15**, 12573–12593; doi:10.3390/s150612573 *Open access article.*)

first-order diamond Raman peak at 1311 cm^{-1}, and two additional features at 520 cm^{-1} and at 950 cm^{-1}, of which the first is the first-order Raman peak of c-Si and that at 950 cm^{-1} is the convolution of two phonons overtones [224].

A broad band at 1218 cm^{-1} dominates the Raman spectrum of the high B doped materials, attributed by Subranian et al. [223] to a Fano-like resonance. The absence of bands in the 1400–1600 cm^{-1} region indicates the absence of sp2 carbon bands, well-evident in the diamond nanowires prepared with CVD processes.

Si wafer   Silicon Nanowires   BDDNF

FIGURE 3.61 Schematic drawing of a dual step process used for the growth of diamond nanowires. The first step is a metal assisted electroless deposition (EMD) of an array of silicon NWs; the second one is the CVD deposition of a diamond layer. (*Reprinted with permission from* D. Luo, L. Wu, J. Zhi (2009) Fabrication of boron-doped diamond nanorod forest electrodes and their application in nonenzymatic amperometric glucose biosensing. *ACS Nano.* 2009; **3**:2121–2128. DOI: 10.1021/nn9003154 Copyright 2009 American Chemical Society.)

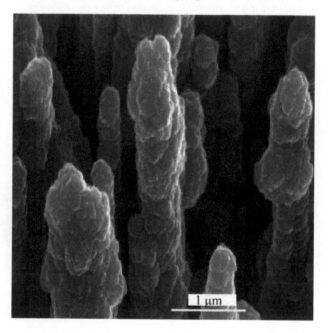

FIGURE 3.62 SEM image of a forest of diamond nanowires grown with the process displayed in Figure 3.61. (*Reprinted with permission from* D. Luo, L., Wu, J. Zhi (2009) Fabrication of boron-doped diamond nanorod forest electrodes and their application in non-enzymatic amperometric glucose biosensing. *ACS Nano.* 2009; **3**:2121–2128. DOI: 10.1021/nn9003154 Copyright 2009 American Chemical Society.)

Apparently, at low and moderate B-doping the main Raman feature of the diamond nanowires is the first-order Raman peak of diamond, indicating the rather good quality of the wires.

CVD process for DNWs are instead carried out by deposition of a thin microcrystalline diamond layer on templates consisting of silicon nanowires [225], as it is schematically shown in Figure 3.61, though tungsten, copper, titanium, and silicon carbide NWs have been also used.

The typical morphology of the diamond wires grown with this method is illustrated in the SEM image of a forest of DNWs displayed in Figure 3.62, where the microcrystallinity of the diamond layer is well-evident.

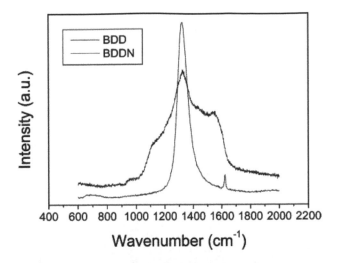

FIGURE 3.63   Raman spectra of boron-doped diamond (BDD) and boron-doped DNWs. (*Reprinted with permission from* D. Luo, L., Wu, J. Zhi (2009) Fabrication of boron-doped diamond nanorod forest electrodes and their application in nonenzymatic amperometric glucose biosensing. *ACS Nano.* 2009; 3:2121–2128. DOI: 10.1021/nn9003154 Copyright 2009 American Chemical Society.)

The Raman spectra of boron-doped diamond and boron-doped diamond nanowires on silicon templates is displayed in Figure 3.63. One can observe that the narrow Raman peak of diamond at 1331 cm^{-1}, well-evident in the boron-doped diamond sample, is associated with a feature at 1580 cm^{-1} and with a shoulder at 1130 cm^{-1}. The feature at 1580 cm^{-1} is attributed to the stretching modes of sp^2-bonded carbon at grain-boundaries of nanocrystalline diamond, which apparently should play a critical role in the electrical properties of DNWs.

The feature at 1130 cm^{-1} might be attributed to the presence of an organic impurity, most probably transpolyacetylene [226], at grain boundaries, formed during the deposition of the layer of nanocrystalline diamond on the silicon template and often observed in nanocrystalline CVD diamond films.

Diamond nanowires were also RIE grown on diamond substrates covered with a thin metallic layer, a technique which recalls that used by Irrera et al. (164, 166–168) for the Au-assisted MACE growth of Si NWs. Here a thin (100 nm) Al layer is DC sputtered onto a single crystal diamond substrate, leading to a coverage of 70% [227]. The fractal distribution of Al at the surface leads to the spontaneous formation of a distribution of Al micromasks, which do allow the growth of the forests of diamond nanowires, after a short (10 min) treatment with inductively coupled plasma (ICP)-RIE shown in Figure 3.64. The different density of NWs depends on the Al micromasks distribution on the diamond surface.

The diameter and the height of these wires is rather large (125–126 nm and 860–880 nm) and almost uniform within dense and sparse areas. Apparently, the diameter of the wires looks constant along the length. No additional information is present on the structural aspects of this material.

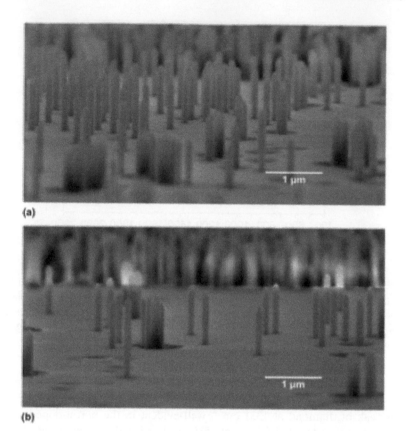

FIGURE 3.64   TEM images of diamond nanowires RIE grown on diamond substrates covered with a thin layer of Al (a) dense area (b) sparse area. (*Reproduced with permission from* K. Wakui, Y. Yonezu, T. Aoki, M. Takeoka, and K. Semba (2017) Simple method for fabrication of diamond nanowires by inductively coupled plasma reactive ion etching *Japan. J. Appl. Phys.* **56** (5) 058005 © 2017 The Japan Society of Applied Physics.)

REFERENCES

1. A. Shah (2007) Thin-film silicon solar cells: The « micromorph » option. *Proc. 22nd European Photovoltaic Solar Energy Conference*, Milano.
2. A. Cavallini, D. Cavalcoli, M. Rossi, A. Tomasi, S. Pizzini, D. Chrastina, G. Isella (2007) Defect analysis of hydrogenated nanocrystalline silicon films. *Physica B* **401–402**, 519.
3. L. Bagolini, A. Mattoni, G. Fugallo, L. Colombo, E. Poliani, S. Sanguinetti, E. Grilli (2010) Quantum confinement by an order-disorder boundary in nanocrystalline silicon. *Phys. Rev. Lett.* **104**, 176803.
4. Y. T. Tan, T. Kamiya, Z. A. K. Durrani, H. Ahmed (2003) Room temperature nanocrystalline silicon single-electron transistors. *J. Appl. Phys.* **94**, 633–637.
5. A. M. Ali (2006) Origin of photoluminescence in nanocrystalline Si:H films. *J. Luminesc.* **127**, 614–622.
6. M. Zacharias, J. Heitmann, R. Scholz, U. Kahler, M. Schmidt, J. Bläsing (2002) Size-controlled highly luminescent silicon nanocrystals: A SiO/SiO$_2$ superlattice approach. *Appl. Phys. Lett.* **80**, 661.

7. T. Z. Lu, M. Alexe, R. Scholz, V. Talelaev. M. Zacharias (2005) Multilevel charge storage in silicon nanocrystals multilayers. *Appl. Phys. Lett.* **87**, 202110.
8. A. Zimina, S. Eisebitt, W. Eberardt, J. Heitmann, M. Zacharias (2006) Electronic structure and chemical environment of silicon nanoclusters embedded in a silicon dioxide matrix. *Appl. Phys. Lett.* **88**, 163103.
9. R-J. Zhang, S-Y. Seo, A. P. Milenin, M. Zacharias (2006) Visible range whispering-gallery mode in microdisk array based on size-controlled Si nanocrystals. *Appl. Phys. Lett.* **88**, 153120.
10. T. Z. Lu, M. Alexe, R. Scholz, V. T. Talalaev, R. J. Zjang, M. Zacharias (2006) Si nanocrystals based memories: Effect of the nanocrystal density (2006). *J. Appl. Phys.* **100**, 014310.
11. G. Di Stefano, D. Sanfilippo, A. Piana, P. G. Fallica, F. Priolo (2007) Light emitting devices based on silicon nanostructures. *Physica E* **38**, 181–187.
12. A. Stesmans, M. Ivanescu, S. Godefroo, M. Zacharias (2008) Paramagnetic point defects at SiO2/nanocrystalline Si interfaces. *Appl. Phys. Lett.* **93**, 023123.
13. M. Jvanescu, A. Stedsmans, M. Zacharias (2008) Inherent paramagnetic defects in layered Si/SiO$_2$ superstructures with Si nanocrystals. *J. Appl. Phys.* **104**, 103518.
14. S. Godefroo, M. Heyne, M. Jvanescu, A. Stesmans, M. Zacharias, O. I. Lebedev, G. Van Tendeloo, V. V. Moshchalkov (2008) Classification and control of the origin of photoluminescence from nanocrystals. *Nat. Nanotechn.* **3**, 174–178.
15. U. Gösele (2008) Shedding new light on silicon. *Nat. Nanotechn.* **3**, 134–135.
16. S-Y. Seo, R-J. Zhang, W. Loffler, H. Kalt, K. J. Kim, M. Zacharias (2009) Far-field observation of the radial profile of visible whispering-gallery modes in a single microdisk based on Si-nanocrystal/SiO$_2$ superlattices. *J. Appl. Phys.* **106**, 123102.
17. L. Pavesi, R. Turan (2010) *Silicon Nanocrystals: Fundamentals, Synthesis and Applications.* J. Wiley & Sons.
18. Z. Ma, X. Liao, G. Kong, J. Chu (1999) Absorption spectra of nanocrystalline silicon embedded in SiO$_2$ matrix. *Appl. Phys. Lett.* **75**, 1857.
19. V. Lu, C. M. Lieber (2006) Semiconductor nanowires. *J. Phys. D. Appl. Phys.* **39**, R387–R406.
20. R. Agarwal, C. M. Lieber (2006) Semiconductor nanowires: Optics and optoelectronics. *Appl. Phys. A* **85**, 209–215.
21. V. Schmidt, J. V. Wittemann, S. Senz, U. Gösele (2009) Silicon nanowires: A review on aspects of their growth and their electrical properties. *Adv. Mat.* **21**, 2681–2782.
22. R. Rurali (2010) Colloquium: Structural, electronic, and transport properties of silicon nanowires. *Rev. Modern Phys.* **82**, 427–449.
23. E. R. Leite, C. Ribeiro (2012) *Crystallization and Growth of Colloidal Nanocrystals*, Springer New York, viii, 95 p.: ill. - (SpringerBriefs in materials). – Incl. bibl. ref. - Ind.: p.93–95. - ISBN 978-1-4614-1307-3; ISSN 2192-1091.
24. Y. Yin, A. P. Alivisatos (2005) Colloidal nanocrystals synthesis and the organic-inorganic interface. *Nature* **437**, 644–669.
25. K. J. van Schooten (2013) Optically active charge traps and chemical defects in semiconducting nanocrystals probed by pulsed optically detected magnetic resonance. PhD Thesis, University of Utah, May 2013.
26. H. Lin, S. Lee, M. Spelling (2017) Clathrate colloidal crystals. *Science* **335**, 931–935.
27. R. D. Tweney (2001) Epistemic artifacts: Michael Faraday's search for the optical effects of gold. *PaviaPaper 2001.*
28. Y. Liu, M. Liu, M. T. Swihart (2017) Plasmonic copper sulphide-based materials: A brief introduction to their synthesis, doping, alloying and applications. *J. Phys. Chem. C* **121**, 13435–13447.
29. Y. Yin, A. P. Alivisatos (2005) Colloidal nanocrystal synthesis and the organic-inorganic interface. *Nature* **437**, 664–670.

30. A. Fojtik, A. Henglein (1994) Luminescent colloidal silicon particles. *Chem. Phys. Lett.* **221**, 363–367.
31. A. Fojtik, A. Henglein (2006) Surface chemistry of luminescent colloidal silicon nanoparticles. *J. Phys. Chem. B.* **110**, 1994–1998.
32. M. C. Beard, K. P. Knutsen, P. Yu, J. M. Luther, Q. Song, W. K. Metzger, R. J. Ellingson, A. J. Nozik (2007) Multiple exciton generation in colloidal silicon nanocrystals. *Nano Lett.* **7**, 2506–2012.
33. Z. Zhou, L. Brus, R. Friesner (2003) Electronic structure and luminescence of 1.1 and 1.4 nm silicon nanocrystals: Oxide shall versus hydrogen passivation. *Nano Lett.* **3**, 163–167.
34. Y. Chao, L. Šiller, S. Krishnamurthy, P. R. Coxon, U. Bangert, M. Gass, L. Kjeldgaard, S. N. Patole, L. H. Lie, N. O'Farrell, T. A. Alsop, A. Houlton, B. R. Horrocks (2007) Evaporation and deposition of alkyl-capped silicon nanocrystals in ultrahigh vacuum. *Nat. Nanotechnol.* **2**(8), 486–489.
35. U. Gösele, V. Lehmann (1995) Light-emitting porous silicon. *Mater. Chem. Phys.* **40**, 253–259.
36. T. Lopez, L. Mangolini (2014) On the nucleation and crystallization of nanoparticles in continuous-flow nonthermal plasma reactors. *J. Vac. Sci. Technol. B* **32**, 061802.
37. L. Mangolini, E. Thimsen, U. Kortshagen (2005) High-yield plasma synthesis of luminescent silicon nanocrystals. *Nano Lett.* **5**, 655–659.
38. L. Mangolini (2013) Synthesis, properties, and applications of silicon nanocrystals. *J. Vac. Sci. Technol. B* **31**, 020801.
39. D. Coleman, T. Lopez, O. Yr. Inceoglu, L. Mangolini (2015) Hollow silicon carbide nanoparticles from a non-thermal plasma process. *J. Appl. Phys.* **117**, 193301.
40. T. Lopez, L. Mangolini (2016) In situ monitoring of hydrogen desorption from silicon nanoparticles dispersed in a nonthermal plasma. *J. Vac. Sci. Technol. B*, **34**, 041206.
41. K. Herynkova, C. Vorkotter, P. Simakova, J. Benedikt, O. Cibulka (2016) Structural and luminescence properties of silicon nanocrystals in colloidal solutions for bio applications. *Phys. Status. Solidi A* **213**, 2873–2878.
42. U. V. Bhandarkar, M. T. Swihart, S. L. Girshick, U. R. Kortshagen (2000) Modeling of silicon hydride clustering in low-pressure silane plasma. *J. Phys. D: Appl. Phys.* **33**, 2731–2746.
43. U. Bhandarkar, U. Kortshagen, S. L. Girshick (2003) Numerical study of the effect of gas temperature on the time for onset of particle nucleation in argon–silane low-pressure plasmas. *J. Phys. D: Appl. Phys.* **36**, 1399–1408.
44. A. Shah, P. Torres, R. Tscharner, N. Wyrsch, H. Keppner (1999) Photovoltaic technology: The case for thin-film solar cells. *Science* **285**, 692–698.
45. A. V. Shah, H. Schade, M. Vanecek, J. Meier, E. Vallat-Sauvain, N. Wyrsch, U. Kroll, C. Droz, J. Bailat (2004) Thin-film silicon solar cell technology. *Prog. Photovoltaics Res. Appl.* **12**, 113–142.
46. B. Yan, G. Yue, X. Xu, J. Yang, S. Guha (2010) High efficiency amorphous and nano-crystalline silicon solar cells. *Phys. Status. Sol A* **207**, 671–677.
47. J. Rath (2012) Thin-film deposition processes. In: *Advanced Silicon Materials for Photovoltaic Applications*, S. Pizzini Ed., J. Wiley & Sons, pp. 235–285.
48. J. Rath (2012) Thin-film silicon solar cells. In: *Advanced Silicon Materials for Photovoltaic Applications*, S. Pizzini Ed., J. Wiley & Sons, pp. 311–353.
49. R. Murri (2013) *Silicon Based Thin Film Solar Cells*, Bentham Science Publishers, Sharjah, United Arab Emirates.
50. K. Ha, E. Jang, S. Jang, J-K. Lee, M. S. Jang, H. Choi, J.-S. Cho, M. Choi (2016) A light-trapping nanoparticle structures. *Nanotechnology* **27**, 055403.
51. S. Ossicini, L. Pavesi, F. Priolo (2003) Light emitting diodes for microphotonics. In: *Springer Tracts in Modern Physics*, Springer Verlag, Berlin, Germany.

52. Z. Yuan, A. Anopochenko, L. Pavesi (2012) Innovative quantum effects in silicon for photovoltaic applications. In: *Advanced Silicon Materials for Photovoltaic Applications*, S. Pizzini Ed., J. Wiley & Sons, pp. 356–391.
53. R. S. Crandall (1991) Defect relaxation in amorphous silicon: Stretched exponentials, the Meyer-Neldel rule, and the Staebler-Wronski effect. *Phys. Rev. B* **43**, 4057–4070.
54. B. Yan, G. Yue, J. M. Owens, J. Yang, S. Guha (2004) Light-induced metastability in hydrogenated nanocrystalline silicon solar cells. *Appl. Phys. Lett.* **85**, 1925.
55. M. R. Black, W. B. Rever, III (2017) Black Silicon: There's more than meets the eye. *PVTech*, Sept. 17.
56. X. Liu, P. R. Coxon, M. Peters, B. Hoex, J. M. Cole, D. J. Fray (2014) Black silicon: Fabrication methods, properties and solar energy applications. *Energy Environ. Sci.* **7**, 3223–3263.
57. B. Ma, Z. Huang, L. Mei, M. Fang, Y. Liu, X. Wu (2015) Investigation on a magnesiothermic reduction process for preparation of nanocrystalline silicon thin film. *J. Surf. Eng.* **32**, 761–765.
58. D. P. Wong, H-T. Lien, Y-T. Chen, K.-H. Chen, L-C. Chen (2012) Patterned growth of nanocrystalline silicon thin films through magnesio-thermic reduction of soda lime glass. *Green Chem.* **14**, 896–900.
59. Y. Tsuboi, S. Ura, K. Takahiro, T. Henmi, A. Okada, T. Wakasugi, K. Kadono (2017) Magnesiothermic reduction of silica glass substrate—Chemical states of silicon in the generated layers. *J. Asian Ceram. Soc.* **5**, 341–349.
60. I. Gutman, L. Klinger, I. Gotman, M. Shapiro (2002) Kinetics and mechanism of periodic structure formation at SiO_2/Mg interface. *Scr. Mater.* **45**, 363–367.
61. H. Soma, T. Uchino (2017) Blue and orange photoluminescence and surface band-gap narrowing in lithium-doped MgO microcrystals. *J. Phys. Chem. C* **121**, 1884–1892.
62. M. Zacharias, P. Streitenberger (2000) Crystallization of amorphous superlattices in the limit of ultrathin films with oxide interfaces. *Phys. Rev. B* **62**, 8391.
63. L. X. Yi, J. Heitmann, R. Scholz, M. Zacharias (2002) Si rings, Si clusters, and Si nanocrystals—different states of ultrathin SiO_x layers. *Appl. Phys. Lett.* **81**, 4248–4266.
64. C. Spinella, S. Lombardo, F. Priolo (1998) Crystal grain nucleation in amorphous silicon. *J. Appl. Phys.* **84**, 5383–5414.
65. G. Franzò, F. Priolo, S. Coffa (1998)Understanding and control of the erbium nonradiative de-excitation processes in silicon. *J. Lumin* **80**, 19–28.
66. D. Pacifici, E. C. Moreira, G. Franzo, V. Martorino, F. Priolo, F. Iacona (2002) Defect production and annealing in ion-irradiated Si nanocrystals. *Phys. Rev B* **65**, 144109.
67. N. Daldosso, G. Das, S. Larcheri, G. Mariotto, G. Dalba, L. Pavesi, A. Irrera (2007) Silicon nanocrystal formation in annealed silicon-rich silicon oxide films prepared by plasma enhanced chemical vapor deposition. *J. Appl. Phys.* **101**, 113510.
68. O. P. Dey, A. Khare (2017) Fabrication of photoluminescent nc-$Si:SiO_2$ thin films prepared by PLD. *Phys. Chem. Chem. Phys.* **19**, 21436–21445.
69. PVTech Newsletters, 17 May 2017.
70. S. Oda (1997) Preparation of nanocrystalline silicon quantum dot structure by a digital plasma process. *Adv. Colloid Interface Sci.* **71–72**, 31–47.
71. F. Iacona, C. Bongiorno, C. Spinella, S. Boninelli, F. Priolo (2004) Formation and evolution of luminescent Si nanoclusters produced by thermal annealing of SiO_x films. *J. Appl. Phys.* **95**, 3723–3732.
72. M. Jamel, F. Karbassian, S. Mohajerzadeh, Y. Abdi, M. D. Robertson, D. Yuill (2007) The preparation of nanocrystalline silicon by plasma-enhanced hydrogenation for the fabrication of light-emitting diodes. *IEEE Electron Device Lett.* **28**, 207–210.
73. F. Priolo, T. Gregorkiewicz, M. Galli, T. F. Krauss (2014) Silicon nanostructures for photonics and photovoltaics. *Nature Nanotechnol.* **9**, 19–32.

74. N. Lin, Y. Han, L. Wang, J. Zhou, J. Zhou, Y. Zhu, Y. i Qian (2015) Preparation of nano-crystalline silicon from SiCl4 at 200°C in molten salt for high-performance anodes for lithium ion batteries. *Angew. Chem.* **54**, 3822–3825.

75. S-C. Lee, J-M. Hur, C-S. Seo (2008) Silicon powder production by electrochemical reduction of SiO_2 in molten $LiCl-Li_2O$. *J. Ind. Eng. Chem.* **14**, 651–654.

76. J. Wang, I. Florea, P. V. Bulkin, J-L. Maurice, E. V. Johnson (2016) Using MDECR-PECVD to study the impact of ion bombardment energy on microstructural proper-ties of µc-Si-H thin film grown from an SiF_4/H_2 chemistry. *Phys. Status. Sol. C* **13**, 782–786.

77. S. Binetti, M. Acciarri, M. Bollani, L. Fumagalli, H. von Känel, S. Pizzini (2005) Nanocrystalline silicon films grown by Low Energy Plasma Enhanced Chemical Vapor Deposition for optoelectronic applications. *Thin Solid Films* **487**, 19–25.

78. R. Amrani, F. Pichot, L. Chahed, Y. Cuminal (2012) Amorphous-nanocrystalline transition in silicon thin films obtained by argon diluted silane PECVD. *Crys. Struct. Theory Appl.* **1**, 57–61. doi:10.4236/csta.2012.13011.

79. V. S. Waman, A. M. Funde, M. M. Kamble, M. R. Pramod, R. R. Hawaldar, D. P. Amalnerkar, V. G. Sathe, S. W. Gosavi, S. R. Jadkar (2011) Hydrogenated nanocrystal-line silicon thin films prepared by hot-wire method with varied process pressure. *J. Nanotechnol.* 10, ID 242398.

80. P. Gogoi, H. S. Jha, P. Agarwal (2011) Variation of microstructure and transport prop-erties with filament temperature of HWCVD prepared silicon thin films. *Thin Solid Films* **519**, 6818–6828.

81. H. Li, R. H. Franken, R. L. Stolk, C. H. M. van der Werf, R. E. I. Schropp, J. K. Rath (2008) Controlling the quality of nanocrystalline silicon made by hot-wire chemical vapor deposition by using a reverse H_2 profiling technique. *J. Non-Crystall. Solids.* **354**, 2087–2091.

82. Y. Wang, X. H. Geng, H. Stiebig, F. Finger (2008) Stability of microcrystalline silicon solar cells with HWCVD buffer layer. *Thin Solid Films* **516**, 733–735.

83. R. E. I. Schropp (2004) Present status of micro- and polycrystalline silicon solar cells made by hot-wire chemical vapor deposition. *Thin Solid Films* **451–452**, 455–465.

84. A. H. Mahan, Y. Xu, B. P. Nelson, R. S. Crandall, J. D. Cohen, K. C. Palinginis, A. C. Gallagher (2001) Saturated defect densities of hydrogenated amorphous silicon grown by hot-wire chemical vapor deposition at rates up to 150 °A/s. *Appl. Phys. Lett.* **78**, 3788–3790.

85. A. H. Mahan (2006) An update on silicon deposition performed by hot wire CVD. *Thin Solid Films* **501**, 3–7.

86. M. Brinza, C. H. M. van der Werf, J. K. Rath, R. E. I. Schropp (2008) Optoelectronic properties of hot-wire silicon layers deposited at 100 oC. *J. Non-Crystall. Solids* **354**, 2248–2252.

87. P. Alpuim, V. Chu, J. P. Conde (2000) Low substrate temperature deposition of amor-phous and microcrystalline silicon films on plastic substrates by hot-wire chemical vapor deposition. *J. Non-Crystall. Solids* **266–269**, 110–114.

88. M. Acciarri, M. Bernasconi, S. Binetti, D. Cavalcoli, A. Cavallini, C. Cavallotti , S. Cereda, D. Chrastina, L. Colombo, E. Grilli, G. Isella, M. Lancin, A. Le Donnne, A. Mattoni, L. Miglio, F. Montalenti , K. Peter, B. Pichaud, S. Pizzini, E. Poliani, M. Rossi, S. Sanguinetti, M. Texier, H. von Känel (2005) Final progress report of the nano-photo program. *Project no. 013944 S. Pizzini Edt.*

89. J. K. Rath (2003) Low temperature polycrystalline silicon: A review on deposition, phys-ical properties and solar cell applications. *Solar Energy Mater. Sol. Cells* **76**, 431–487.

90. S. Binetti, M. Acciarri, M. Bollani, L. Fumagalli, H. von Känel, S. Pizzini (2005) Nanocrystalline silicon films grown by Low Energy Plasma Enhanced Chemical Vapor Deposition for optoelectronic applications. *Thin Solid Films* **487**, 19–25.

91. K. Bhattacharya, D. Das (2008) Nanocrystalline silicon prepared at high growth rate using helium dilution. *Bull. Mater. Sci.* **31**, 467–471.

92. G. Cicala, G. Bruno, P. Capezzuto, L. Sciavulli, V. Capozzi, G. Perna (1997) Deposition of photoluminescent nanocrystalline silicon films by SiF_4,SiH_4,H_2 plasmas. *Conference Paper in MRS Online Proceeding Library Archive November 1997.* doi:10.1557/PROC-452-809.

93. M. Kummer, C. Rosenblad, A. Dommann, T. Hackbarth, G. Höck, M. Zeuner, E. Müller, H. von Känel (2002) Low energy plasma enhanced chemical vapor deposition. *Mater. Sci. Eng., B* **89**, 288–295.

94. M. Acciarri, S. Binetti, M. Bollani, A. Comotti, L. Fumagalli, S. Pizzini, H. von Känel (2005) Nanocrystalline silicon film grown by LEPECVD for photovoltaic applications. *Sol. Energy Mater. Sol. Cells* **87**, 11–24.

95. S. Pizzini, M. Acciarri, S. Binetti, D. Cavalcoli, A. Cavallini, D. Chrastina, L. Colombo, E. Grilli, G. Isella, M. Lancin, A. Le Donnne, A. Mattoni, K. Peter, B. Pichaud, E. Poliani, M. Rossi, S. Sanguinetti, M. Texier, H. von Känel (2006) Nanocrystalline silicon films as multifunctional materials for optoelectronic and photovoltaic applications. *Mater. Sci. Eng. B* **134**, 118–124.

96. A. Mattoni, L. Colombo (2007) Nonuniform growth of embedded silicon nanocrystals in an amorphous matrix. *Phys. Rev. Lett.* **99**, 205501.

97. M. Rondanini, S. Cereda, F. Montalenti, L. Miglio, C. Cavallotti (2007) A multiscale model of the plasma assisted deposition of crystalline silicon. *Surf. Coat. Technol.* **201**, 8863.

98. S. Cereda, F. Montalenti, L. Miglio (2007) Interaction of SiHx precursors with hydrogen-covered Si surfaces: Impact dynamics and adsorption sites. *Surf. Sci.* **601**, 3970.

99. S. Cereda, M. Ceriotti, F. Montalenti, M. Bernasconi, L. Miglio (2007) Quantitative estimate of H abstraction by thermal SiH_3 on hydrogenated Si(001)(2x1). *Phys. Rev. B* **75**, 235311.

100. A. Cavallini, D. Cavalcoli, M. Rossi, A. Tommasi, S. Pizzini, D. Chrastina, G. Isella (2007) Defect analysis of hydrogenated nanocrystalline silicon films. *Physica B* **401–402**, 519–522.

101. A. Le Donne, S. Binetti, G. Isella, S. Pizzini (2008) Structural homogeneity of nc-Si films grown by low-energy PECVD. *Electrochem. Solid State Lett.* **11** P5–P7.

102. A. Le Donne, S. Binetti, G. Isella, B. Pichaud, M. Texier, M. Acciarri, S. Pizzini (2008) Structural characterization of nc-Si films grown by low-energy PECVD on different substrates. *Appl. Surf. Sci.* **254**, 2804–2808.

103. A. Le Donne, S. Binetti, G. Isella, B. Pichaud, M. Texier, M. Acciarri, S. Pizzini (2008) Advances in structural characterization of thin film nanocrystalline silicon for photovoltaic applications. *Solid State Phenom.* **131–133**, 33–38.

104. D. Cavalcoli, M. Rossi, A. Tomasi, A. Cavallini, D. Chrastina, G. Isella (2008) Hydrogenated nanocrystalline silicon thin film studied by scanning force spectroscopy. *Solid State Phenom.* **131–133**, 547–552.

105. A. Cavallini, D. Cavalcoli (2008) Nanostructures in silicon investigated by atomic force microscopy and surface photovoltage spectroscopy. *Scanning* **30**(4), 358–363.

106. A. Mattoni, L. Colombo (2008) Crystallization kinetics of mixed amorphous-crystalline nanosystems. *Phys. Rev. B* **78**, 075408.

107. M. Rondanini, C. Cavallotti, D. Ricci, D. Chrastina, G. Isella, T. Moiseev, H. Kanel (2008) An experimental and theoretical investigation of a magnetically confined dc plasma discharge. *J. Appl. Phys.* **104**, 013304.

108. S. Cereda, F. Zipoli, M. Bernasconi, L. Miglio, F. Montalenti (2008) Thermal-hydrogen promoted selective desorption and enhanced mobility of adsorbed radicals in silicon film growth. *Phys. Rev. Lett.* **100**, 046105.

109. P. L. Novikov, A. Le Donne, S. Cereda, Leo Miglio, S. Pizzini, S. Binetti, M. Rondanini, C. Cavallotti, D. Chrastina, T. Moiseev, H. von Kaenel, G. Isella, F. Montalenti (2009) Crystallinity and microstructure in PECVD-grown Si films: A simple atomic-scale model validated by experiments. *Appl. Phys. Lett.* **94**, 051904.

110. E. Poliani, C. Somaschini, S. Sanguinetti, E. Grilli, M. Guzzi, A. Le Donne, S. Binetti, S. Pizzini, D. Chrastina, G. Isella (2009) Tuning by means of laser annealing of electronic and structural properties of n-Si/a-Si-H. *Mater. Sci. Eng. B* **159–160**, 31–33.

111. G. Micard, G. Hahn, B. Terheiden, D. Chrastina, G. Isella, T. Moiseev, D. Cavalcoli, A. Cavallini, S. Binetti, M. Acciarri, A. Le Donne, M. Texier, B. Pichaud (2010) Electrical and structural properties of *p*-type nanocrystalline silicon grown by LEPECVD for photovoltaic applications. *Phys. Status. Sol. C* **7**, 712.

112. A. Moiseev, D. Chrastina, G. Isella, C. Cavallotti (2009) Plasma composition and kinetic reaction rates in a LEPECVD Ar-SiH4-H2 plasma during nc-Si films deposition for photovoltaic applications. *ECS Trans.* **25**, 1065–1072.

113. J. Ramm. E. Beck, A. Zuger, R. E. Pixley (1992) Low-temperature *in situ* cleaning of silicon wafers with an ultra high vacuum compatible plasma source. *Thin Solid Films* **222**, 126–131.

114. C. Rosenblad, H. R. Deller, A. Dommann, T. Meyer, P. Schroeter, H. von Känel (1998) Silicon epitaxy by low-energy plasma enhanced chemical vapor deposition. *J. Vac. Sci. Technol. A* **16**, 2785–2790.

115. K. H. Chung (2010) Silicon -based epitaxy by chemical vapor deposition using novel precursor neopentasilane. PhD Thesis, Princeton university, Dept. Electrical Eng, p. 54.

116. M. Kummer, C. Rosenblad, A. Dommann, T. Hackbarth, G. Höck, M. Zeuner, E. Müller, H. von Känel (2002) Low energy plasma enhanced chemical vapor deposition. *Mater. Sci. Engin. B* **89**, 288–295.

117. S. Cereda, F. Montalenti, D. Branduardi, M. Radny, P. Smith, Leo Miglio (2006) Binding sites for SiH$_2$/Si(001): A combined ab initio, tight binding, and classical investigation. *Surf. Sci.* **600**, 4445.

118. S. Izumi, Y. Sato, S. Hara, S. Sakai, (2004) Development of a molecular dynamics potential for Si–H systems and its application to CVD reaction processes. *Surf. Sci.* **560**, 1–11.

119. A. Mattoni, L. Colombo (2008) Crystallization kinetics of mixed amorphous-crystalline nanosystems. *Phys. Rev. B* **78**, 075408.

120. D. J. Lockwood, A. Wang, B. Bryskiewicz (1994) Optical absorption evidence for quantum confinement effects in porous silicon. *Solid State Commun.* **89**, 587–589.

121. I. H. Campbell, P. M. Fauchet (1986) The effects of microcrystal size and shape on the one phonon Raman spectra of crystalline semiconductors. *Solid State Commun.* **58**, 739–741.

122. M. Avrami (1939) Kinetics of phase change. I. General theory. *J. Chem. Phys.* **7**, 1103.

123. M. Avrami (1940) Kinetics of phase change. II transformation-time relations for random distribution of nuclei. *J. Chem. Phys.* **8**, 212.

124. Y. Zhang, H. Wang, S. Jiang (2018) Evolution of structural topology of forming nanocrystalline silicon film by atomic-scale-mechanism-driven model based on realistic network. *AIP Adv.* **8**, 095321.

125. L. Mishnaevsky, M. Tsapatsis (2016) Hierarchical materials: Background and perspectives. *MRS Bull.* **41**, 661–664.

126. Md. N. Islam, A. Pradhan, S. Kumar (2005) Effects of crystallite size distribution on the Raman- scattering profiles of silicon nanostructures. *J. Appl. Phys.* **98**, 024309.

127. V. Schmidt, J. V. Wittemann, S. Senz, U. Gösele (2009) Silicon nanowires: A review on aspects of their growth and their electrical properties. *Adv. Mat.* **21**, 2681–2782.

128. R. Rurali (2010) *Colloquium*: Structural, electronic, and transport properties of silicon nanowires. *Rev. Modern Phys.* **82**, 427–449.

129. A. I. Hochbaum, P. Yang (2010) Semiconductor nanowires for energy conversion. *Chem. Rev.* **110**, 527–546.
130. T. J. Kempa, R. W. Day, S.-K. Kim, H.-G. Park, C. M. Lieber (2013) Semiconductor nanowires: A platform for exploring limits and concepts for nano-enabled solar cells. *Energy Environ. Sci.* **6**, 719–733.
131. K. A. Dick (2008) A review of nanowire growth promoted by alloys and non-alloying elements with emphasis on Au-assisted III–V nanowires. *Prog. Cryst. Growth Charact. Mater.* **54**(3), 138–173.
132. S. N. Mohammad (2011) General hypothesis for nanowire synthesis. I. Extended principles and evidential (experimental and theoretical) demonstration. *J. Appl. Phys.* **110**, 054311; General hypothesis for nanowire synthesis. II: Universality *J. Appl. Phys.* **110**, 054312.
133. R. S. Wagner, W. C. Ellis (1964) Vapor-Liquid-Solid mechanism of single crystal growth. *Appl. Phys. Lett.* **4**, 89.
134. R. S. Wagner, W. C. Ellis (1965) The vapour-liquid-solid mechanism of crystal growth and its application to silicon. *Trans. Met. Soc. AIME* **233**, 1053.
135. V. Schmidt, J. V. Wittemann, S. Senz, U. Gösele (2009) Silicon nanowires: A review on aspects of their growth and their electrical properties. *Adv. Mat.* **21**, 2681–2782.
136. V. Schmidt, J. V. Wittemann, U. Gösele (2010) Growth, thermodynamics, and electrical properties of silicon nanowires. *Chem. Rev.* **110**, 361–388.
137. J. Westwater, D. P. Gosain, S. Tomiya, S. Usui (1997) Growth of silicon nanowires via gold/silane vapor–liquid–solid reaction. *J. Vac. Sci. Technol.* B **15**, 554–557.
138. A. Sarikov (2011) Metal induced crystallization mechanism of the metal catalyzed growth of silicon wire-like crystals. *Appl. Phys. Lett.* **99**, 143102.
139. F. Fabbri, E. Rotunno, L. Lazzarini, D. Cavalcoli, A. Castaldini, N. Fukata, K. Sato, G. Salviati, A. Cavallini (2013) Preparing the way for doping wurtzite silicon nanowires while retaining the phase. *Nano Lett.* **13**, 5900–5906.
140. D. Hourlier, P. Lefebre-Legry, P. Perrot (2009) Preparation of silicon-based nanowires and the thermochemistry of the process. *JEEP 00002 EDP Science.*
141. D. Hourlier, P. Perrot (2010) Au-Si and Au-Ge phases diagrams for nanosytems. *Mater. Sci. Forum* **653**, 77–85. Trans Tech Publications, Switzerland.
142. B. J. Kim, J. Tersoff, C.-Y. Wen, M. C. Reuter, E. A. Stach, F. M. Ross (2009) Determination of size effects during the phase transition of a nanoscale Au-Si eutectic. *Phys. Rev. Lett.* **103**, 155701.
143. Z. J. Yanfeng, Z. Yamin (2010) Influence of gold particles on melting temperature of VLS grown silicon nanowires. *J. Semicond.* **31**(1), 012002-1/5.
144. X. Li, J. Ni, R. Zhang (2017) A thermodynamic model of diameter- and temperature-dependent semiconductor nanowire growth. *Sci. Rep.* **7**, 15029.
145. T. Xu, J. P. Nys, A. Addad, O. I. Lebedev, A. Urbieta, B. Salhi, M. Berthe, B. Grandidier, D. Stiévenard (2010) Facetted sidewalls of silicon nanowires: Au-induced structural reconstructions and electronic properties. *Phys. Rev.* B **81**, 115403.
146. H. Jeong, T. E. Park, H. K. Seong, H. J. Choi (2009) Growth kinetics of silicon nanowires by platinum assisted vapour-liquid-solid mechanism. *Chem. Phys. Lett.* **467**(4), 331–334.
147. A. M. Beers, J. Bloem (1982) Temperature dependence of the growth rate of silicon prepared through chemical vapor deposition from silane. *Appl. Phys. Lett.* **41**, 153.
148. K. Sato, A. Castaldini, N. Fukata, A. Cavallini (2012) Electronic level scheme in boron- and phosphorus-doped silicon nanowires. *Nano Lett.* **12**, 3012–3017.
149. R. Jones, A. Resende, S. Öberg, P. R. Briddon (1999) The electronic properties of transition metal hydrogen complexes in silicon. *Mater. Sci. Eng.* B **58**, 113–117.
150. F M Ross, J. Tersoff, S. Kodambaka, M. C. Reuter (2005) Growth and surface structure of silicon nanowires observed in real time in the electron microscope. *Microsc. Semiconducting Mater. Springer Proc. Phys.* **107**, 283–286. Springer, Berlin, Heidelberg.

151. M. F. Hainey, J. M. Redwing (2016) Aluminum-catalyzed silicon nanowires: Growth methods, properties and applications. *Appl. Phys. Rev.* **3**, 040806.
152. Y. Wang, V. Schmidt, S. Senz, U. Gösele (2006) Epitaxial growth of silicon nanowires using an aluminium catalyst. *Nature Nanotechn.* **1**, 186–189.
153. R. L. Boatright, J. O. McCaldin (1976) Solid-state growth of Si to produce planar layers. *J. Appl. Phys.* **47**, 2260–2262.
154. V. I. Levitas, K. Samani (2011) Coherent solid/liquid interface with stress relaxation in a phase-field approach to the melting/solidification transition. *Phys. Rev. B* **84**, 140103.
155. S. L. Lai, J. R. A. Carlsson, L. H. Allen (1998) Melting point depression of Al clusters generated during the early stages of film growth: Nanocalorimetry measurements. *Appl. Phys. Lett.* **72**, 1098.
156. M. Rodot, J. E. Burree, A. Mesli, G. Revel, R. Kishore, S. Pizzini (1987) Al-related recombination centres in polycrystalline silicon. *J. Appl. Phys.* **62**, 2556.
157. B.-S. Kim, T.-W. Koo, J.-H. Lee, D. S. Kim, Y. C. Jung, S. W. Hwang, B. L. Choi, E. K. Lee, J- M. Kim, D. Whang (2009) Catalyst-free growth of single-crystal silicon and germanium nanowires. *Nano Lett.* **9**, 864–869.
158. T. Ishiyama, S. Nakagawa, T. Wakamatsu (2016) Growth of epitaxial silicon nanowires on a Si substrate by a metal-catalyst-free process. *Sci. Rep.* **6**, 30608. doi:10.1038/srep30608.
159. Gurdeep Raj (2004) *Advanced Inorganic Chemistry*, Vol. 2 p. 220, Goel Publ. House, Dehli (India).
160. B. Fazio, P. Artoni, M. A. Latì, C. D'Andrea, M. J. Lo Faro, S. Del Sorbo, S. Pirotta, P. G. Gucciardi, P. Musumeci, C. S. Vasi, R. Saija, M. Galli, F. Priolo, A. Irrera (2016) Strongly enhanced light trapping in a two-dimensional silicon nanowire random fractal array. *Light Sci. Applic.* **5**, e16062.
161. X. Li, P. W. Bohn (2000) Metal-assisted chemical etching in HF/H_2O_2 produces porous silicon. *Appl. Phys. Lett.* **77**, 2572.
162. R. L. Smith, S. D. Collins (1992) Porous silicon formation mechanisms. *J. Appl. Phys*, **71**, R1.
163. R. L. Smith, S.-F. Chuang, S. D. Collins (1988) A theoretical model of the formation morphologies of porous silicon. *J. Electron. Mater.* **17**, 533.
164. A. Irrera, J. Lo Faro, C. D'Andrea, A. A. Leonardi, P, Artoni, B. Fazio, R. A. Picca, N. Cioffi, S. Trusso, G. Franzò, P. Musumeci, F. Priolo, F. Iacona (2017) Light-emitting silicon nanowires obtained by metal-assisted chemical etching. *Semicond. Sci. Technol.* **32**, 043004.
165. A. Pal, R. Ghosh, P. K. Giri (2015) Early stages of growth of Si nanowires by metal assisted chemical etching: A scaling study. *Appl. Phys. Lett.* **107**, 072104.
166. A. Irrera, P. Artoni, R. Saija, P. G. Gucciardi, M. A. Iatì, F. Borghese, P. Denti, F. Iacona, F. Priolo, O. M. Maragò (2011) Size-scaling in optical trapping of silicon nanowires. *Nano Lett.* **11**, 4879–4884.
167. A. Irrera, P. Artoni, F. Iacona, E. F. Pecora, G. Franzo`, M. Galli , B. Fazio, S. Boninelli, F. Priolo (2012) Quantum confinement and electroluminescence in ultrathin silicon nanowires fabricated by a maskless etching technique. *Nanotechnology* **23**, 075204 (7pp).
168. B. Fazio, P. Artoni, M. A. Latì, C. D'Andrea, M. J. Lo Faro, S. Del Sorbo, S. Pirotta, P. G. Gucciardi, P. Musumeci, C. S. Vasi, R. Saija, M. Galli, F. Priolo, A. Irrera (2016) Strongly enhanced light trapping in a two-dimensional silicon nanowire random fractal array. *Light Sci. Appl.* **5**, e16062.
169. S. Carapezzi, A. Cavallini (2019) The importance of design in nanoarchitectonics: Multifractality in MACE silicon nanowires. *Beilstein J. Nanotechnol.* **10**, 2094–2102.
170. A. M. Morales, M. Lieber (1998) A laser ablation method for the synthesis of crystalline semiconductor nanowires. *Science* **279**(5348), 208–211.

171. F. Zhang, Y. H Tang, N. Wang, D. P. Yu, C. S. Lee, I. Bello, S. T. Lee (1998) Silicon nanowires prepared by laser ablation at high temperature. *Appl. Phys. Lett.* **72**, 1835.

172. W. Shi, H. Peng, Y. Zheng, N. Wang, N. Shang, Z. Pan, C. Lee, S. Lee (2000) Synthesis of large areas of highly oriented, very long silicon nanowires. *Adv. Mater.* **12**, 1343–1345.

173. R. J. Barsotti, Jr., J. E. Fischer, C. H. Lee, J. Mahmood, C. K. W. Adu, P. C. Eklund (2002) Imaging, structural, and chemical analysis of silicon nanowires. *Appl. Phys. Lett.* **81**, 2866.

174. S. Bhattacharya, D. Banerjee, K. W. Adu, S. Samui, S. Bhattacharyya (2004) Confinement in silicon nanowires: Optical properties. *Appl. Phys. Lett.* **85**, 2008.

175. S-L. Zhang, W. Ding, Y. Yan, J. Qu, B. Li, L-Yu. Li, K. T. Yue, D. Yu (2002) Variation of the Raman feature on excitation wavelength of silicon nanowires. *Appl. Phys. Lett* **81**, 4446.

176. B. Ceccaroli, S. Pizzini (2012) Processes. In: *Advanced Silicon Materials for Photovoltaic Applications*, S. Pizzini Ed., Wiley, p. 62.

177. D. Yu, C. S. Lee, I. Bello, X. S. Sun, Y. H. Tang, G. W. Zhou, Z. G. Bai, Z. Zhang, S. Q. Feng (1998) Syntesis of nano-scale silicon wires by excimer laser ablation. *Solid State Commun.* **105**, 403.

178. S. L. Zhang personal communication.

179. B. Li, D. Yu, S.-L. Zhang (1999) Raman spectral study of silicon nanowires. *Phys. Rev. B* **59**, 1654–1648.

180. F. Fabbri, E. Rotunno, L. Lazzarini, N. Fukata, G. Salviati (2014) Visible and infra-red light emission in boron-doped wurtzite silicon nanowires. *Sci. Rep.* **4**, 3603. Nature Publishing Group.

181. A. Foncuberta i Morral, J. Arbiol, J. D. Prades, A. Cirera, J. R. Morante (2007) Synthesis of silicon nanowires with wurtzite crystalline structure by using standard chemical vapor deposition *Adv. Mater* **19**, 1347–1351.

182. G. S. Doerk, C. Carraro, R. Maboudian (2009) Temperature dependence of Raman spectra for individual silicon nanowires. *Phys. Rev. B* **80**, 073306.

183. Y. Wang, J. Zhang, J. Wu, J. L. Coffer, Z. Lin, S. V. Sinogeikin, W. Yang, Y. Zhao (2008) Phase transition and compressibility of silicon nanowires. *Nano Lett.* **8**, 2891–2895.

184. P. B. Sorokin, P. V. Avramov, V. A. Demin, L. A. Chernozatonski (2010) Metallic beta-phase silicon nanowires: Structure and electronic properties. *JETP Lett.* **92**, 352.

185. Y. Zhang, Z. Iqbal, S. Vijayalakshmi, H. Grebel (1999) Stable hexagonal-wurtzite silicon phase by laser ablation. *Appl. Phys. Lett.* **75**, 2758.

186. D. B. Zhang, M. Hua, T. Dumitrica (2008) Stability of polycrystalline and wurtzite nanowires via symmetry-adapted tight-binding objective molecular dynamics. *J. Chem. Phys.* **128**, 084104.

187. C. C. Yang, Q. Jiang (2005) Effect of pressure on melting temperature of silicon and germanium. *Mater. Sci. Forum* **475–479**, 1893–1896.

188. S.-Y. Lee, G.-S. Kim, J. Lim, S. Han, B. Li, J. T. L. Thong, Y.-G. Yoon, S.-K. Lee (2014) Control of surface morphology and crystal structure of silicon nanowires and their coherent phonon transport characteristics. *Acta Mater.* **64**, 62–71.

189. X. Li, J. Ni, R. Zhang (2017) A thermodynamic model of diameter- and temperature-dependent semiconductor nanowire growth. *Sci. Rep.* **7**, 15029.

190. Y. Zhao, B. I. Yacobson (2003) What is the ground-state structure of the thinnest Si nanowires? *Phys. Rev. Lett.* **91**, 035501.

191. R.-P. Wang, G.-W. Zhou, Y-L. Liu, S-H. Pan, H.-Z. Zhang, D.-P. Yu, Z. Zhang (2000) Raman spectral study of silicon nanowires: High-order scattering and phonon confinement effects. *Phys. Rev. B* **61**, 16827.

192. S. Piscanec, M. Cantoro, A. C. Ferrari, J. A. Zapien, Y. Lifshiz, S. T. Lee, S. Hoffmann, J. Robertson (2003) Raman spectroscopy of silicon nanowires. *Phys. Rev. B* **68**, 241312.

193. S. Piscanec, A. C. Ferrari, M. Cantoro, S. Hoffmnn, J. A. Zapien, Y. Lifshiz, S. T. Lee, J. Robertson (2003) Raman spectrum of silicon nanowires. *Mater. Sci. Engin. C* **23**, 931–934.

194. H. Richter, Z. P. Wang, L. Ley (1981) The one phonon Raman spectrum in microcrystalline silicon. *Solid State Commun.* **39**, 625–629.

195. I. H. Campbell, P. M .Fauchet (1986) The effects of microcrystal size and shape on the one phonon Raman spectra of crystalline semiconductors. *Solid State Commun.* **58**, 739–741.

196. S-L. Zhang, W. Ding, Y. Yan, J. Qu, B. Li, L.-Y. Li, K. T. Yue, D. Yu (2002) Variation of Raman feature on excitation wavelength in silicon nanowires. *Appl. Phys. Lett.* **81**, 4446.

197. A. Antidormi, X. Cartoixà, L. Colombo (2018) Nature of microscopic heat carriers in nanoporous silicon. *Phys. Rev. Mater.* **2**, 056001.

198. X. Cartoixà, R. Dettori, C. Melis, L. Colombo, R. Rurali (2016) Thermal transport in porous Si nanowires from approach-to-equilibrium molecular dynamics calculations. *Appl. Phys. Lett.* **109**, 013107.

199. A. Giri, B. F. Donovan, P. E. Hopkins (2018) Localization of vibrational modes leads to reduced thermal conductivity of amorphous heterostructures. *Phys. Rev. Materials* **2**. 056002.

200. D. Banerjee, C. Trudeau, L. F. Gerlein, S. G. Cloutier (2016) Phonon processes in vertically aligned silicon nanowire arrays produced by low-cost all-solution galvanic displacement method. *Appl. Phys. Lett.* **108**, 113109.

201. N. Daldosso, M. Luppi, S. Ossicini, E. Degoli, R. Magri, G. Dalba, P. Fornasini, R. Grisenti, F. Rocca, L. Pavesi, S. Boninelli, F. Priolo, C. Spinella, F. Iacona (2003) Role of interface region on the optoelectronic properties of silicon nanocrystals embedded in SiO$_2$. *Phys. Rev. B* **68**, 085327.

202. Z. Ma, X. Liao, J. He, W. Cheng, G. Yue, Y. Wang, G. Kong (1998) Annealing behaviour of the luminescence from SiOx:H. *J. Appl. Phys.* **83**, 7934–7939.

203. B. G. Gribov, K. V. Zinov'ev, O. N. Kalashnik, N. N. Gerasimenko, D. I. Smirnov, V. N. Sukhanov (2012) Structure and phase composition of silicon monoxide. *Semiconductors* **46**, 1576–1579.

204. J. A. Salazar, R. L. Estopier, E. Q. Gonzales, A. M. Sanchez, J. P. Chavez, I. E. Z. Huerta, M. A. Mijares (2016) *Silicon-Rich Oxide Obtained by Low-Pressure CVD to Develop Silicon Light Sources.* INTEC.

205. T. Hanrath, B. A. Korgel (2002) Nucleation and growth of germanium nanowires seeded by organic monolayer-coated gold nanocrystals. *J. Am. Chem. Soc.* **124**, 1424–1429.

206. Z. Qi, H. Sun, M. Luo, Y. Jung, D. Nam (2018) Strained germanium nanowire optoelectronic devices for photonic-integrated circuits. *J. Phys.: Cond. Matter* **30**, 334004.

207. D. S. Sukhdeo, D. Nam, J-H. Kang, M. L. Brongersma, K. C. Saraswat (2014) Direct bandgap germanium-on-silicon inferred from 5.7% (1 00) uniaxial tensile strain. *Photonics Res.* **2**, A8–13.

208. C. O'Regan, S. Biswas, N. Petkov, J. D. Holmes (2014) Recent advances in the growth of germanium nanowires: Synthesis, growth dynamics and morphology control. *J. Mater. Chem. C*, **2**, 14–33.

209. P. Periwal, T. Baron, P. Gentile, F. Bassani (2014) Growth strategies to control tapering in Ge nanowires. *APL Mater.* **2**, 46105(1-8).

210. C. Li, H. Mizuta, S. Oda (2011) Growth and characterisation of Ge nanowires by chemical vapour deposition. In: *Nanowires - Implementations and Applications*, Abbass Hashim Ed., Intechopen, London, UK, pp. 487–508.

211. S. Kodambaka, J. Tersoff, M. C. Reuter, F. M. Ross (2007) Germanium nanowire growth below the eutectic temperature. *Science* **316**(5825), 729–732.

212. A. R. Phani, V. Grossi, M. Passacantando, L. Ottaviano, Sandro Santucci (2006) Growth of Ge nanowires by chemical vapour deposition technique. Conference: 2006 NSTI Nanotechnology Conference and Trade Show - NSTI Nanotech 2006 Technical Proceedings.

213. T. Hanrath, B. A. Korgel (2002) Nucleation and growth of germanium nanowires seeded by organic monolayer-coated gold nanocrystals. *J. Am. Chem. Soc.* **124**, 1424–1429.

214. Y. Sierra-Sastre, S. Choi, S. Picraux, C. A. Batt (2008) Vertical growth of Ge nanowires from biotemplated Au nanoparticle catalysts. *J. Am. Chem. Soc.* **130**, 10488–10489.

215. Z. Zhu, Y. Song, Z. Zhang, H. Sun, Yi. Han, Y. Li, L. Zhang, Z. Xue, Z. Di, S. Wang (2017) The vapor-solid-solid growth of Ge nanowires on Ge (110) by Molecular Beam Epitaxy. Available at: https://arxiv.org/ftp/arxiv/papers/1706/1706.01605.pdf.

216. S. J. Rezvani, N. Pinto, L. Boarino (2016) Rapid formation of single crystalline Ge nanowires by anodic metal assisted etching. *Cryst. Eng. Comm.* **18**, 7843–7848.

217. C. Fang, H. Foll, J. Carstensen (2006) Long germanium nanowires prepared by electrochemical etching. *Nano Lett.* **67**, 1578–1580.

218. S. Szunerits, Y. Coffinier, E. Galopin, J. Brenner, R. Boukherroub (2010) Preparation of boron-doped diamond nanowires and their application for sensitive electrochemical detection of tryptophan. *Electrochem. Comm.* **12**(3), 438–441.

219. S. Szunerits, Y. Coffinier, R. Boukherroub (2015) Diamond nanowires: A novel platform for electrochemistry and matrix-free mass spectrometry. *Sensors* **15**, 12573–12593. doi:10.3390/s150612573.

220. X. Peng, J. Chu, L. Wang, S. Duan, P. Feng (2017) Boron-doped diamond nanowires for CO gas sensing application (2017). *Sens. Actuators B: Chemical* **241**, 383–389.

221. B. J. M. Hausmann, M. Khan, Y. Zhang, T. M. Babinec, K. Martinick, M. McCutcheon, P. R. Hemmer, M. Loncar (2010) Fabrication of diamond nanowires for quantum information processing applications. *Diamond Relat. Mater.* **19**, 621–629.

222. M. Shellaiah, K. W. Sun (2019) *Diamond Nanowire Synthesis, Properties and Applications.* doi:10.5772/intechopen.78794.

223. P. Subramanian, S. Kolagatla, S. Szunerits, Y. Coffinier, W. Siang Yeap, K. Haenen, R. Boukherroub, A. Schechter (2017) Atomic force microscopic and Raman investigation of boron-doped diamond nanowire electrodes and their activity toward oxygen reduction. *J. Phys. Chem. C.* **121**, 3397–3403.

224. I. Iatsunskyi, S. Jurga, V. Smyntyna, M. Pavlenko, V. Myndrul, A. Zaleska (2014) Raman spectroscopy of nanostructured silicon fabricated by metal-assisted chemical etching. *Proc. SPIE*, **9132**, 913217-1.

225. D. Luo, L. Wu, J. Zhi (2009) Fabrication of boron-doped diamond nanorod forest electrodes and their application in nonenzymatic amperometric glucose biosensing. *ACS Nano* **3**, 2121–2128. doi:10.1021/nn9003154.

226. S. Prawer, R. J. Nemanich (2004) Raman spectroscopy of diamond and doped diamond. *Philos. Trans. R. Soc. London. Ser A* **362**, 2537–2565.

227. K. Wakui, Y. Yonezu, T. Aoki, M. Takeoka, K. Semba (2017) Simple method for fabrication of diamond nanowires by inductively coupled plasma reactive ion etching. *Jpn J. Appl. Physics* **56**(5), 058005.

4 Preparation, Structural, and Physical Properties of Nanocrystalline Dots, Films, and Nanowires of Compound Semiconductors

4.1 BACKGROUND CONCEPTS

Unlike homopolar semiconductors diamond, Si, and Ge, bonding in compound semiconductors is partially ionic, and the ionicity increases along the sequence of IV–IV, III–V, II–VI, and I–VII compounds as shown by Christensen et al. [1].

In turn, the ionicity f_i is defined as

$$f_i = \frac{\left(\Delta E_{sp^3}\right)^2}{E_G^2} \tag{4.1}$$

where ΔE_{sp^3} is the offset of the anions and cations hybrid energy levels and E_G is the energy separation between sp^3 hybrid bonds and antibonding states, which in turn is the sum of two terms

$$E_G = \left[\left(\Delta E_{sp^3}\right)^2 + (2h)^2\right] \tag{4.2}$$

of which the second term $E_h = 2h$ is a hybridization contribution.

In homopolar semiconductors the first term vanishes, and only the hybridization term E_h contributes to E_G. In compound semiconductors, as can be seen in Table 4.1, both terms contribute and there is a continuous increase of the ionicity f_i from SiC to CdSe.

Ionicity in compound semiconductors induces polarity effects, with the presence of an effective electrical charge at their *polar* surfaces, terminated either with a group III or group V element in the case of III–V compounds.

Polarity is, therefore, an intrinsic feature of bulk compound semiconductors [2–5], which influences their electrical, optical, and photophysical properties as well as those of compound semiconductor 1D nanostructures [6–8].

TABLE 4.1

Calculated Values of Ionicity f for Homopolar and Covalent Semiconductors [1]

Compound	d(Å)	ΔE_{sp^3} (eV)	E_h (eV)	f_i
CdSe	2.63	8.35	3.62	0.841
CdS	2.53	5.90	4.67	0.794
CdTe	2.81	6.35	3.77	0.739
InAs	2.62	5.06	0.357	0.553
AlP	2.36	5.16	6.04	0.421
GaP	2.36	5.00	6.66	0.361
GaAs	2.45	4.29	6.40	0.310
SiC	1.88	7.47	9.27	0.177
C	1.54	0.00	13.31	0.000
Si	2.35	0.00	6.82	0.000
Ge	2.45	0.00	6.38	0.000

Note: d is the lattice constant

Ionicity influences also the structural properties of II–VI and III–V compound semiconductors, which crystallize either with the zincblende and rock-salt structure or the zincblende and wurtzite structure, respectively, depending on their ionicities. In both cases, zincblende is the equilibrium structure for low-ionicity compounds.

Compounds crystallizing with the zincblende structure present also high-pressure polytypes with the rock-salt structure, and the pressure at which the transition from the zincblende to the rock-salt structure occurs decreases with their ionicity, as is shown in Figure 4.1 [1]. Apparently, compounds with the largest ionicity (ZnSe, ZnTe, CdTe) tend to crystallize with the rock-salt structure, that is, eventually, the equilibrium structure of MgS.

Point defects (vacancies and self-interstitials) are equilibrium properties of compound semiconductors and control their doping efficiency. It should be remarked that in bulk compound semiconductors point defects are generated at high temperatures, during their growth process, in equilibrium with the gaseous components of the material, i.e. Ga vapors and As vapors in the case of GaAs.

Therefore the formation of As vacancies in GaAs is given by the following equation

$$Ga_{Ga} \rightleftharpoons Ga(g) + V_{As} \tag{4.3}$$

showing that their concentration depends on the partial pressure of Ga in the growth atmosphere, and their equilibrium concentration is

$$c_{V,As} = Kp_{Ga}; \; K = -\exp\frac{\Delta G_f^V}{kT} \tag{4.4}$$

where ΔG_f^V is the Gibbs free energy of formation of As vacancies.

Similar equations hold for Ga vacancies V_{Ga}.

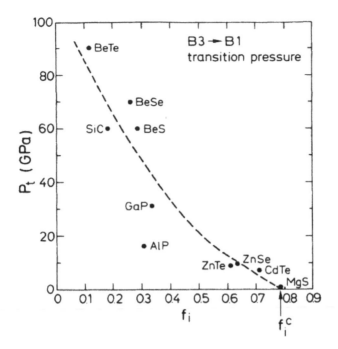

FIGURE 4.1   Calculated dependence of the transition pressure P_t from the zincblende (B3) to the rocksalt structure for a family of compound semiconductors. (*Reprinted with permission from N.E. Christensen, S. Sapaty, Z. Pawlowska (1987) Bonding and ionicity in semiconductors Phys.Rev. B* **36**(2) *1032–1050. Copyright 1987 American Physical Society License Number: RNP/19/SEP/019036 License date: Sep 26, 2019.*)

Their properties have been reviewed by Hurle [9] for the case of GaAs and by Newmark [10] for the II–VI compounds, and equilibrium thermodynamic models [9] are shown to describe accurately the doping and the self-compensation features arising from dopants incorporated in both sublattices and are a powerful guide for the optimization of growth and doping processes.

Non-stoichiometry is another intrinsic feature of compound semiconductors, as shown by Rudolph [11] and Wenzl et al. [12]. It leads to the set-up of a wide homogeneity region of their solid phases in their $x - T$ domain, well-illustrated by Rudolph [11] for the case of PbTe, CdTe, GaAs, and InP, and by Wentzl et al. [12] for the case of GaAs, where the homogeneity region is almost entirely located in the As-rich side.

Non-stoichiometry is common to all compound semiconductors, including SiC [13], which can be grown non-stoichiometric under form of thin films.

Non-stoichiometry shifts the congruent melting point composition from that of the stoichiometric phase, and the amount of stoichiometric deviation $\Delta\delta$ in the temperature range of stability of the homogeneity phase varies with the temperature, with a strong decrease at low temperatures, but the deviations from stoichiometry are small but remain consistent even at low temperatures.

TABLE 4.2

Maximum Width of the Existence Region for Some Compound Semiconductors [11]

Material	InP	GaAs	CdTe	CdS	CdSe	PbTe	SnTe
$\Delta\delta_{max}$	$\approx 5 \cdot 10^{-5}$	$\approx 2 \cdot 10^{-4}$	$\approx 1 \cdot 10^{-4}$	$\approx 1 \cdot 10^{-4}$	$\approx 5 \cdot 10^{-4}$	$\approx 1 \cdot 10^{-3}$	$\approx 1 \cdot 10^{-2}$
Region of max excess	In-rich	As-rich	Te-rich	Cd-rich	Cd-rich	Te-rich	Te-rich

As can be seen in Table 4.2, there is an increase of the maximum of stoichiometric deviation $\Delta\delta_{max}$ from InP to SnTe. Considering that the growth temperatures of compound semiconductor nanowires range in the 400–800°C interval, nanowires could also suffer from stoichiometry offsets, small but not-negligible in the case of GaAs NWs, important in the case of II–VI compounds.

We will see in the following sections how far ionicity, polarity, non-stoichiometry, and polytypism do influence the properties of compound semiconductor nanowires.

4.2 SILICON CARBIDE NANOWIRES

SiC NWs have been intensively studied over the last few decades, [14] in order to develop performant growth and doping processes and to optimize their morphological and optoelectronic properties.

The interest in SiC nanostructures arises from several, unique properties of their parent material, such as its wide band gap, its excellent thermal conductivity, its chemical and physical stability, and superior mechanical properties, which favor the use of bulk SiC in specific electronic device applications, operating in extreme conditions or/and harsh environments [15].

The polytypism is another typical feature of SiC [16, 17]. It presents more than 200 different polytypes, which result from different periodic stacking sequences of the hexagonally packed double layers of Si and C along the cubic [111] or hexagonal [0001] direction. The most common polytypes of SiC are the cubic (3C-SiC or βSiC) one and the three hexagonal ones (2H, 4H, and 6H), which present slightly different band gaps, ranging between 2.417 eV for the cubic material and 3.26 eV for the 4H phase and 3.02 eV for the 6H phase. The hexagonal polytypes are the most thermodynamically stable, but the cubic one is most stable at low growth temperatures.

Given the wide band gaps of SiC polytypes, nanocrystalline SiC or SiC nanowires have been envisaged, like porous Si and Nc-Si, as materials potentially capable of shifting their optical emission in the blue. We will see in Chapter 5 that surface defects and wire diameters, definitely larger than the Bohr radius of 3C-SiC (≈ 2.7 nm), seem to limit or preclude completely quantum confinement effects [18, 19]. Still, it will be seen in the next section that 4H-SiC nanowires and multi-twinned 3C-SiC nanowires do present favorable opportunities, due to the presence of twin boundaries located at the right distance.

Twins,* stacking faults (SF), and grain boundaries (GB)† are the most common native defects in SiC, seeded by stress-induced dislocations, whose density is enhanced by temperature [20].

Most of the physical properties of bulk SiC are common to SiC nanowires, with some typical differences.

As an example, common to all 1D nanostructures, size and surface effects do unfavorably influence their doping processes, and the ionization energy of dopants, which increases with the increase of the diameter of the nanostructures [21].

Furthermore, unlike bulk SiC, which can be grown either with the cubic or the hexagonal structures, the cubic structure is found the most thermodynamically stable under the process conditions available for the growth of SiC NWs, though also 6H-SiC nanowires were successfully grown [22, 23] using Al as the catalyst. Since 6H-SiC nanowires present favorable opportunities for their use as blue or ultraviolet LED applications, their optimized growth is a real challenge.

Experimental results and theoretical calculation demonstrate, also, that the mechanical properties of 1D SiC nanostructures are enhanced with respect to bulk SiC [24], but are strongly influenced by twins, stacking faults, and amorphous layers [21].

Amorphous layer coatings, typical of core-shell structures, in fact, can induce brittle to ductile transition in SiC nanowires, and cause a decrease in the critical yield stress and Young's modulus, whereas the critical strain of the nanowires can be enhanced by twins and stacking faults. As an example, defective SiC NWs with a high density of stacking faults experience superior plasticity behavior at low temperatures (80°C), with a 200% elongation before fracture [25].

The mechanical properties of SiC NWs are also influenced by the size-dependent density of structural defects (SF) [26]. It was, in fact, demonstrated that while the fracture strength was strongly enhanced (from 8.1 to 25.3 GPa) with the decrease of the wire diameter from 45 to 17 nm, approaching the fracture strength of bulk 3C SiC (28.5 GPa), the density of stacking faults does substantially decrease with the decrease of the diameter down to 17 nm.

Structural defects not only influence the mechanical properties of SiC NWs, but also their opto-electronic properties. In fact, the periodically twinned 3C SiC nanowires prepared by Wang et al. [27] exhibit a PL emission spectrum that manifests quantum confinement effects, due to the fact that the NW is a periodical sequence of nanocells, with a 3D confinement.

4.2.1 SYNTHESIS, MORPHOLOGY, AND STRUCTURAL FEATURES OF SiC NANOWIRES: THERMAL REACTION PROCESSES

Thermal reaction processes were systematically used, often with excellent results, for the synthesis of SiC NWs.

As a first example, Chen et al. [28] demonstrated that the reaction of graphite with Si vapors in a graphite crucible at 1550°C for about three hours might be successfully used to synthetize SiC NWs. Due to some oxygen contamination of the argon

---

* Twins are low-energy stacking faults.
† In the case of polycrystalline SiC nanowires.

used as cover gas, the nanowires present a core-shell morphology, with a 3C-SiC core (from XRD measurements) and an amorphous SiO_2 shell. XRD measurements show also that the predominant orientation is [111] along the wire axis.

TEM measurements carried out on these wires show that they are straight and randomly oriented and present smooth surfaces. Their average diameter is 60 nm, with an amorphous silica layer 5–15 nm thick, but HRTEM measurements show also the presence of a high density of stacking faults and planar defects, which are the typical defects of all the NWs prepared with thermal reaction processes, as we will see in the further examples. It is also demonstrated that the growth of SiC nanowires is a spontaneous process, which occurs without the use of a catalyst, when a graphitic material is used as the source of carbon.

Similar results were obtained by Cheong and Lockman [29] by the reaction of activated carbon powder (8.5% O) with silicon vapors sublimated from a silicon wafer at a maximum temperature of 1300°C under vacuum in a graphite crucible. The reaction product is the mesh of randomly oriented twinned nanowires seen in Figure 4.2,

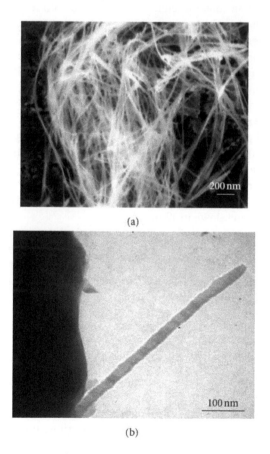

(a)

(b)

FIGURE 4.2 SEM (a) and TEM (b) images of SiC nanowires. (*After* K.Y. Cheong, Z. Lockman (2009) Growth mechanism of cubic-silicon carbide nanowires *J. Nanomat.* 2009, January 2009 Article No.31 Hindawi Open Access Journal.)

with a diameter within 10 and 30 nm and a length of several micrometers. XRD measurements established the 3C-SiC nature of the phase, confirmed by the Raman peak at 796 cm^{-1}, typical of SiC [30].

We will see in the following section that the formation of arrays of randomly oriented nanowires is the common output of high-temperature processes, with a few exceptions given by the use of carbon-based templates.

Contextually to the formation of SiC NWs at the surface of the carbon powder, the surface of the silicon wafer placed on top of the graphite crucible was shown to be heavily corroded, with the formation of faceted SiC micropillars shown in Figure 4.3.

It seems obvious to conclude that the silicon wafer during sublimation is also chemically etched and reacted by the CO atmosphere generated by the oxygen contamination of the carbon with the formation of SiC nanopillars, and that this process belongs to the family of the dry etching processes discussed in Section 4.2.4.

A simple thermal process carried out at 1700°C in flushing argon on a mixture of SiC powder and Fe powder, this last working as the catalyst, was used by Wu et al. [31] to prepare bunches of needle-like nanowires, with diameters ranging between 20 and 50 nm and length of 1–2 μm, growing from a metal particle working as the seed that remains on the tip of the wires.

The wires are shown by TEM to be single crystalline, [100] oriented along the wire axis, and their Raman spectrum peak at 796 cm^{-1} indicates the presence of the cubic 3C-SiC phase [30].

Although the authors of ref. 30 do not provide a suggestion about the physics of the process occurring, we suppose that a supersaturated solution of SiC in liquid iron droplets is formed at the process temperature, from which the SiC wire growth process occurs during the cooling process.

It is however apparent that Fe is not a true catalyst, but an active ingredient of the mixture, since Fe–Si alloys are excellent solvents for SiC [32] and already at 1723 K a liquid Si–C–Fe phase in equilibrium with SiC is thermodynamically stable as is shown in Figure 4.4.

A similar process using Fe as the catalyst was carried out by Wang et al. [33], who used as reactants a micrometric powder of graphene sheets and a micrometric

FIGURE 4.3 Faceted SiC micropillars formed at the silicon surface during a thermal treatment at 1300–1350°C in the presence of traces of CO. (*After* K.Y. Cheong, Z. Lockman (2009) Growth mechanism of cubic-silicon carbide nanowires *Journal of Nanomaterials*, January 2009 Article No.31 Hindawi Open Access Journal.)

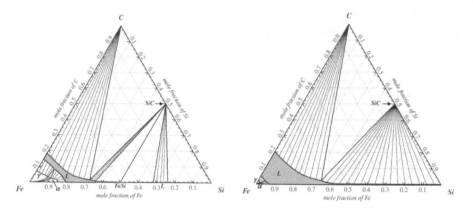

FIGURE 4.4 Ternary phase diagram of the system Fe–Si–C at 1523 K (left) and 1723 K (right). (*Reproduced with permission of the Japan Institute of Metals and Materials* from S. Kawanishi, T. Yoshikava, T. Tanaka (2009) Equilibrium phase relationship between SiC and liquid phase in the Fe-Si-C system at 1523–1723 K *Mat. Trans.* **50** (4) 806–813 *Aug 28, 2019.*)

powder of silicon and iron, all of them of high purity grade. The process was carried out under high vacuum conditions at 1573 K, very close to the temperature of the equilibrium phase diagram of Figure 4.4 on the left. After cooling, the graphene sheets are shown to be covered by meshes of SiC NWs, whose features lead the author to suggest that the reaction occurs at the graphene surface, and that NWs with an average length of 10 μm and a diameter of 60 nm segregate from a supersaturated Si–C–Fe alloy, like in the former case. XRD measurements demonstrate the cubic βSiC structure of these wires, and high-magnification TEM images of these wires not only show that straight and twisted sections are present but also that that the tips of these wires are metallic, confirming that these wires segregate from a liquid metallic alloy.

Several authors successfully used the siliconization of carbonaceous materials for the production of SiC NWs without intentionally added catalysts, as was done as an example by Kim et al. [34] who employed the reaction of silica powder with carbon black at 1300–1600°C, temperatures at which the reduction of silica by carbon

$$2C + SiO_2 \rightleftharpoons Si + 2CO \tag{4.5}$$

is thermodynamically allowed (see Figure 4.5). It was, however, shown that SiC NWs present a metallic (Fe) cap on the tips of the wires, that suggests that a liquid drop of the Fe–Si–C alloy does form at the surface of the iron-contaminated SiO$_2$ powder.*

In fact, see again Figure 4.5, the reduction of Fe$_2$O$_3$ to metallic iron does occur at temperatures above 1000 K in a CO atmosphere.

Similar results were obtained by Chiu et al. [35], who reacted high-purity silica powder and high-purity graphite in an arc discharge reactor. Dense arrays of βSiC randomly oriented nanowires, with diameters in the range of 3–15 nm and a length

---

* I suppose that the SiO$_2$ powder is contaminated by iron oxides.

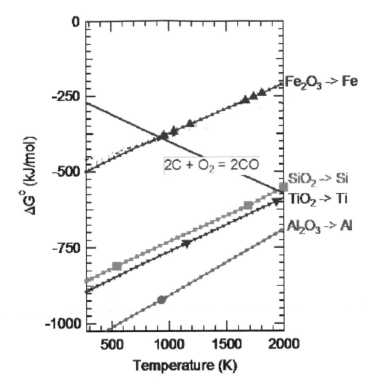

FIGURE 4.5 Ellingham diagram of the Si/SiO_2 system, containing different impurities, in equilibrium with carbon.

of a few µm are the product of the reaction, similarly to the case of Je et al. [36], who used silica powder and Si and graphite as the Si and C sources and iron powder as the catalyst to grow 3C-SiC NWs at a process temperature of 960°C. The SiC nanowires grown with this process have a metallic cap on their tips, a diameter of about 20–80 nm, and an amorphous shell coating. As in most of the cases of SiC NWs, dense arrays of SFs orthogonal to the growth axis are observed in the HRTEM images.

In another recent work, Hua et al. [37] obtained results very close to those obtained by Wang [27] and Kim [34] by carrying out a different process, which consisted of heating under vacuum, in a silicate tube, a powder of graphene sheets contained in an alumina crucible. A simple heating cycle at 1400°C followed by a slow cooling to room temperature causes the growth of twisted SiC nanowires, whose aspect at different magnifications is shown in the SEM images of Figure 4.6.

The XRD spectra measured on these wires confirm that they consist of SiC. However, the very nature of the process occurring is unknown, although the authors suppose that SiO vapors evolved from the refractory tube during high-temperature heating in a CO atmosphere

$$CO + SiO_2 \rightleftharpoons SiO + CO_2 \qquad (4.6)$$

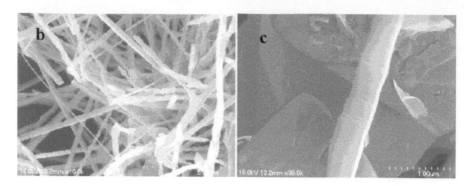

FIGURE 4.6   SEM images of SiC nanowires grown from graphene. (*From* Z. Hua, Y. Li, Y. Zhan, P. Huan, X. Wang, and C. Chen (2018) Fabricating SiC Nanowire by Vacuum Heating Multilayer Graphene in Silicate Refractory Tube *IOP Conf. Ser.: Mater. Sci. Eng.* 381 012107 Open access article.)

would react with CO

$$CO + SiO \rightleftharpoons SiC + O_2 \qquad (4.7)$$

or directly with graphene

$$SiO(g) + C \rightleftharpoons SiC + \frac{1}{2}O_2 \qquad (4.8)$$

leading to the formation of SiC.

Carbon nanotubes as templates and their reaction with SiO were used by Pan et al. [38] and by Sun et al. [39] to grow arrays of oriented, cubic silicon carbide (βSiC) nanowires.*

Schematically, SiO powder is placed on the bottom of a graphite crucible, and then covered with a molybdenum grid, which works as the support of the carbon nanotubes array stripped from their original silica substrate. The crucible is heated in flushed argon and held at 1400°C for two hours, to allow the sublimation of SiO and its reaction with the carbon nanotubes, which are converted in an array of SiC nanowires, with similar diameters (10–40 nm), lengths (2 mm), and spacings ($\approx$ 100 nm) to the original carbon nanotubes array.

Selected area electron diffraction (SAED) measurements on these wires demonstrate that their structure is cubic, single crystalline, and [111] oriented along the wire axis. TEM investigations show the presence of a high density of planar defects perpendicular the wire axis.

The success of this process for the synthesis of SiC nanowires suggests its utilization for other types of carbide and nitrides NW arrays. Arrays of SiC NWS have been used as cathodic materials for field emission measurements, showing excellent electrical and morphological stability, which is promising for future device applications.

The reaction of SiO(g) with carbon nanotubes[†] was also used by Zhu and Fan [40] to grow shell- core 3C-SiC nanowires, with a shell of impure and non-stoichiometric amorphous SiC and a core of SiC.

---

* The Pal paper reports details about the synthesis of carbon nanotubes.

† The authors do not give indications about the reaction temperature (800°C?) nor about process details.

FIGURE 4.7   TEM image of a SiC nanowire presenting a continuous sequence of stacking faults. An electron diffraction image is displayed on the right top of the image. (*Reproduced from* J. Zhu, S. Fan (1999) Nanostructure of GaN and SiC Nanowires Based on Carbon Nanotubes *J. Mat.Res.14*(4)1175–1177 *with permission of Cambridge University Press, License Number 4654830510839, Aug 23, 2019.*)

Like the case of Pan et al. [38] discussed before, TEM investigations carried out on these SiC NWs show (see Figure 4.7) [40] the presence of a continuous sequence of stacking faults, that seem, in fact, a constant of this process.

Heavily twisted SiC NWs were also grown at 1200°C by Wang et al. [41] using low-grade (98.5%) silicon powder, petroleum coke, and lignin-phenolic resin, in the absence of deliberately added iron as the catalyst.

On the base of the composition of a spherical cap present on the tip of the wires, the authors suppose that a VLS process does occur, which involves SiO in the vapor phase and carbon, following Equation 4.7, while a liquid sodium and calcium silicate phase, formed by impurities present in the raw materials used, works as catalyst.

Similar results were obtained by Hu et al. [42] with the reaction of silicon powder with a phenolic resin at 1450°C in argon atmosphere. The several hundred μm-long SiC nanowires, see Figure 4.8 present a shell-core morphology with a βSiC core with diameters in the range of 140–560 nm and an amorphous shell about 2 nm thick. Also here the core is heavily damaged by stacking faults.

The systematic absence of a metallic cap over the tip of the wires was assumed as an indication that the growth occurred in the absence of a catalyst.

Silicon-based templates, instead of carbon-based templates, were used by Pavlikov et al. [43] to grow SiC nanowires. In fact, they employed porous silicon or silicon nanowires, which were converted to 40–50 nm thick SiC nanowires by reaction at

FIGURE 4.8 SEM image of the SiC nanowires grown at 1480°C by reaction of silicon powder with a phenolic resin. (*After* P. Hu, S. Dong, X. Zhang, K. Gui, G. Chen, Z. Hu (2017) Synthesis and characterization of ultralong SiC nanowires with unique optical properties, excellent thermal stability and flexible nanomechanical properties *Scientific Reports* **7** *3011* | DOI:10.1038/s41598–017-03588-x 1 Open Access Journal.)

1360°C with a mixture of hydrocarbons, growing onto the porous silicon substrate, which works as seed, not as a template, as is seen in Figure 4.9.

The chemical composition of the nanowires could be inferred from their Raman spectra, which include a peak at 796 cm^{-1} typical of SiC, while XRD measurements demonstrate the 3C-SiC structure of the wires.

The reaction of Si with CCl_4 in the presence of sodium as the reductant was used by Hu et al. [44], following the praxis successfully adopted for the growth of diamond powder. The process was carried out in autoclave at 700°C, with the formation of carbon powder, SiC NWs (Figure 4.10a) and carbon nanorods (see Figure 4.10f). At the end of the process the residual silicon was etched off with a HF-HNO_3 solution, while the residual carbon was burned off at 600°C in air, with partial oxidation of the SiC material, which was submitted to a final cleaning with HF to get the final sample of straight and [111] oriented SiC filaments (Figure 4.10a, b), with an average diameter of 15–20 nm and a length of 5–15 μm, with a 3C-SiC structure resulting from XRD measurements. Raman spectra carried out on this material confirm the SiC nature, in view of the presence of a narrow peak at 776 cm^{-1}.

The HRTEM image of Figure 4.10c shows that stacking faults and planar defects are absent, unlike the NWs grown with high-temperature processes, previously discussed.

Some of the wires (Figure 4.10d) present a hollow morphology, with walls heavily damaged by stacking faults (Figure 4.10e). These hollow SiC NWs presumably are the product of silicization of carbon nanorods, acting as templates.

SiC NWs were also grown using the laser ablation technique by Shi et al. [24]. The process consists of the ablation with an excimer KrF laser a SiC target held in the center of an alumina tube under an Ar–H_2 atmosphere (0.01 torr). The SiC target

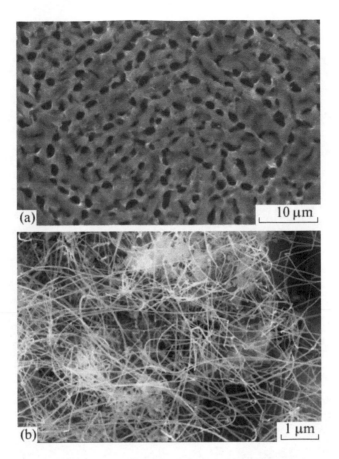

FIGURE 4.9 SEM images of a porous silicon substrate before (a) and after reaction (b) with hydrocarbons at 1360°C. (*Reproduced with permission of Springer Nature from* A.V. Pavlikov, V. Latukhina, V.I. Chepurnov, V. Yu. Timoshenko (2017) Structural and optical properties of silicon-carbide nanowires produced by the high-temperature carbonization of silicon nanostructures *Semiconductors* **51** (3) 402–406 *License number 4655440576890, License data Aug 24, 2019.*)

is held at 1100°C. The ablation products are collected on an iron-catalyzed* graphite substrate sitting at the periphery of the alumina tube.

An array of straight and curved SiC[†] nanowires, with an average diameter of 80 nm, was formed onto the graphite substrate after two hours of laser annealing. The nanowires have a core-shell structure, with a core of SiC and a shell of amorphous SiO_x. The average diameter of the core is 55 nm and the thickness of the shell amounts to 17 nm.

---

* The substrate was preliminarily dipped in an iron nitrate solution and dried, which reduces to metallic iron in the hydrogen atmosphere.
[†] From micro Raman measurements.

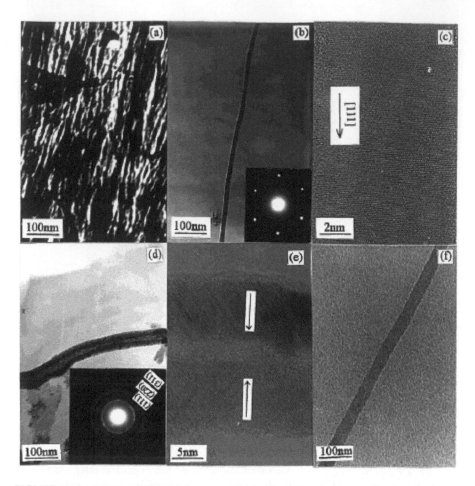

FIGURE 4.10   TEM and HRTEM images of SiC NWS and carbon nanotubes. (a) SEM image of an array of SiC NWs (b) TEM image of an oriented SiC filament (c) HRTEM image of an SiC filament (d) TEM image of an hollow filament (e) TEM image of the shell of a hollow wire (f) TEM image of a carbon nanorod. (*Reprinted with permission from* J.Q. Hu, Q.Y. Lu, K.B. Tang, B. Deng, R.R. Jiang, Y.T. Qian, W.C. Yu, G.E. Zhou, X.M. Liu, J.X. Wu (2000) Synthesis and characterization of SiC nanowires through a reduction–carburization route *J. Phys. Chem. B* **104** (22) 5251–5254, *Copyright 2000 American Chemical Society Dec 2, 2019.*)

Typical of SiC nanowires grown at high temperature, the core is damaged by numerous SFs parallel to the wire axis, that is, [100] oriented in the present case.

Like in the most of the previously discussed growth processes occurring in the presence of iron as the SiC solvent, a spherical, metallic nanoparticle was found on top of the nanowires, whose chemical components were iron, silicon, carbon (and oxygen). It is therefore possible to suppose that the growth process involves the sublimation of SiC under laser ablation, and the further dissolution of SiC in liquid droplets of an Fe–C alloy (see Figure 4.11) of eutectic composition (4.30 wt%),* melting

---

* View of SP. The authors of the paper do not consider the nature of the liquid alloy.

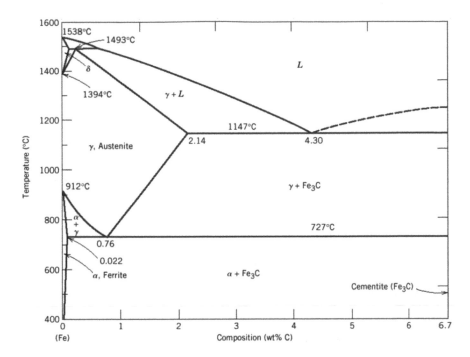

FIGURE 4.11   Phase diagram of the system Fe–C. (*From MSE 300, Materials Laboratory procedures, University of Tennessee, Dept. Materials Science and Technology.*)

at 1147°C, with the formation of a ternary, supersaturated solution, from which SiC nanowires do segregate. The temperature of the experiment is slightly below the bulk eutectic temperature, but a size-dependent decrease of the eutectic temperature might be expected, given the diameter of the wires.

To conclude the analysis of thermal reaction processes addressed at the growth of SiC nanowires, it is interesting to mention the almost unique case of a process suitable for growing 6H-SiC nanowires developed by Wei et al. [23], who used a simple microwave furnace (800 W) as the heat source, capable of reaching a maximum temperature of 1200°C on an SiC support, and a xerogel based on sucrose, tetra-ethyl-orthosilicate (TEOS), and oxalic acid* as the precursor material, mixed with ultrafine Al powder. The process was carried out with 3 g of the xerogel enclosed in a crucible, using flowing Ar as the protective atmosphere, and an heating time of 30 min. After cooling to room temperature, the product was heat treated in air at 700°C to oxidize the residual carbon and then etched in a HF solution to remove silica. The SEM analysis of the final product shows that it consists of an array of nanowires with an average diameter of 80 nm and a length of a few μm, and the dispersive X-ray (EDX) analysis shows that the chemical composition of the nanowires consists of Si, C, and Al with traces of oxygen, with Al homogeneously distributed in the material. All the XRD diffractions of the material can be indexed as pertaining to a 6H-SiC

---

* Details on the preparation of the xerogel and of the ultrafine Al powder are omitted but available in the Wei paper.

phase, though the resulting lattice parameters are smaller than the standard one, possibly because of the 1.7% of Al dissolved in the 6H phase. It will be shown, later, that in the absence of Al nanopowder, 6H-SiC wires could not be grown.

Further TEM and HRTEM investigations show that the morphology of the wires is of the core-shell type, with a very thin SiO_x shell. The core is heavily damaged by multiple twins and SFs, these last leading to the formation of single crystal cells with size of the order of a few nm. We will see in Chapter 5 that the optical properties of these NWs could be, in fact, discussed by assuming the presence of quantum confinement effects, a unique phenomenon for SiC nanowires. Eventually, it was also found that 20% of the material produced consists of ultrathin nanowires with a diameter less than 10 nm. The tips of the wires are naked, without the presence of a metal cap, unlike the case of iron-catalyzed growths.

The chemistry of the growth process implies the oxidation and decomposition of the xerogel, with the intermediate formation of CO and SiO, which are dissolved in liquid Al nano-droplets to form an SiC-saturated Al–Si–C solution, from which Al-doped SiC nanowires segregate, see phase diagram of Figure 4.12 [45].

The fact that Al-doping represents the necessary condition to grow hexagonal SiC nanowires, as supposed by Li et al. [22], is confirmed by the results of a work of Wang et al. [27], who succeeded in growing 3C-SiC nanowires by thermal decomposition at 1300°C of the same xerogel, but only in the presence of iron nitrate as the catalyst.

The diameter of the NWs ranges within 50 and 300 nm, and the length within tens and hundreds of micrometers, with a zig-zag morphology and hexagonal cross sections.

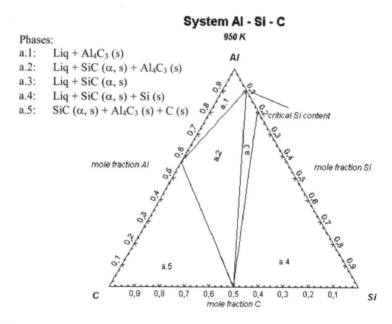

FIGURE 4.12 Ternary phase diagram of the Al–C–Si system at 950 K. (*After* M.S. Yaghmaee, G. Kaptay (2001) On the stability of SiC in the ternary liquid Al-Si-Mg Alloy Material's World (e-Journal ISSN 1586–0140 http://materialworld.fw.hu (open access).)

Periodical twinning is at the origin of the unique morphology of these wires, which leads not only to the set-up of an ordered sequence of parallel twin segments distributed along the length of the wire, but to a thickness of these twin segments which linearly increases with the increase of the wire diameter, starting from nano cells of 20 nm.

After etching the wires in a mixture of HF and HNO_3, one can TEM image the wires, that show clearly the positions of the twin boundaries, thanks to their improved resistance to the chemical attack.

As already noted in the introduction of this section, twinning in correspondence with small wire diameters leads to the formation of nano cells confined in 3D, where twin boundaries behave as barriers, at the origin of the photoluminescence spectra of these nanowires, with three peaks at 404 nm (3.07 eV), 434 nm(2.86 eV), and 547 nm (2.27 eV), of which the first shows a blue-shift of 0.5 eV vs. the main emission of 3C-SiC at 2.417 eV .

4.2.2 SYNTHESIS, MORPHOLOGY, AND STRUCTURAL FEATURES OF SiC NANOWIRES: CVD PROCESSES

Lagonegro et al. [46] carried out the growth of cubic SiC nanowires on [111] silicon substrates coated with a thin (2 nm) film of sputtered iron or nickel, using purified hydrogen as the carrier gas and a mixture of silane and propane (both at 3% in H_2) as the source gas at a temperature of 1250°C. Before introducing the source gases, a short dewetting* stage was carried out, to obtain an optimized surface distribution of the catalyst.

Considering only the case of iron catalyst, at the process temperature a binary Fe–Si alloy does form (the eutectic temperature of Fe–Si is 1203°C) and, thus, a uniform distribution of liquid Fe–Si droplets with diameters in the range 30–100 nm. With the introduction of the source gases the Fe–Si droplets reach the Si–C saturation concentration (see Figure 4.4) that does allow the spontaneous SiC segregation under form of an array of nanowires,† randomly distributed on the wafer surface, like in the cases discussed in the previous section.

The SiC NWs grown with this process are cubic and single crystalline, 30–100 nm in diameter, and tens of μm long, with a [111] growth direction. HRTEM images of the wires show the presence of dense arrays of stacking faults lying orthogonally or at 20° to the wire axis, though regions free of SFs are also observed.

Nickel-catalyzed NWs grown at 1100°C in the same growth atmosphere were further studied by Carapezzi et al. [47]. Scanning TEM investigations of these 3C SiC NWs (see images in Figure 4.13) show the presence of tapering and zig-zag faceting, as well as of SFs orthogonal and 20° to the wire axis, like in the previous case.

Some of the previous authors [48] carried out also the CVD growth of arrays of core-shell SiC–SiO$_x$ nanowires (see Figure 4.14a, b) using CO as the carbon source and nickel or iron as the catalysts, deposited on the silicon substrate as ethanolic solutions of Fe (or Ni) nitrates with the addition of a surfactant.‡ As in the case of

---

* Dewetting means here the aggregation of the metal in metallic islands.
† The paternity of the explanation is of the book's author.
‡ Which favors the homogeneous distribution of the catalyst on the silicon surface.

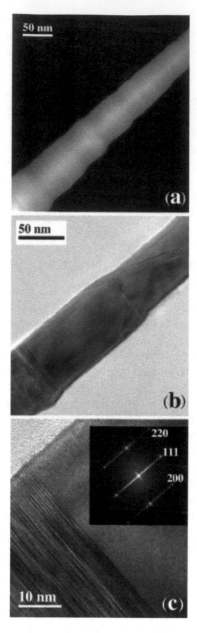

FIGURE 4.13 Scanning TEM images of Ni-catalyzed grown SC NWs. (*Reproduced with permission of the Royal Society of Chemistry, after S*. Carapezzi, A. Castaldini, F- Fabbri, F. Rossi, M. Negri, G.F. Salviati, A. Cavallini (2016) Cold field electron emission of large-area arrays of SiC nanowires: Photo-enhancement and saturation effects *J. Mater. Chem. C***4** (35)8226–8234 *Confirmation Number: 11847084 Order Date: Aug 31, 2019.*)

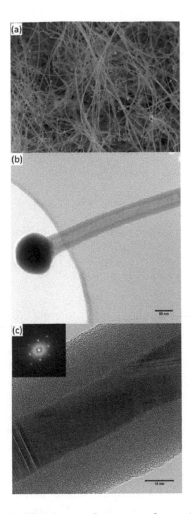

FIGURE 4.14 Top view (a) SEM image of an array of core-shell SiC NWs grown with an iron-catalyzed process, (b) false color TEM image of an SiC core-shell nanowire with a metallic cap on the top (green is SiC, violet is SiOx), (c) HRTEM image of an SiC nanowire. (*Reproduced with permission of the Royal Society of Chemistry from* M. Negri, S.C. Dhanabalan, G. Attolini, P. Lagonegro, M. Campanini, M. Bosi, F. Fabbri, G. Salviati (2015) Tuning the radial structure of core-shell silicon carbide nanowires *CrystEngComm* **17**,1258– 1263 *Confirmation Number: 11846154; Order Date: Aug 28, 2019.*)

metal-assisted growth processes of Si NWs (see Chapter 3, Section 3.4.4) an electro-less reaction does occur in correspondence with the interface of the silicon wafer with the nitrate solution

$$Si + 2Fe^{3+} \rightleftharpoons Si^{4+} + 2Fe \qquad (4.9)$$

leading to the deposition of metal drops on the silicon surface.

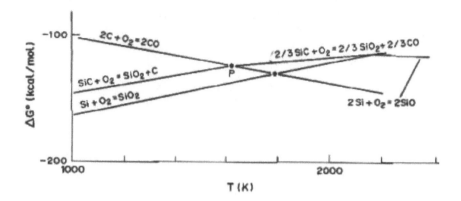

FIGURE 4.15　Ellingham diagram of the oxidation of silicon, carbon, and SiC (P = 0.1 MPa). (*From* K. Ishizaki, K. Matsumaru (2011) Thermodynamics of gas-solid reaction under hot isostatic pressing conditions *AZO Journal Mater.* DOI: 10.2240/azojomo0309 by *with the courtesy of AZO materials Aug 30, 2019 open access.*)

The growth was carried out using nitrogen as the carrier gas, with 4% of CO, at temperatures ranging between 1050 and 1150°C (1323–1423 K), lower than the melting temperature of the Si–Fe eutectic* and lower than the temperature at which a liquid ternary low-carbon Fe–Si–C phase[†] is still stable (see Figure 4.4). It was experimentally observed that only at 1100°C are core-shell morphologies stable, while at lower and higher temperatures only oxygen-deficient SiO_x nanowires could be grown.[‡]

The effect of CO on the chemical nature of the wire could be discussed accounting for the thermodynamics of the Si–C–O system [49, 50], as is also well-illustrated in the Ellingham diagram of Figure 4.15, holding for a gas total pressure of 0.1 MPa (1 bar), which shows that above 1000 K the SiO_2 phase is always the most stable, and that up to 1600 K the SiC should easily convert to SiO_2 in the presence of an excess of oxygen. It is therefore reasonable that in a CO atmosphere SiC would remain the stable phase. At temperatures above 1600°C in the presence of a CO atmosphere we expect, instead, the thermodynamic stability of the SiC phase and the decomposition of SiO_2. Since a temperature of 1100°C does not seem a critical (thermodynamic) temperature, the CO pressure and kinetic factors influencing the rates of the reactions

$$SiC + O_2 \rightleftharpoons SiO_2 + C \qquad (4.10)$$

$$Si + O_2 \rightleftharpoons SiO_2 \qquad (4.11)$$

are responsible for the peculiar conditions observed.

---

* See Section 3.4.1 where the CVD growth at hypoeutectic temperatures is discussed.
[†] It is apparent from the phase diagrams of Figure 4.3 that the ternary phase is a low-carbon phase.
[‡] No comment on the thermodynamics of the system is given by the authors.

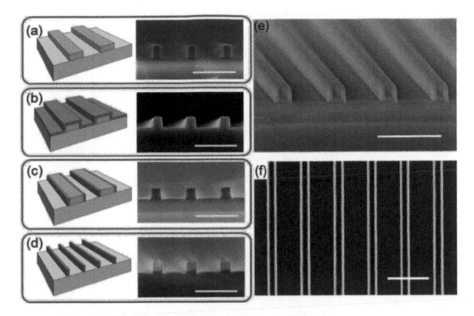

FIGURE 4.16 Schematic diagram of the nanofabrication steps for the growth of ultra-thin self-aligned SiC nanowires (scale bar-500 nm) (a), the silicon wafer with the preformed ribbon array (b), ultrathin conformal deposit of SiC over the ribbon array (c), the array after etching (d), the NWs array after removal of the ribbon (e), and SEM images of the wire arrays (f). (*After* B.N. Tabassum, M. Kotha, V. Kaushik, B. Ford, S. Dey, E. Crawford, V. Nikas, S. Gallis (2018) On-demand CMOS-compatible fabrication of ultrathin self-aligned SiC nanowire arrays *Nanomaterials* **8**, 906 Open Access Journal.)

Considering that the EDX spectra of the metallic cap on the tip of the core-shell SiC wire (see Figure 4.14b) manifest the presence of Fe, Si, and traces of C, one could suppose, also, that the growth actually occurred from a low-carbon ternary Fe–Si–C phase, whose melting temperature is lowered by size effect.

It is, eventually, important to observe, see Figure 4.14c, that also in the case of CVD growth from a CO gas, stacking faults orthogonal to the growth direction and at 20° to the growth axis are present, demonstrating the invariability of their presence in SiC NWs, independently of the growth temperature.

Ultrathin SiC nanowires were CVD-grown by Tabassum et al. [51] by CVD-growing ultrathin (10–40 nm) layers of SiC on a HSQ (hydrogen silsequioxane) ribbon array, deposited on [100] silicon substrates. The conformal synthesis was carried out at 800°C using as Si and C single precursor [$C_6H_{16}Si_2$ (1,1,3,3-tetramethyl-1,3-disilacyclobutane)], diluted in forming gas (5% H_2, 95% N_2);* see Figure 4.16a, b, c where details of the process are schematically drawn.

After synthesis, the undesired material was RIE etched off, obtaining the ultra-thin wires array shown in Figure 4.16d, which were eventually submitted to a final thermal annealing at 900–1200°C.

* See details in the original paper.

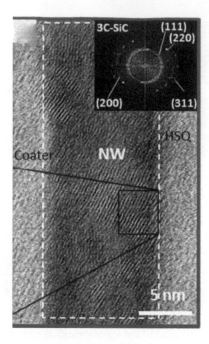

FIGURE 4.17 TEM image of the microstructure of a CVD-grown SiC nanowire: in blue a single crystal domain. (*After* B.N. Tabassum, M. Kotha, V. Kaushik, B. Ford, S. Dey, E. Crawford, V. Nikas, S. Gallis (2018) On-demand CMOS-compatible fabrication of utrathin self-Aligned SiC nanowire arrays *Nanomaterials* **8**, 906 Open Access Journal.)

FTIR spectroscopy measurements carried out on the as-grown material showed that it is a mixture of amorphous and crystalline SiC, but that a complete conversion to crystalline 3C-SiC does occur after annealing at 1100°C.

The XPS analysis of the material confirms the sole presence of C and Si, and optical absorption measurements reveal a Tauc's gap of 2.4 eV, slightly above the energy gap of 3C-SiC.

Since HR-STEM measurements show that the NWs are nanocrystalline, see Figure 4.17, with grains of an average size of 5 nm, one would expect quantum confinement effects, and a blue-shift of the energy gap, that optical absorption measurements do not put in evidence. Due to the use of the oxygen-free precursor and of the very reducing atmosphere, grain boundaries are clean and, apparently, do not behave as confinement barriers [52].

4.2.3 SYNTHESIS OF SiC NANOPILLARS: DRY-ETCHING PROCESSES

Dry-etching processes are employed for the synthesis of SiC nanopillars, with the advantage that pillars with cubic and hexagonal crystal structures could be generated by making use of the appropriate single crystal substrates. Furthermore, while NWs grown with CVD or thermal reaction processes suffer from high densities of structural defects, as we have seen in the previous sections, dry etching would maintain the structural identity and defectivity of the substrate used.

An inductively coupled plasma, with SF_6/O_2 as the etching components and a Ti/Cr mask was used by Khan and Adesida [53] to deep dry etch (up to 7.5 um) 6H-SiC substrates.

The profiles were shown by SEM imaging to be highly anisotropic, with highly smooth surfaces.

Using the same etching mixture, and a SiO_2 patterned mask, Jang and Cheung [54] investigated the surface contamination of etched samples of 4H-SiC by X-ray photoelectron spectroscopy. It was found that while Si–Si and Si–F bonds are absent, various C–F bonds could be detected, whose concentrations are shown to depend on the etch conditions, with a substantial enhancement with the decrease of the oxygen concentration in the plasma phase.

Choi et al. [55, 56] used as well an SF_6/O_2 plasma in a commercial high-density plasma etching chamber to dry etch 3C-SiC, 4H[0001] and 6H-SiC [0001] substrates. The etch mask, consisting of an array of circular Ni patterns of 300 nm in diameter and 110 nm in thickness, is suitably created using electron beam lithography for the resist and Ni e-beam evaporation for the mask. Ni was selected as the mask material as it was shown to work better than other metals, but it is laterally eroded during the etching toward a complete erosion after 840 sec of normal operation.

Consequently, pseudo-conical shapes are expected to develop during the etching.

Figure 4.18a displays the pillar heights as a function of the etching times and shows that the etching efficiency is almost independent of the nature of the bulk phase. Figure 4.18b shows, instead, the morphology of the pillars after 700 sec of etching time, i.e. before the complete erosion of the Ni mask. It is apparent that due to the lateral erosion of the mask, the diameter shrunk below 100 nm, from the original diameter of 500 nm, but the diameter remains substantially constant on top of the pillars, favoring their application in nano FET applications [56]. It appears, also, that the shape of the pillars depends on the crystal structure and orientation of the substrate, as can be better envisaged looking at the images of Figure 4.19.

Ou et al. [57] used a self-assembled, nanopatterned Au mask to dry etch with SF_6/O_2 plasma SiC substrates of unmentioned structure. The nanopatterning technique is just the same as used by Irrera et al. [58–63] for the chemical etching of Si substrates, as discussed in Chapter 3.

As in the previous cases, an array of pseudo-conical pillars is obtained, displayed in Figure 4.20, whose quality and shape depend on the etching conditions.

Eventually, a different approach was used by Hsu et al. [64], who developed a process for the synthesis of ordered arrays of nanotips grown on various substrates (Figure 4.21), which includes the preliminary self-assembly of masks of SiC caps and the further dry etching of the substrate.

A CH_4 and SiH_4 plasma is used to create SiC clusters in the plasma phase, which deposit on the surface of the substrate as an ordered self-assembly of caps, which work as a protective mask for the successive dry-etching process, carried out with an argon and hydrogen plasma. So far, the process has been used for the synthesis of well-aligned and uniformly distributed nanotips of Si, GaN, GaP, sapphire, and aluminum, materials which are softer than SiC, but the process could be obviously applied to SiC.

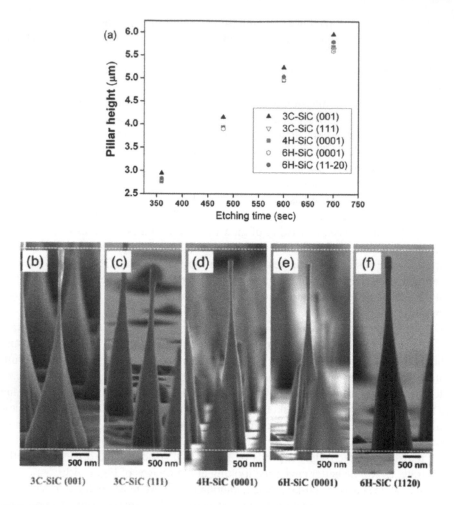

FIGURE 4.18 (a) Dependence of the pillar heights on the etching time of a SiC substrate with a SF_6/O_2 plasma. (b) Morphology of the SiC pillars after 700 sec of etching. (*After* J.H. Choi, L. Latu-Romain, E. Bano, A. Henry, W.J. Lee, T. Chevolleau and T. Baron (2012) Comparative study on dry etching of alpha- and beta-SiC nano-pillars, Materials letters (General ed.), (87), 9–12 *Linkoping University Post Prints, Open Access in DiVA*.)

However, it is well-evident that also in this case the lateral erosion of the hard SiC cap leads to a conical morphology of the pillars, as can be seen in Figure 4.21.

It could be concluded that, almost independent of the mask materials and reactive species, only pseudo-conical pillars could be obtained by dry etching. Nevertheless, a Ti/Cr mask seems to work better than Au, Ni, and SiO_2 [53] despite the strongly anisotropic pillars, still present also in this case.

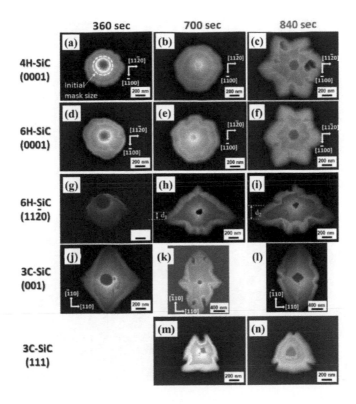

FIGURE 4.19 Top view SEM images of SiC pillars after different etching times. (*After* J.H. Choi, L. Latu-Romain, E. Bano, A. Henry, W.J. Lee, T. Chevolleau and T. Baron (2012) Comparative study on dry etching of alpha- and beta-SiC nano-pillars, *Materials letters (General ed.)*, 87 9–12 Linkoping University Post Prints, *Open Access in DiVA*.)

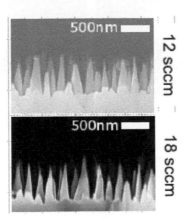

FIGURE 4.20 SEM images of arrays of SiC nanopillars after 10 min of etching time, for two different SF_6 flow rates in the etching chamber. (*After* Y. Ou, I. Aijaz, V. Jokubavicius, R. Yakimova, M. Syvaijarvi, H. Ou (2013) Broadband antireflection silicon carbide surface by self-assembled nanopatterned reactive-ion etching *Optical Materials Express* **3** (1) 86–93 Open Access Journal.)

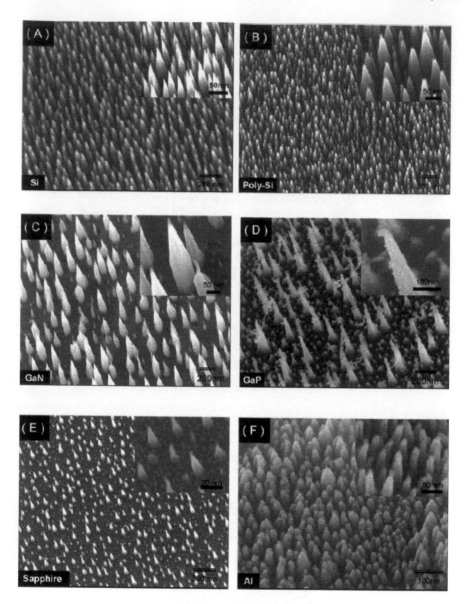

FIGURE 4.21 HRSEM images of SiC nanotip arrays fabricated with a self-masked process on a variety of substrates. (*Reproduced with permission from* C.H. Hsu, H-C Lo, C-F. Chen, C. T. Wu, J-S. Hwang, D. Das, J. Tsai, L-C. Chen, K-H. Chen (2004) Generally applicable self-Masked Dry Etching Technique for Nanotip Array Fabrication *Nano Lett.* **4** (3) 471–475 *Copyright 2004 American Chemical Society Sep 15, 2019.*)

4.3   GaAs AND OTHER III–V COMPOUND SEMICONDUCTORS NANOSTRUCTURES

Similar to the case of SiC nanowires, with which we dealt in the last section, the interest in GaAs nanowires is based on the potentialities of their parent bulk material [65], on which a vast amount of research has been carried out since the 1950s, which opened the door to the development of modern optoelectronics, high-frequency microelectronics, and nano-optoelectronics.

Unlike Si and SiC, GaAs is, in fact, a direct gap semiconductor, with an energy gap of 1.42 eV,* that leads to an intense PL emission at 870 nm (1.42 eV), which should be blue-shifted by size effects in nanostructured GaAs, as suggested by Duan et al. [66], offering several advantages over nanostructured silicon carbide, though twins and stacking faults are also common extended defects in GaAs, where twins interrupt the cubic stacking sequence [67] in GaAs NWs.

Depending on the application, the growth of bulk, single crystal GaAs is carried out with the liquid encapsulated (LEC) CZ technique for microwave applications, which require quasi-undoped and semi-insulating material, or with the horizontal or vertical Bridgman technique for optoelectronic applications [68], taking care of the delicate phase relations in GaAs crystal growth illustrated by Wentzl et al. [12], which preclude, as an example, GaAs growth from As-rich solutions, and of the thermodynamics of GaAs doping [69].

III–V bulk compounds crystallize either with the cubic zincblende or the hexagonal (wurtzite) structures, and the ionicity of the compound (see Table 4.1) determines the preferred structure [67]. In fact, GaP and GaAs crystallize with the cubic zincblende structure, as originally shown by Goldschmidt [70], while GaN crystallizes with the wurtzite structure.

The zincblende–wurtzite polytypism in semiconductors has been discussed by Yeh et al. [71] in terms of the calculated total energy difference ΔE_{W-ZB} (meV/atom)[†] and displayed as a function of Pauling's electronegativity and of the difference in tetrahedral radii; see Figure 4.22. It is interesting to observe that the distribution of the ΔE_{W-ZB} values follows a linear relationship in both cases, which indicates that phase stability is governed by basic chemical factors, and it could also be expected that bi-stability would be a feature of semiconductors with $\Delta E_{W-ZB} \to 0$, as is the case of GaAs for which $\Delta E_{W-ZB} \approx 24$ meV/pair, or 2.5 KJ/mol, or of CdS, for which $\Delta E_{W-ZB} \simeq 0$.

Polytypism is also a common feature of all III–V nanowires, whose actual structure is, however, substantially influenced by their diameter, as shown by the Monte Carlo simulations of Akiyama et al. [72], who found that III–V nanowires of diameters smaller than 12 nm grow with the wurtzite structure, while thicker nanowires, up to 32 nm in diameter, were found to be bi-stable, growing with both the wurtzite and zincblende structure.

Bi-stability is also a feature of GaAs NWs, as shown Pengfei et al. [73] who studied, using first-principles DFT, the effect of surface dangling bonds on the cohesive

---

* And two indirect energy gaps at 1.71 and 1.90 eV.
† Using the local density formalism (LDF).

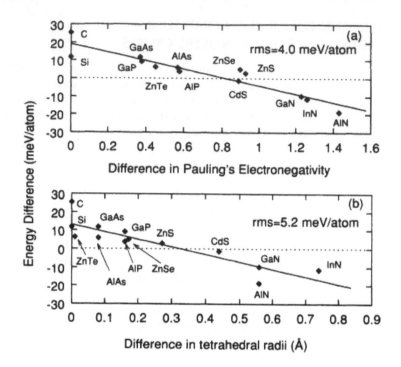

FIGURE 4.22 Calculated total energy differences $\Delta E_{W\text{-}ZB}$ between the zincblende (diamond) and wurtzite structures of some common semiconductors, displayed in function of their ionicity and tetrahedral radii. (*Reproduced with permission from C-Y.Yeh, Z.W.Lu, S.Froyen, A.Zunger (1992) Zinkblende-wurtzite polytypism in semiconductors Phys.Rev.B* **46**(16) *10086–10097 Copyright 1992 American Physical Society License Number: RNP/19/ SEP/019109 License date: Sep 29, 2019.*)

energy of ZB and WZ GaAs nanowires. They found that due to the larger number of dangling bonds at ZB facets, where both two-coordinated and three-coordinated atoms do exist, WZ-NWs are more stable than ZB-NWs up to a diameter of 90 Å. Above this diameter the cohesive energy of ZB and WZ nanowires converge to the bulk value and GaAs NWs are structurally bi-stable.

Also the growth rate was demonstrated by Zardo et al. [74] to have a role in the structure of GaAs NWs, since low growth rates favor, as an example, the increase of the density of the wurtzite phase in mixed zincblende–wurtzite GaAs NWs.*

Growth rate and minute growth temperature differences have been, also, seen at the origin of structural changes of In-assisted growth of InP nanostructures by Pozuelo et al. [75]. They found that at 385°C pure zincblende, [111] oriented InP nanostructures, MOCVD grown, are observed, while at 400°C, [0001] oriented pure wurtzite pillars could only be grown. The explanation given is a temperature-induced lowering of the nucleation barrier for the growth of the wurtzite phase with a vapor phase process.

Eventually, Jiang et al. [76] and Zhou et al. [77] demonstrated a weak influence of the size on the transition pressure from the semiconducting cubic phase to the

---

* Details on the growth procedure are given in the next section.

orthorhombic, metallic phase of GaAs. It does occur, in fact, at a pressure of 17 GPa for bulk GaAs vs. a pressure of 20 GPa for nanostructured GaAs.

As already discussed in the introduction, partial ionic bonding and surface termination of a single element of group III or V are at the background of polarity and of the polar faces of bulk compound semiconductors [2–5].

These features are also common to compound semiconductor nanostructures [6–8] leading to the presence of polar facets at anisotropically grown, single-crystal nanowires, as is the case of the [100] A or [100] B surfaces of III–V nanowires crystallizing with the zincblende structure or of the [1000] A or [1000] B wurtzite faces, where A and B are elements of the III group and V group, respectively.

Polarity in III–V compound semiconductors arises from the partial ionicity of the A–B bonds, that leads to a distribution of dipoles in the volume of the material, with the consequent set-up of an electrostatic field. It has been demonstrated with basic electrostatic arguments by Nosker et al. [2] that a monopole distribution on the polar faces sets up to compensate the electric field generated by the dipole distribution in the bulk. In other terms, the compensation of the inner electrostatic field is obtained by spontaneous arrangement of the polar faces, involving a proper arrangement of the surface double layer [2], which leads to the set-up of a net excess surface charge.

It is apparent that the polarities of surfaces of III–V semiconductors imply a specific chemical reactivity towards impurities, which is not present in homopolar semiconductors.

As already discussed in the introduction of this chapter, melt-grown ($T_m = 1513$ K) bulk GaAs, see Figure 4.2, presents a wide region of non-stoichiometry, eminently located in the As-rich side, which extends down to less than 600 K [11].

A variety of native point defects [vacancies (V_{Ga}, V_{As}), interstitials (Ga_i, As_i), and antisites (As_{Ga}, Ga_{As})] is also present in GaAs [9, 78, 79]. From the values of the equilibrium constants tabulated by Hurle [9], the enthalpies of formation of Ga- and As-vacancies are estimate to yield 1.86 and 0.49 eV. Interaction between vacancies and impurities leads to point defect complexes as is the case of the vacancy–oxygen pair studied by Son et al. [80].

Direct gap, small deviations from the stoichiometry at low temperatures, zinc-blende structure, and relativity low impact of extended defects (stacking faults and twins) are all bulk GaAs features common to GaAs NWs, which favor their use in a wide range of optoelectronic applications [81, 82] including photovoltaic cells [83].

A limiting factor to the use of GaAs NWs could be the unfavorable high sensitivity of GaAs to surface states and the ease of oxidation of GaAs surfaces, that causes the formation of an oxide layer, which does not passivate the material, but sets up acceptor-like levels, which would trap free carriers and induce carrier depletion.

Concerning GaAs surface states, they pin the Fermi level close to mid-gap [84], leading to surface-depleted layers even with high doping levels [85] and to an extremely high surface recombination velocity, three orders of magnitude higher than of other III–V compounds, which can limit the performance of GaAs-based optoelectronic devices, especially in the case of GaAs NWs due to their high surface to volume ratio [86].

4.3.1 Thermodynamics of Growth and Doping of GaAs Nanowires

Vapor-liquid-solid (VLS) processes [87], carried out by CVD, molecular beam epitaxy (MBE) [88–90], or metal organic chemical vapor deposition (MOCVD), are widely used to grow GaAs nanowires with the assistance of a transition metal (M = Au, Ag, Cu), though GaAs NWs and InP NWs could also be grown with a self-assisted growth, as shown by Fontcuberta i Morral [84], Zamani et al. [88–90], and Zardo et al. [74] for self-assisted GaAs NWs growth and Pozuelo et al. [75] for self-assisted growth of InP nanowires.

Looking first to the VSL growth of GaAs NWs with the assistance of a transition metal M, the basic thermodynamic conditions for the growth imply the formation of a supersaturated solution of GaAs in a liquid metal droplet, from suitable precursors present in the vapor phase which are catalytically decomposed to Ga and As.* Eventually, from the supersaturated solution GaAs segregation does occur under form of a wire, whose radius depends on the size of the metal droplet.

Metal-assisted epitaxial growth does occur using as a substrate a single crystal wafer of GaAs.

The knowledge of the ternary phase diagrams of Ga–As–M and the temperature range of stability of the liquid ternary phase, or at least knowledge of the binaries is the prerequisite for the synthesis of GaAs nanowires† from gas phase processes using metal assistance [91, 92]. As shown in Chapter 2 the melting temperature of metal nanoparticles decreases with the decrease of their size. That of Au, mostly used as metallic catalyst for the growth of GaAs nanowires, drops to 1250 K for nanoparticles of 10 nm, and below 700 K for nanoparticles of 1 nm, according to a recent work of Font and Myers [93].

It could be, therefore, supposed that since the growth temperature of the Au-assisted growth processes is held within 400 and 700°C, the Au catalyst is stable in the liquid phase.

In the case of the ternary Ga–As–Au system [94], one sees that the $Au_l - GaAs_l$‡ equilibrium does occur under the condition

$$\mu_{GaAs}^{Au} = \mu_{GaAs}^{GaAs} \tag{4.12}$$

where μ_{GaAs}^{Au} is the chemical potential of GaAs dissolved in liquid Au, and μ_{GaAs}^{GaAs} is the chemical potential of GaAs in liquid GaAs, with a pseudo-eutectic temperature at 620°C; see the pseudo-binary phase diagram of the system Au–GaAs in Figure 4.23.

Conditions for GaAs to grow from a liquid Au droplet saturated in GaAs have been calculated by Ghasemi and Johansson [95]. The calculated phase diagram shows that GaAs growth could occur in a wide range of temperatures, above the calculated GaAs–Au eutectic temperature at 874.42 K, that compares well to the pseudo-eutectic temperature of Panish [94] reported above.

Against the common belief that Au-assisted growth of nanowires of homopolar or heteropolar semiconductor NWs does always occur with a VLS process from a

* With MBE Ga and As vapors are already present.
† The same conditions apply to the other III–V compounds.
‡ With l as the symbol of a liquid phase.

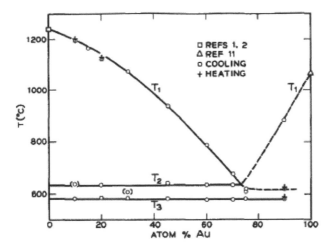

FIGURE 4.23   Pseudo-binary phase diagram of the system Au–GaAs. (*Reprinted with permission of the Electrochemical Society after* M.B. Panish (1967) Ternary Condensed Phase Systems of Gallium and Arsenic with Group IB Elements *J. Electrochem. Soc.* **114** 516–521 *Order detail ID:72042476; Order License Id:4686561223692 License date Oct 12, 2019.*)

liquid Au droplet, Dick [87] suggests that also a VSS process could occur from a solid Au droplet, as demonstrated by the concentration of As and Ga measured in the Au droplet which was insufficient to form a liquid solution, and notes that melting temperatures of eutectics are size-dependent, a concept we already mentioned.

When self-assistance conditions are selected, as formerly mentioned for the case of GaAs NWs [74, 82] and InP NWs [75], only the knowledge of the binary Ga–As and In–P is needed, see Figure 4.24, from which the thermodynamic conditions for the liquid phase growth of both GaAs and InP could be deduced.

It has been already mentioned that the ZB structure is experimentally shown to be the most stable for large-diameter nanowires, and that the critical radius for the wurtzite structure to be stable lies within 5 and 25.5 nm NWs [74]. Moreover, low growth rates do increase the fraction of WZ NWs in mixed ZB-WZ GaAs NW samples.

It is also experimentally demonstrated that due to the effective strain relaxation granted by thin nanowires, the quality of GaAs NWs, heteroepitaxially grown on a silicon substrate, should not be drastically compromised by structural defects, as shown in Figure 4.25 [85].

The doping efficiency of GaAs NWs is strongly influenced by the nature of the dopants, by the type and temperature of the growth process adopted [96], and by the nature of the substrate.

As an example, the role of the solubility and of the segregation coefficients of a dopant i in liquid Au, where $\kappa_i = \dfrac{x_i^{Au}}{x_i^{GaAs}}$ is the segregation coefficient, in Au-assisted heteroepitaxial growth was studied by Haggren et al. [97] and by Mäntynen [85], using a metal organic vapor phase epitaxy (MOVPE) process and silicon substrates.

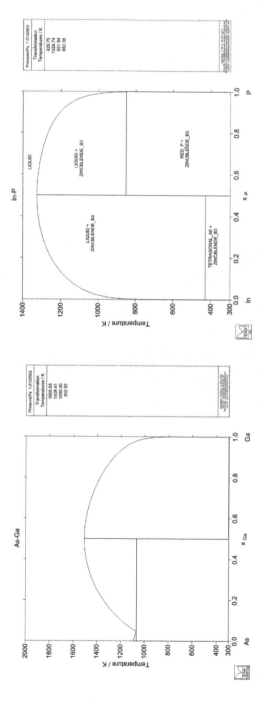

FIGURE 4.24 Phase diagrams of the binaries Ga–As and In–P. With the courtesy of the National Physical Laboratory MTDATA (MP Ga = 302.76 K, MP In = 429 K MP As 1090 K MP P 317.15 K).

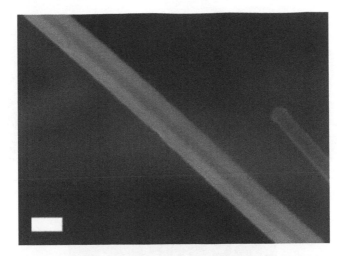

FIGURE 4.25 SEM image of an undoped GaAs nanowire grown with the Au-assisted MOVPE process. (The scale bar is 200 nm.) (*Reproduced with permission of Prof. Harri Lipsanen and Henrik Mäntynen after* H. Mäntynen (2016) Doping of Vapor–Liquid–Solid Grown Gallium Arsenide Nanowires Thesis Aalto University, School of Electrical Engineering, Finland.)

As dopants, Zn and Sn were used for p-type and n-type doping, both of which display high solubility in liquid Au and small values of κ_i.

Regardless of differences in solubility and segregation coefficients of the dopants, it is supposed that the diffusion of dopants from the liquid seed through the Au/GaAs interface is fast and does not represent a kinetic hinderance.

Instead, twinning, twisting, and kinking are observed in SEM images of GaAs nanowires grown with high Zn/Ga ratios (see Figure 4.26), supposedly due to phase changing in the Au seed particles during the growth, unlike the case of Sn, where no additional structural defects are observed as a function of the increase of the Sn/Ga ratio during the growth, except for some tapering effects.

Doping effects in core and shell GaAs nanowires heteroepitaxially grown on Si substrates with the self-assisted method were, instead, considered by Goktas et al. [98], who used an MBE process for the growth and the post-growth doping. Shell doping was also carried out on a undoped core-shell wire. As dopants, tellurium for n-type doping and beryllium for p-type doping were used. The wires grow with a hexagonal size, and with an average diameter of 80 nm and a length of 1.5 μm. The doping features of planarized Be- and Te-doped GaAs NWs were compared to those of Be- and Te-doped GaAs thin films (TF) grown as 400 nm thick epilayers on (nominally) undoped GaAs. The growth temperature of GaAs TF samples was 600°C, while that of NWs was 500°C.

The main results of SIMS measurements carried out on doped GaAs TF and GaAs NW samples was that the concentrations of both dopants remain constant along the entire length of the wire, that the measured concentration of Be is well-comparable with that of the reference TF sample, while that of Te is one order of magnitude less than that measured on the reference TF samples.

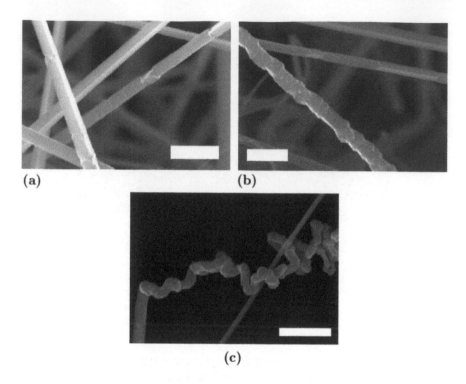

FIGURE 4.26 SEM images displaying the effect of the Zn/Ga ratio on the defectivity of the GaAs nanowires (a) 0.020 (b) 0.042 (c) 0.082 (the scale bar is 1 μm). (*Reproduced with permission of Prof. Harri Lipsanen and Henrik Mäntynen after* H. Mäntynen (2016) Doping of Vapor–Liquid–Solid Grown Gallium Arsenide Nanowires Thesis Aalto University, School of Electrical Engineering, Finland.)

The authors suggest that the higher growth temperature of NWs with respect to that used for TFs deposition and the higher volatility of Te, with respect to that of Be, could explain the lower concentration of Te in GaAs NWs.

As could be qualitatively understood, doping GaAs with amphoteric dopants, like Si, should induce carrier compensation effects, arising from Si impurities incorporated both in substitutional Ga and As sites in the GaAs lattice, with the formation of donors Si_{Ga} and Si_{As} acceptors, as well as of $Si_{Ga}^+ - Si_{As}^-$ neutral pairs. Doping temperature strongly influences the Si substitution in Ga and As sites. In fact, at low temperatures ($\approx 600°C$) Si substitutes mostly As sites, while Ga sites are substituted at higher temperatures.

The physico-chemical grounds underneath could not be understood only by considering the values of the covalent radii of Si (117.3pm), Ga (122.5 pm), and As (122.5 pm) [99].

A reasonable conclusion arises by considering that the enthalpy of formation of Ga vacancies is much higher (1.86 eV) than the enthalpy of formation of As vacancies (0.46 eV), as shown in the former section. Teramoto [100] and Fontcuberta i Morral et al. [101, 102] have, eventually, considered the thermodynamic aspects of the doping process.

Teramoto [100] considered the thermodynamics of Si doping from a Ga liquid phase solution and showed that the thermodynamic driving force for Si doping of GaAs is the difference between the enthalpies of solution of Si in the different Ga and As sublattices, influenced by the strain energy involved, leading to the spontaneous formation of a solution with Si in As sites.

Fontcuberta i Morral et al. [101, 102] have shown, in addition, that silicon doping of a GaAs NW is guided by the segregation coefficients of Si at the interface between the source, a Ga droplet, and the GaAs substrate during the doping cycle.

They used Raman spectroscopy to detect the dopant incorporation in Ga or As sites and to study the compensation effects associated with Si-doping of GaAs NWs in a wide range of Si doping ($1{\cdot}4\ 10^{18}$–$4{\cdot}10^{19}$ cm^{-3}) by measuring the intensity and frequency of the local vibration modes (LVM) of Si in GaAs.

Considering the biphasic equilibrium between a Si-doped Ga droplet supersaturated in GaAs and a solid GaAs nanowire

$$\mu_{Si}^{Ga/GaAs} = \mu_{Si}^{GaAsNW} \qquad (4.13)$$

it could be demonstrated* that at 630°C the segregation coefficients of silicon to As sites $\kappa_{Ga/GaAs}^{Si/As} = 0.1$, larger than the Si segregation coefficient to Ga sites, $\kappa_{Ga/GaAs}^{Si/Ga} = 0.06$. Instead, $\kappa_{Ga/GaAs}^{Si/Ga} > \kappa_{Ga/GaAs}^{Si/As}$ at 750°C.

They could also show an important effect of dopant concentration, with the set-up of a consistent concentration of neutral $Si_{Ga}^{+} - Si_{As}^{-}$ pairs at a total Si dopant concentration of $4{\cdot}10^{19}$ cm^{-3}.

4.3.1.1 Morphology and Defects of Metal-Assisted Grown GaAs Nanowires

Metal-assisted growth of GaAs nanowires is generally carried out with Au-assistance, but Fe [103], Ni, Ag, and Pt have also been used.

As shown by Wang et al. [83], the quality of the NWs arrays CVD grown† with the use of a metallic catalyst depend on the size, composition, and phase of the metal seed, which could be a solid Ga–Me alloy for VSS growth.

As an example, taking a Ni-Ga solid alloy as the catalyst, one has the minimum lattice misfit between the catalyst surface and that of the GaAs NW, leading to conditions of effective epitaxial growth. Therefore, catalysts engineering could be used to tune the crystal phase and orientation of NWs.

Also supersaturation conditions of Ga in the Au seed in an NW growth process have a large impact on the morphology and defects of grown GaNWs. In fact, low-Ga supersaturation leads to low growth rates, and to polycrystalline NWs containing a high density of defects, while high-Ga supersaturation, with the formation of low-melting Ga_mAu_n eutectics, leads to the unidirectional growth of defect-free single crystalline nanowires crystallized with the zincblende structure, using Au nanocrystals with a diameter of 40 nm.

A study of the role of Ga_mAu_n eutectics in the morphology of GaAs nanowires grown with Au-assistance was carried out recently by Chen et al. [104].

---

* Using MBE as the growth technique and Ga self-assistance.
† With a solid source sublimated in a flux of hydrogen.

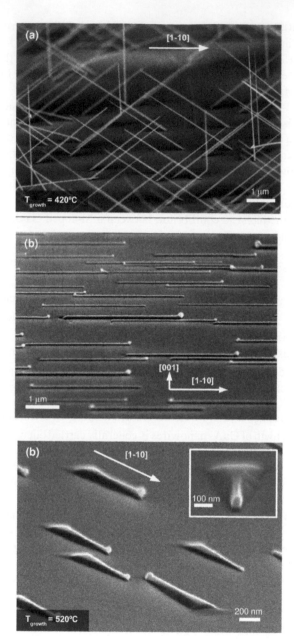

FIGURE 4.27 Effect of the growth temperature on the orientation and morphology of GaAs nanowires grown using an MOCVD process (a) predominantly [111] oriented nanowires grown at 420°C (b) planar <110> oriented NWs grown at 460°C (c) tapered and well aligned <110> nanowires grown at 520°C. (*Reprinted with permission from* S.A. Fortuna , J. Wen, I.S. Chun , X. Li (2008) Planar GaAs nanowires on GaAs (100) substrates: Self-aligned, nearly twin-defect free, and transfer-printable. *Nano Lett.* **8**(12) 4421–4427 *Copyright 2008 American Chemical Society.*)

The importance has been, eventually, shown of adopting a two-step process, consisting of a high-temperature nucleation of the catalytic alloy at 640–660°C followed by the true growth process carried out at 600°C. The merit of a two-step process is to improve the morphology as well as the diameter and the growth orientation distribution of GaAs nanowires.

The process temperature was shown to rule the orientation and morphology of planar, defect-free GaAs nanowires* grown at growth temperatures of 420–520°C, using an MOCVD process with trimethyl gallium and AsH$_3$ (arsine) as precursors [105]. Single crystal (100) GaAs were used as substrates and gold nanoparticles as catalysts.

One can see from the SEM images displayed in Figure 4.27 that at 460°C planar arrays of half cylindrical-shaped GaAs nanowires, <110> oriented, are grown, while at 420°C the planar distribution completely disappears, and at 520°C the wires are tapered, though maintaining the <110> orientation.

4.3.1.2 Morphology and Defects of GaAs Nanowires Grown with a Self-Assisted Growth

A Ga-assisted, epitaxial growth process of GaAs nanowires was developed by the A. Foncuberta i Morral group. The process is carried out in the temperature range 640–720°C, using [100] GaAs single crystal substrates, preliminarily spin-coated with a 3.2 nm thick, nano-porous SiO$_x$ layer, in an MBE chamber brought to an initial pressure of $5 \cdot 10^{-10}$ torr, using for the growth As/Ga ratios varying in the range 2–13[†] [74, 88, 89]. The growth precursors are atomic gallium and As$_4$ molecules, and the growth of the wire is driven by the As$_4$ flux. In turn, the nano-porous SiO$_x$ layer works as a mask and provides a self-assembled network of nano-holes for the localization of liquid Ga droplets at the very beginning of the process. When the nano-holes are passing through the entire thickness of the layer, they promote the epitaxial growth of GaAs NWs, since the growth starts with the Ga droplet wetting the single crystal substrate. Considering that not all the holes are passing through, only a fraction of nanowires will grow epitaxially, as will be shown in the SEM images displayed in Figures 4.28, 4.29, and 4.30.

One of the problems of this technique[‡] is that the process is carried out at a constant As/Ga flux, thus not providing the best conditions for the initial deposition of a liquid Ga droplet in the mask holes. Probably this is the reason why the process temperature should be dropped to 640°C in order to grow zincblende GaAs nanowires with a Ga-assisted process.[§]

Figure 4.28 is an SEM image of an assembly of epitaxial GaAs nanowires, where some nanowires grow vertically oriented (B), while some others (A) grow with an angle of 35.3° with respect to the actual substrate surface. The different orientation of GaAs nanowires with respect to the single crystal substrate demonstrates, in fact, the occurrence of an epitaxial process, and gives also the indication of their polarity

---

* With the zincblende structure.
† As the ratio of the pressures of As and Ga in the MBE chamber.
‡ The author of the book takes the responsibility for this opinion.
§ GaAs NWs are also grown at higher temperatures with an unspecified assistance.

FIGURE 4.28   SEM images of GaAs nanowires grown without metal catalyst. The droplet shape changes for B polar NWs (vertical) and A polar NWs (horizontal). (*After* M. Zamani (2017) Growth and Characterization of High Quality GaAs Nanowires, Thesis, Politecnico di Milano, *courtesy of* Prof. Anna Fontcuberta I Morral a Laboratoire des Matériaux Semiconducteurs, École Polytechnique Fédérale de Lausanne, and Prof. Carlo Spartaco Casari Micro and Nanostructured Materials Lab, Energy Dept. Politecnico di Milano.)

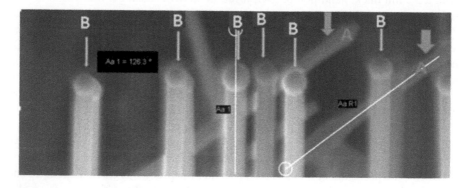

FIGURE 4.29   SEM image of polar [111]A and [111]B GaAs nanowires. (*After* M. Zamani (2017) Growth and Characterization of High Quality GaAs Nanowires Grown without Metal Catalyst, Thesis, Politecnico di Milano, courtesy of Prof. Anna Fontcuberta I Morral a Laboratoire des Matériaux Semiconducteurs, École Polytechnique Fédérale de Lausanne, and Prof. Carlo Spartaco Casari Micro and Nanostructured Materials Lab, Energy Dept. Politecnico di Milano.)

[89], such that B-type wires present As-terminated polar facets and A-type wires present Ga-terminated polar facets, belonging to the <110> family.

Figure 4.29 additionally shows that also the shape of the tip of the wires depends on the polarity of the wire, and presents a collar for A-polar NWs (horizontal), as is also apparent in Figure 4.30 (left panel), while Figure 4.30 (right panel) shows that type-A wires are almost defect-free (only a twin is clearly visible) while several twins perpendicular to the growth axis are present in type-B wires.

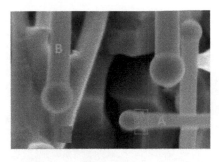

FIGURE 4.30 SEM images of polar A and B GaAs nanowires. (*After* M. Zamani (2017) Growth and Characterization of High Quality GaAs Nanowires Grown without Metal Catalyst, Thesis, Politecnico di Milano, courtesy of Prof. Anna Fontcuberta I Morral aLaboratoire des Matériaux Semiconducteurs, École Polytechnique Fédérale de Lausanne, and Prof. Carlo Spartaco Casari Micro and Nanostructured Materials Lab, Energy Dept. Politecnico di Milano.)

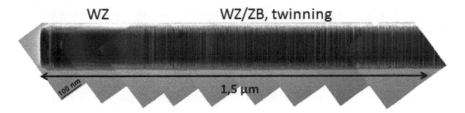

FIGURE 4.31 Bright field TEM image of a GaAs wire with 30% of wurtzite. (*Reproduced with permission from* I. Zardo, S. Conesa-Boj, F. Peiro, J.R. Morante, J. Arbiol, E. Uccelli, G. Abstreiter, A. Fontcuberta i Morral (2009) Raman spectroscopy of wurtzite and zincblende GaAs nanowires: Polarization dependence, selection rules, and strain effects *Physical Rev. B,* **80** 245324 *Copyright 2009 American Physical Society, License number RNP/19/ OCT/019231; License date Oct 05, 2019.*)

ZB nanowires exhibit hexagonal sections, with facets of the <110> family and are characterized by a Raman spectrum with a single peak at 266.7 cm^{-1}, only slightly below (267.2 cm^{-1}) that of a reference sample of GaAs, showing the excellent quality of the wires.

The bright-field, HRTEM image of a GaAs nanowire MBE-grown at low growth rates to get 30% of wurtzite content is displayed in Figure 4.31. The image shows that the NW radius ($\approx$ 60 nm) is larger than the critical thickness and, thus, prevents the stability of a pure WZ phase along the entire length of the wire, that, in fact, is only present on the left side of the wire (A zone)* in correspondence with the tip.

Away from the top of the wire (zone A) on the left, one observes an intermediate region characterized by the presence of a mixture of WZ-ZB GaAs phases, where the ZB content increases with the distance from the zone A, and of a high density of ZB twins orthogonal to the wire axis, whose density decreases with the distance from the zone and with the increase of the amount of the ZB phase. Equilibrium conditions are

---

* The nature of the phase is determined by electron diffraction spectra.

FIGURE 4.32 Morphology of GaAs NWs grown with Au-assistance (a) and with self-assistance (b). (*Reproduced with permission of American Chemical Society from S.* Breuer, C. Pfüller, T. Flissikowski, O. Brandt, H.T. Grahn, L. Geelhaar, H. Riechert (2011) Suitability of Au- and self-assisted GaAs nanowires for optoelectronic applications *Nano Lett.* **11** (3) 1276–1279 *Nov 03, 2019.*)

eventually established when the wire grows with the pure ZB structure. It is interesting to observe that the establishment of equilibrium conditions is dominated by an intermediate step, where the ZB phase is nucleated together with the WZ phase, giving rise to segments of the ZB phase encapsulated within the WZ phase.

An optimization of this growth procedure was recently suggested by Küpers et al. [106], who developed a two-step process, where the first is carried out at low As-fluxes to deposit Ga in the holes of the SiO_x mask and to nucleate the NW, and the second to control the shape and the elongation of the wire, with excellent results and development of full untapered morphology.

A question remains about the effect of Au contamination on the electronic and optical quality of GaAs nanowires grown with Au-assistance. This issue was considered by Breuer et al. [83] and by Ahtapodov et al. [107], who compared the properties of Au-assisted and self-assisted GaAs NWs, with conflicting evidence.

The nanowires considered by Brauer [83] were MBE-grown at 500°C for the Au-assisted case and at 580°C for the self-assisted case on oxidized <111> silicon substrates. To reduce the dominant recombination at GaAs surfaces the GaAs core is covered with a thin (Al–Ga)As shell in both cases. For the Au-assisted growth, 10 nm Au nanocrystals were used as seeds.

The Au-assisted grown NWs are thin, see Figure 4.32a, with a constant diameter of 68 ± 12 nm up to 3/4 of the length (5 ± 1 μm), and tapered on top. Their cores are 48 ± 7 nm. The self-assisted nanowires, see Figure 4.32b, are thicker, with a core of 106 ± 18 nm and a constant diameter of 150 ± 5 nm along the entire length. Self-assisted NWs crystallize with the ZB structure, while Au-assisted NWs crystallize with the wurtzite structure.

Minority carrier measurements carried out by means of micro-luminescence transients on self-assisted NWs and Au-assisted nanowires show a systematic difference of their lifetimes τ_s, which hold 2.5 ± 0.1 ns for the self-assisted NWs and 9 ± 1

ps for the Au-assisted NWs, leading to the conclusion that the more than hundredfold lifetime depression occurring in Au-assisted NWs is due to Au-contamination.

The wires prepared by Ahtapodov et al. [107] were Au-assisted MBE-grown, with an (Al–Ga)As sheet, and a core consisting of a pure wurtzite phase, as in the case of the Brauer [83] ones. A single NW was studied, selected for its excellent morphological quality. The wire was 2.4 µm long, with only 17 stacking faults over the entire length of the wire, separated by an average distance of 8 nm. No ZB segments were individuated along the wire.

The optical quality of the wire was checked by micro-luminescence and time-resolved micro-luminescence measurements, as done by Brauer [83], with results which indicate a room temperature carrier lifetime $\tau_s = 0.8$ ns, a factor of ten higher than that measured by Breuer on Au-assisted NWs and a factor of three lower than that measured on self-assisted NWs.

The excellent morphological quality of the wire, with the virtual absence of extended defects (SFs) which could behave as gettering and recombination sites for Au, is presumably at the origin of the very limited detrimental effect of gold observed in Ahtapodov's wires.

4.4 II–VI SEMICONDUCTORS, CdS, CdSe, AND CdTe NANOWIRES

Among the wide family of direct gap II–VI semiconductors, only the structural and physico-chemical properties of CdS, CdSe, and CdTe NWs [108] will be discussed here, in view of the well-known applications of their bulk counterparts in a variety of electronic and photonic devices, including solar cells* and radiation detectors, and of the known lasing potentialities of CdS and CdSe NWs [109, 110].

CdSe, CdTe, and CdS quantum dots [111] would, however, also merit attention, since they present quantum confinement features, and find applications also in domains far outside the optoelectronic one [112].

II–VI semiconductors, see Figure 4.22, share their polytypism [73] with III–V semiconductors, discussed in Section 4.2, since they crystallize with the zincblende or the wurtzite structure, with the thermodynamic stability of the wurtzite structure that does increase with the increase of their ionicities.

In close agreement with prediction based on their ionicities (see Table 4.1 and Figure 4.22), in fact, bulk CdTe crystallizes with the zincblende structure, while bulk CdS and CdSe are bi-stable [113], though wurtzite structure is the most stable for both of them.

In turn, CdTe NWs crystallize with the zincblende structure, while CdSe and CdS nanowires typically grow with the wurtzite phase [111, 114]. An exception is presented by CdTe NWs synthesized in an aqueous dispersion of CdTe nanodots [115] and by CdS NWs prepared by a wet process [116].

II–VI compound semiconductors share with III–V compounds also non-stoichiometry, with potentially severe implications concerning the growth and doping of nanowires using the VSS process, as we will see below.

---

* Thanks to their room temperature energy gaps (E_G(CdSe) = 1.74 eV; E_G(CdTe) = 1.56 eV) and their optical absorption features, which allow the main part of the optical spectrum of the sun to be covered.

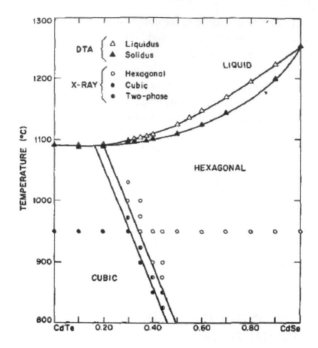

FIGURE 4.33 Pseudo-binary phase diagram of the system CdSe–CdTe. (*Reproduced with permission from A.J. Srauss, J. Steiniger (1970) Phase diagram of the CdTe-CdSe pseudobinary system J.Electrochem.Soc 117 (11) 1420–4426 Copyright 1970 Electrochemical Society, Order Date 29-Oct-2019 Order license ID1001222-1ISSN0013-4651.*)

As expected, the properties of II–VI nanomaterials could be tailored by size selection to favor the set-up of quantum confinement effects [117], but also by alloying, as is the case of CdTe–CdSe alloys; see the phase diagram of a CdTe–CdSe system in Figure 4.33 [118], which indicates the existence of a range of CdTe-rich solid solutions crystallizing with the zincblende structure, and a wide range of CdSe-rich solutions crystallizing with the wurtzite structure, where the extension of both ranges depends on the temperature.

Gas-phase synthesis of II–VI nanowires could be carried out by physical vapor deposition (PVD) processes, using the direct sublimation of precursors, which sublimate without decomposition, or by a variety of CVD methods as in the case of III–V compounds.

PVD processes based on the thermal sublimation of granular precursors are systematically carried out at the industrial scale for the fabrication of II–VI solar cells with so-called close-spaced sublimation (CCS) [119, 120] and are used also for the growth of II–VI compound nanowires.

The deposition temperature of II–VI compound nanostructures should be carefully selected, since their melting temperatures strongly depend on size, see Figure 2.5 for the case of CdS and Figure 4.34 for CdSe, though only modest effects are expected for sizes above a few nanometers.

CdS, CdTe, and CdSe nanowires could be PVD-grown without and with a catalyst.

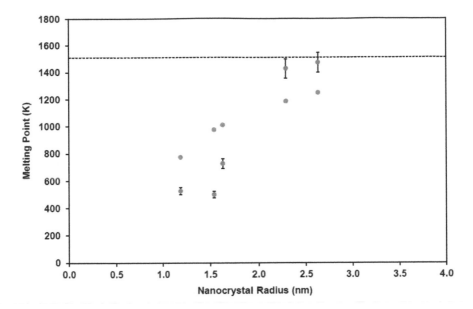

FIGURE 4.34   Size dependence of melting point of CdSe nanocrystals. The dotted line displays the melting point of bulk CdSe (1541 K) (*After* A.D. Dukes III, C.D. Pitts, A.B. Kapingidza, D.E. Gardner, R.C. Layland (2018) Comparison of the observed size-dependent melting point of CdSe nanocrystals to theoretical predictions *Europ. J. Chem.* **9** (1) 39–43 Open Access Journal.)

A typical example of a PVD process carried out without a metal catalyst, successfully employed for the growth of CdS, CdTe, and CdSe nanowire arrays, is given by the work of Utama et al. [121], who carried out the sublimation under vacuum (50 torr) of high-purity powders of CdS (T=750°C), of CdSe (T=775°C), and of CdTe (Ts=650°C). The deposition temperature was the same as in the sublimation process. Van der Waals epitaxy [122] was successfully used to carry out the process as a catalyst-free process,* employing cleaved muscovite mica layers as the deposition substrates.

Vertically aligned arrays of single crystal nanowires of CdTe, CdSe, and CdS could be grown, see Figure 4.35, with an average diameter of 100 nm. Their crystal structure, measured with XRD, is zincblende for CdTe NWs, while it is wurtzite for CdS and CdSe NWs, in agreement with our previous considerations. TEM measurements carried out on these wires show that CdS and CdSe are grown with the <0001> growth direction, while the growth direction of CdTe NWS is <111>; these are, in fact, the thermodynamically favorite directions for wurtzite and zincblende structures, respectively.

In CdTe wires several twin defects could be observed very on top of the wire, while CdS and CdSe wires are defect-free.

The lasing capability of vertically aligned CdS and CdSe nanowires, demonstrated by Pan et al. [109] and Chen et al. [110], will be discussed in Chapter 5, Section 5.5.3.

---

* It was experimentally shown that the use of a catalyst is unnecessary using muscovite as the support.

FIGURE 4.35   SEM images of as-grown CdS (a), CdSe (b), and CdTe (c) nanowire samples epitaxially grown on muscovite substrates. (Scale bars correspond to 100 nm.) *(Reproduced with permission from* M.I.B. Utama, Z. Peng, R. Chen, B. Peng, X. Xu, Y. Dong, L.M. Wong, S.J. Wang, H.D. Sun, Q. Xiong (2011) Vertically aligned cadmium chalcogenide nanowire arrays on Muscovite mica: A demonstration of epitaxial growth strategy *Nano Lett.* **11**(8) 3051–3057 *Copyright 2011 American Chemical Society Oct 26, 2019.)*

Tang et al. [115] demonstrated that CdTe nanowires could be obtained also by spontaneous self-assembly at room temperature of an aqueous mono-dispersed distribution of CdTe dots [115], with diameters ranging between 2.5 and 5.4 nm.

The nanowire diameter was shown to correspond almost exactly to the diameter of the dots, and the growth is associated with a phase change, since the crystalline CdTe nanodots have the cubic zincblende structure and the wires crystallize with the wurtzite structure, with their long axis parallel to the <001> direction of the wurtzite.

The necklace pearl appearance of the wires does explain the growth mechanism, consisting of the spontaneous mechanical aggregation of dots, that change their structure during aggregation, driven by dipole–dipole attraction, which is strong and long-range, and by a thermodynamic driving force, associated with the difference between the free energy of nanowires and that of individual nanodots.

The critical features and deterministic or non-deterministic effects inherent to vapor phase depositions of II–VI semiconductor nanowires carried out with and without metal catalysts were studied by Colli et al. [123] and by Fasoli et al. [108, 124].

Colli et al. [123] preliminarily argue that the inconveniencies observed with the conventional use of a single-zone furnace reactor for the synthesis of nanocrystals and nanowires with a vapor transport process, i.e. the presence of a wide distribution of nanocrystal shapes or of NW diameters and lengths, depends on the absence of a shutter, which could insulate the growth zone during furnace ramps at the beginning and at the end of the process. He demonstrates that deterministic shape selective synthesis of CdSe nanowires could be instead obtained by a proper control of the furnace pressure, which works as a vapor flow shutter during the process of deposition, consisting of ramping the precursor powder from room temperature to $T_p = 700°C$ and the Au-coated silicon substrate to $T_s = 600°C$, chosen as the CdSe deposition temperature, followed by a constant temperature cycle and a cooling-down cycle to room temperature.

In fact, they show that by carrying out the whole process at a constant pressure of 50 torr, CdSe nanostructures of different shape and orientation grow on the Au catalyst already during the heating ramp, serving as nuclei for further growth at constant temperature. Instead, no crystal nucleation occurs if the heating ramp is carried out at a pressure of 1 bar of argon.

They also demonstrate that the morphology of the deposits depends on a proper selection of the T_p and T_s values. As an example, at low sublimation temperatures ($T_p = 550°C$) no nucleation occurs on an Au-covered substrate at $T_s > 470°C$, while untapered, defect -free CdSe NWs 50 nm in diameter grow at T_s of 420–470°C, and nano-saws are obtained for $T_s < 430°C$.

In order to obtain a more general empirical scheme connecting the morphology of CdSe nanostructures and the T_p and T_s values, Fasoli et al. [124] carried out a systematic study both for the case of CdSe-assisted or self-assisted growth and of the Au-assisted growth of CdSe nanostructures at substrate temperatures T_s between 450 and 750°C. CdSe seeds 5–10 nm in diameter were used or Au-patterned (0.5–1.5 nm thick) oxidized <100> silicon substrates (where the Au pattern splits in nano-islands having an average diameter of 10 nm during the heating ramp). For self-assisted growth a bare oxidized silicon substrate was used.

The main concept beyond the morphology selectivity typical of the CdSe vapor growth is that it depends on the balance of sticking of CdSe vapors on different surfaces and on the growth kinetics of various nanostructures on CdSe nuclei.

Since the sticking coefficient of CdSe on Au is high, CdSe NWs could be grown easily at low substrate temperatures ($T_s < 500°C$), without parasitic deposition on NW sidewalls or on the Au-uncoated areas of the substrate.

Self-seeded growth of CdSe NWs or growth of CdSe nanorods assisted by CdSe nuclei does, instead, occur at high powder temperatures and low substrate temperatures, when the substrate temperature is not high enough to desorb CdTe from the uncoated substrate surface. In fact, at $T_p = 525°C$ and $T_s = 750°C$, spontaneous CdSe nucleation does occur on bare oxidized substrates, on which the further growth of thin NWs does occur. At $T_p = 585$ and $T_s = 750°C$, on substrates whose surfaces were previously decorated with a distribution of CdSe colloidal nuclei, the growth of nanorods does, instead, occur. In both last cases growth kinetics dominates the nanowire morphology.

Eventually, branched structures do grow at high temperatures and low substrate temperatures ($T_p = 485°C$ and $T_s = 700°C$) on Au-coated substrates.

It is well-evident that very modest temperature variations do provide relevant morphology changes.

REFERENCES

1. N. E. Christensen, S. Sapaty, Z. Pawlowska (1987) Bonding and ionicity in semiconductors. *Phys. Rev. B* **36**(2), 1032–1050.
2. R .W. Nosker, P. Mark, J. D. Levine (1970) Polar surfaces of wurtzite and zincblende lattices. *Surf. Sci.* **19**(2), 291–317.
3. C. B. Duke (1996) Semiconductor surface reconstruction: The structural chemistry of two-dimensional surface compounds. *Chem. Rev.* **96**, 1237–1259.
4. P. D. C. King, T. D. Veal, C. F. McConville, F. Fuchs, J. Furthmüller, F. Bechstedt P. Schley, R. Goldhahn J. Schörmann, D. J. As, K. Lischka, D. Muto, H. Naoi, Y. Nanishi, Hai Luc ,W. J. Schaff (2007) Universality of electron accumulation at wurtzite c- and a-plane and zincblende InN surfaces. *Appl. Phys. Lett.* **91**, 092101.

5. Y. Zhang, J. Zhang, J. Zhu (2018) Stability of wurtzite semi-polar surfaces: Algorithms and practices. *arXiv.* https://arxiv.org.
6. I. Sohn, W-Ki Hong, S. Lee, S. Lee, J. Y. Ku, Y. J. Park, J. Hong, S. Hwang, K. H. Park, J. H. Warner, S. N. Cha, J. M. Kim (2014) Surface energy-mediated construction of anisotropic semiconductor wires with selective crystallographic polarity. *Sci. Rep.* **4**, 5680.
7. M. Hjort, S. Lehmann, J. Knutsson, A. A. Zakharov, Y. A. Du, S. Sakong, R. Timm, G. Nylund, E. Lundgren, P. Kratzer, K. A. Dick, A. Mikkelsen (2014) Electronic and structural differences between wurtzite and zinc blende InAs nanowire surfaces: Experiment and theory. *ACS Nano* **8**(12), 12346–12355.
8. A. Senichev, P. Corfdir, O. Brandt, M. Ramsteiner, S. Breuer, J. Schilling, L. Geelhaar, P. Werner (2018) Electronic properties of wurtzite GaAs: A correlated structural, optical, and theoretical analysis of the same polytypic GaAs nanowire. *Nano Res.* **11**(9), 4708–4721.
9. D. T. J. Hurle (1999) A comprehensive thermodynamic analysis of native point defect and dopant solubilities in gallium arsenide. *J. Appl. Phys.* **85**, 6957–7018.
10. G. F. Newmark (1997) Defects in wide band gap II-VI crystals. *Material Sci. Eng. R* **21**(1), 1–46.
11. P. Rudolph (2003) Non-stoichiometry related defects at the melt growth of semiconductor compound crystals – a review. *Cryst. Res. Technol.* **38**(7–8), 531–739.
12. H. Wentzl, A. Dahlen, A. Fattah, S. Petersen, K. Mika, D. Henkel (1991) Phase relations in GaAs crystal growth. *J. Cryst. Growth* **109**, 191–204.
13. A. Majid (2018) A perspective on non-stoichiometry in silicon carbide. *Ceram. Intern.* 44(2), 1277–1283.
14. K. Zekentes, K. Rogdakis (2011) SiC nanowires: Material and devices. *J. Phys. D: Appl. Phys.* **44**(13), 133001.
15. M. J. Choyke, H. Matsumami, G. Pensl (2003) *Silicon Carbide: Recent Major Advances*, Springer.
16. F. Bechstaedt, J. Furthmüller, H. Grossner, C. Raffy (2003) Zero-and two-dimensional native defects. In: *Silicon Carbide: Recent Major Advances*, M. J. Choyke, H. Matsumami, G. Pensl Eds., Springer.
17. V. A. Izhevskyi, L. A. Genova, J. C. Bressiani, A. H. A. Bressiani (2000) Review article: Silicon carbide: Structure, properties and processing. *Ceramica* **46**, 297. Print version ISSN 0366-6913On-line version ISSN 1678-4553.
18. X. L. Wu, J. Y. Fan, T. Qiu, X. Yang, G. G. Siu, P. K. Chu (2005) Experimental evidence for the quantum confinement effect in 3C-SiC nanocrystallites. *Phys. Rev. Lett.* **94**, 026102.
19. D. Beke, Z. Szekrenyes, Z. Czigany, K. Kamaras, A. Gali (2015) Dominant luminescence is not due to quantum confinement in molecular-sized silicon carbide nanocrystals. *Nanoscale* **7**, 10982.
20. K. Konishi, S. Yamamoto, S. Nakata, Y. Nakamura, Y. Nakanishi, T. Tanaka, Y. Mitani, N. Tomita, Y. Toyoda, S. Yamakawa (2013) Stacking fault expansion from basal plane dislocations converted into threading edge dislocations in 4H-SiC epilayers under high current stress. *J. Appl. Phys.* **114**, 014504.
21. Z. Wang, F. Gao (2012) Defects and doping in one-dimensional SiC nanostructures. *J. Comput. Theor. Nanosci.* **9**(11), 1967–1974.
22. Y. B. Li, S. S. Xie, W. Y. Zhou, L. J Ci, Y. Bando (2002) Cone-shaped hexagonal 6H–SiC nanorods. *Chem. Phys. Lett.* **356**, 325.
23. G. Wei, G. Qin, W. Gang, J. Sun, J. Lin, R. Kim, D. Zhang, K. Zheng (2008) The synthesis and ultraviolet photoluminescence of 6H–SiC nanowires by microwave method. *J. Phys. D: Appl. Phys.* **41**, 235102.

24. W. Shi, Y. Zheng, H. Peng, N. Wang, C. S. Lee, S. T. Lee (2000) Laser ablation synthesis and optical characterization of silicon carbide nanowires. *J. Amer. Ceram. Soc.* **83**(12), 3228–3230.

25. Y. F. Zhang, X. D Han, K. Zheng, Z. Zhang, X. N. Zhang, J. Y. Fu, Y. Ji, Y. J. Hao, X. Y. Guo, Z. L. Wang (2007) Direct observation of super-plasticity of beta-SiC nanowires at low temperature. *Adv. Funct. Mater.* **17**, 3435–3440.

26. G. Cheng, T-H. Chang, Q. Qin, H. Huang, Y. Zhu (2014) Mechanical properties of silicon carbide nanowires: Effect of size-dependent defect density. *Nano Lett.* **14**(2), 754–758.

27. D-H. Wang, D. Xu, Q. Wang, Y-J- Hao, G-Q. Jin, X-Y. Guo, K. N. Tu (2008) Periodically twinned SiC nanowires. *Nanotechnology* **19**, 215602.

28. K. Chen, Z. Huang J. Huang, M. Fang, Y.-G. Liu, H. Li, L. Yin (2013) Synthesis of SiC nanowires by thermal evaporation method without catalyst assistant. Ceram. Int. **39**(2), 1957–1962.

29. K. Y. Cheong, Z. Lockman (2009) Growth mechanism of cubic-silicon carbide nanowires. *J. Nanomat.* Article No.31 Hindawi.

30. G. Chikvaidze, N. Mironova-Ulmane, A. Plaude, O. Sergeev (2014) Investigation of silicon carbide polytypes by Raman spectroscopy. *Latv. J. Phys. Techn. Sci* **3**, 51–56.

31. Z. S. Wu, S. Z. Deng, N. S. Xu, Jian Chen, J. Zhou, J. Chen (2002) Needle-shaped silicon carbide nanowires: Synthesis and field electron emission properties. *Appl. Phys. Lett.* **80**, 3829.

32. S. Kawanishi, T. Yoshikava, T. Tanaka (2009) Equilibrium phase relationship between SiC and liquid phase in the Fe-Si-C system at 1523–1723 K. *Mater. Trans.* **50**(4), 806–813.

33. D. Wang, C. Xue, H. Bai, N. Jiang (2015) Silicon carbide nanowires grown on graphene sheets. *Ceram. Int.* **41**(4), 5473–5477.

34. J-H. Kim, S-C. Choi (2018) Characteristics of silicon carbide nanowires synthesized on porous body by carbothermal reduction. *J. Korean Ceram. Soc.* **55**(3), 285–289.

35. S-C. Chiu, C-W. Huang, Y-Y. Li (2007) Synthesis of high-purity silicon carbide nanowires by a catalyst-free arc-discharge method. *J. Phys. Chem. C* **111**(28), 10294–10297.

36. N. Jie, L. Zhengcao, Z. Zhengjun (2007) Synthesis of silicon carbide nanowires by solid phase source chemical vapor deposition. *Front. Mater. Sci. China* **1**(3), 304–308.

37. Z. Hua, Y. Li, Y. Zhan, P. Huan, X. Wang, C. Chen (2018) Fabricating SiC nanowire by vacuum heating multilayer graphene in silicate refractory tube. *IOP Conf. Ser.: Mater. Sci. Eng.* **381**, 012107.

38. Z. Pan, H.-L. Lai, F. C. K. Au, X. Duan, W. Zhou, W. Shi, N. Wang, C.-S. Lee, N.-B. Wong, S.-T. Lee, S. Xie (2000) Oriented silicon carbide nanowires: Synthesis and field emission properties. *Adv. Mater.* **12**(16), 1186–1190.

39. X-H. Sun, C-P. Li, W-K. Wong, N.-B. Wong, C.-S. Lee, S.-T. Lee, B.-K. Teo (2002) Formation of silicon carbide nanotubes and nanowires via reaction of silicon (from disproportionation of silicon monoxide) with carbon nanotubes. *J. Am. Chem. Soc.* **124**(48), 14464–14471.

40. J. Zhu, S. Fan (1999) Nanostructure of GaN and SiC nanowires based on carbon nanotubes. *J. Mat. Res.* **14**(4), 1175–1177.

41. F-C. Wang, L. Zhao, W. Fang, X. He, F. Liang, H. Chen, X. Du (2015) Synthesis and characterization of silicon carbide nanowires from lignin-phenolic resin and silicon powder with an *in-situ* formed molten salt as catalyst. *New Carbon Mater.* **30**(3), 222–229.

42. P. Hu, S. Dong, X. Zhang, K. Gui, G. Chen, Z. Hu (2017) Synthesis and characterization of ultralong SiC nanowires with unique optical properties, excellent thermal stability and flexible nanomechanical properties. *Sci. Rep.* **7**, 3011.

43. A. V. Pavlikov, V. Latukhina, V. I. Chepurnov, V. Yu. Timoshenko (2017) Structural and optical properties of silicon-carbide nanowires produced by the high-temperature carbonization of silicon nanostructures. *Semiconductors* **51**(3), 402–406.

44. J. Q. Hu, Q. Y. Lu, K. B. Tang, B. Deng, R. R. Jiang, Y. T. Qian, W. C. Yu, G. E. Zhou, X. M. Liu, J. X. Wu (2000) Synthesis and characterization of SiC nanowires through a reduction–carburization route. *J. Phys. Chem. B* **104**(22), 5251–5254.

45. M. S. Yaghmaee, G. Kaptay (2001) On the stability of SiC in the ternary liquid Al-Si-Mg alloy. *Mater. World* (e-Journal ISSN 1586–0140 http://materialworld.fw.hu).

46. P. Lagonegro, M. Bosi, G. Attolini, M. Negri, S. C. Dhanabalan, F. Rossi, F. Boschi, P. P. Lupo, T. Besagni, G. Salviati (2015) SiC NWs grown on silicon substrate using Fe as catalyst. *Mater. Sci. Forum* **806**, 39–42.

47. S. Carapezzi, A. Castaldini, F. Fabbri, F. Rossi, M. Negri, G. F. Salviati, A. Cavallini (2016) Cold field electron emission of large-area arrays of SiC nanowires: Photo-enhancement and saturation effects. *J. Mater. Chem. C* **4**(35), 8226–8234.

48. M. Negri, S. C. Dhanabalan, G. Attolini, P. Lagonegro, M. Campanini, M. Bosi, F. Fabbri, G. Salviati (2015) Tuning the radial structure of core-shell silicon carbide nanowires. *Cryst. Eng. Comm* **17**, 1258–1263.

49. M. Nagamori, I. Malinsky, A. Claveau (1986) Thermodynamics of the Si-C-O system for the production of silicon carbide and metallic silicon. *Metall. Trans. B* **17**(3), 503–514.

50. C. Varadachari, R. Bhowmick, K. Ghosh (2012) Thermodynamics and oxidation behaviour of crystalline silicon carbide (3C) with atomic oxygen and ozone. *Int. Scholarly Res. Network ISRN, Thermodyn.* 2012. Article ID 108781. doi:10.5402/2012/108781.

51. B. N. Tabassum, M. Kotha, V. Kaushik, B. Ford, S. Dey, E. Crawford, V. Nikas, S. Gallis (2018) On-Demand CMOS-compatible fabrication of ultrathin self-aligned SiC nanowire arrays. *Nanomaterials* **8**, 906.

52. M. Luppi, S. Ossicini (2005) Ab-initio study on oxidized silicon clusters and silicon nanocrystals embedded in SiO2: Beyond the quantum confinement effect. *Phys. Rev. B* **71**, 035340.

53. F. A. Khan, I. Adesida (1999) High rate etching of SiC using inductively coupled plasma reactive ion etching in SF_6-based gas mixtures. *Appl. Phys. Lett.* **75**, 2268–2270.

54. L. Jiang, R. Cheung (2003) Inductively coupled plasma etching of SiC in SF_6/O_2 and etch-induced surface chemical bonding modifications. *J. Appl. Phys.* **93**, 1376–1383.

55. J. H. Choi, L. Latu-Romain, E. Bano, A. Henry, W. J. Lee, T. Chevolleau, T. Baron (2012) Comparative study on dry etching of alpha- and beta-SiC nano-pillars. *Mater. Lett. (General ed.)* **87**, 9–12.

56. J. Choi (2013) Silicon carbide nanowires: From fabrication to related devices. PhD Thesis, École Doctorale EEATS, Universitè du Grenoble(France).

57. Y. Ou, I. Aijaz, V. Jokubavicius, R. Yakimova, M. Syvaijarvi, H. Ou (2013) Broadband antireflection silicon carbide surface by self-assembled nanopatterned reactive-ion etching. *Opt. Mater. Express* **3**(1), 86–93.

58. A. Irrera, J. Lo Faro, C. D'Andrea, A. A. Leonardi, P. Artoni, B. Fazio, R. A. Picca, N. Cioffi, S. Trusso, G. Franzò, P. Musumeci, F. Priolo, F. Iacona (2017) Light-emitting silicon nanowires obtained by metal-assisted chemical etching. *Semicond. Sci. Technol.* **32**, 043004.

59. A. Irrera, P. Artoni, R. Saija, P. G. Gucciardi, M. A. Iatì, F. Borghese, P. Denti, F. Iacona, F. Priolo, O. M. Maragò (2011) Size-Scaling in optical trapping of silicon nanowires. *Nano Lett.* **11**, 4879–4884.

60. A. Irrera, P. Artoni, F. Iacona, E. F. Pecora, G. Franzò, M. Galli, B. Fazio, S. Boninelli, F. Priolo (2012) Quantum confinement and electroluminescence in ultrathin silicon nanowires fabricated by a maskless etching technique. *Nanotechnology* **23**, 075204.

61. B. Fazio, P. Artoni, M. A. Latì, C. D'Andrea, M. J. Lo Faro, S. Del Sorbo, S. Pirotta, P. G. Gucciardi, P. Musumeci, C. S. Vasi, R. Saija, M. Galli, F. Priolo, A. Irrera (2016) Strongly enhanced light trapping in a two-dimensional silicon nanowire random fractal array. *Light Sci. Appl.* **5**, e16062.

62. A. Irrera, A. A. Leonardi, C. Di Franco, M. J. Lo Faro, G. Palazzo, C. D'Andrea, K. Manoli, G. Franzò, P. Musumeci, B. Fazio, L. Torsi, F. Priolo (2018) New generation of ultrasensitive label-free optical si nanowire-based biosensors. *ACS Photonics*, **5**(2), 471–479.

63. M. J. Lo Faro, C. D'Andrea, E. Messina, B. Fazio, P. Musumeci, G. Franzo', P. G. Gucciardi, C. Vasi, F. Priolo, F. Iacona, A. Irrera (2016) A room temperature light source based on silicon nanowires. *Thin Solid Films* **613**, 59–63.

64. C. H. Hsu, H-C Lo, C-F. Chen, C. T. Wu, J-S. Hwang, D. Das, J. Tsai, L-C. Chen, K-H. Chen (2004) Generally applicable self-masked dry etching technique for nanotip array fabrication. *Nano Lett.* **4**(3), 471–475.

65. J. S. Blakemore (1982) Semiconducting and other major properties of gallium arsenide. *J. Appl. Phys.* **53**, R123–R181.

66. X. Duan, J. Wang, C. M. Lieber (2000) Synthesis and optical properties of gallium arsenide nanowires. *Appl. Phys. Lett.* **76**, 1116–1118.

67. L. S. Karlsson (2007) Transmission electron microscopy of III-V nanowires and nanotrees. Doctoral Thesis, Lund University Division of Polymer & Materials Chemistry Department of Chemistry Sweden.

68. P. Rudolph, M. Jurisch (1999) Bulk growth of GaAs: An overview. *J. Cryst. Growth* **198/199**, 325–335.

69. W. A. Oates, H. Wenzl (1998) Foreign atom thermodynamics in liquid gallium arsenide. *J. Cryst Growth* **191**, 303–312.

70. M. Goldschmidt (1929) Crystal structure and constitution. *Trans. Faraday Soc.* **25**, 253–283.

71. C-Y. Yeh, Z. W. Lu, S. Froyen, A. Zunger (1992) Zinkblende-wurtzite polytypism in semiconductors. *Phys. Rev. B* **46**(16), 10086–10097.

72. T. Akiyama, K. Sano, K. Nakamura, T. Ito (2006) An empirical potential approach to wurtzite-zinc blende polytypism in group III-V semiconductor nanowires. *Jpn. J. Appl. Phys.* **45**, L275–L278.

73. L. Pengfei, C. Huawei, Z. Xianlong, Y. Zhongyuan, C. Ningning, G. Tao, W. Shumin (2013) Structural properties and energetics of GaAs nanowires. *Physica E* **52**, 34–39.

74. I. Zardo, S. Conesa-Boj, F. Peiro, J. R. Morante, J. Arbiol, E. Uccelli, G. Abstreiter, A. Fontcuberta i Morral (2009) Raman spectroscopy of wurtzite and zinc-blende GaAs nanowires: Polarization dependence, selection rules, and strain effects. *Phys. Rev. B* **80**, 245324.

75. M. Pozuelo, S. V. Prikhodko, R. Grantab, H. Zhou, L. Gao, S. D. Sitzman, V. Gambin, V. B. Shenoy, R. F. Hicks, S. Kodambaka (2010) Zincblende to wurtzite transition during the self-catalyzed growth of InP nanostructures. *J. Cryst. Growth* **312**, 2305–2309.

76. J. Z. Jiang, J. Staun Olsen, L. Gerward, S. Steenstrup (2002 X-ray diffraction study on pressure-induced phase transformation in nanocrystalline GaAs. *J. High Pressure Res.* **22**(2), 395–398.

77. W. Zhou, X-J. Chen, J-B. Zhang, X-H. Li, Y-Q. Wang, A. F. Goncharov (2014) Vibrational, electronic and structural properties of wurtzite GaAs nanowires under hydrostatic pressure. *Sci. Rep.* **4**, 6472.

78. X. Liu, A. Prasad, J. Nishio, E. R. Weber (1995) Native point defects in low-temperature-grown GaAs. *Appl. Phys. Lett.* **67**, 279.

79. G. A. Baraff, M. Schlüter (1985) Electronic structure, total energies, and abundances of the elementary point defects in GaAs. *Phys. Rev. Lett.* **55**, 1327.

80. N. T. Son, C. G. Hemmingsson, T. Paskova, K. R. Evans, A. Usui, N. Morishita, T. Ohshima, J. Isoya, B. Monemar, E. Janzén (2009) Identification of the gallium vacancy–oxygen pair defect in GaN. *Phys. Rev. B* **80**, 153202.
81. S. Breuer, C. Pfüller, T. Flissikowski, O. Brandt, H. T. Grahn, L. Geelhaar, H. Riechert (2011) Suitability of Au- and self-assisted GaAs nanowires for optoelectronic applications. *Nano Lett.* **113**, 1276–1279.
82. A. Fontcuberta i Morral (2011) Gold-free GaAs nanowire synthesis and optical properties. *IEEE J. Sel. Top. Quantum Electron* 17, 819–828.
83. Y. Wang, X. Zhou, Z. Yang, F. Wang, N. Han, Y. Chen, J. C. Ho (2018) GaAs nanowires grown by catalyst epitaxy for high performance photovoltaics. *Crystals* **8**(9), 347.
84. E. F. Schubert (1990) Delta doping of III-V compound semiconductors: Fundamentals and device applications. *J. Vac. Sci Techn. A* **8**(3), 2980–2990.
85. H. Mäntynen (2016) Doping of vapor–liquid–solid grown gallium arsenide nanowires. Thesis, Aalto University, Finland.
86. H. J. Joyce, Q. Gao, Y. Kim, H. H. Tan, C. Jagadish, X. Zhang, Y. Guo, J. Zou, M. A. Fickenscher, S. Perera, T. B. Hoang, L. M. Smith, H. E. Jackson, J. M. Yarrison-Rice (2008) Growth, structural and optical properties of high quality GaAs nanowires for optoelectronics. *Conference Paper* September 2008. doi:10.1109/NANO.2008.25. Source: IEEE Xplore.
87. K. A. Dick (2008) A review of nanowire growth promoted by alloys and non-alloying elements with emphasis on Au-assisted III–V nanowires. *Prog. Cryst. Growth Charact. Mater.* **54**(3), 138–173.
88. M. Zamani (2017) Growth and characterization of high quality. GaAs Nanowires Thesis, Politecnico di Milano.
89. M. Zamani, G. Tütüncüoglu, S. Martí-Sánchez, L. Francaviglia, L. Güniat, L. Ghisalberti, H. Potts, M. Friedl, E. Markov, W. Kim, J-B. Leran, V. G. Dubrovskii, J. Arbiol, A. Fontcuberta i Morral (2018) Optimizing the yield of A-polar GaAs nanowires to achieve defect-free zinc blende structure and enhanced optical functionality. *Nanoscale* **10**, 17080–17091.
90. D. L. Dheeraj, H. L. Zhou, A. F. Moses, T. B. Hoang, A. T. J. van Helvoort, B. O. Fimland, H. Weman (2010) Growth of heterostructured III-V nanowires by molecular beam epitaxy for photonic applications. *Proc. Spie* **7608**, 76081. 1-6 Quantum sensing and Nanophonic devices VII.
91. A. S. Ameruddin (2015) Growth and characterisation of gold-seeded indium gallium arsenide nanowires for optoelectronic. Applications Thesis. The Australian National University.
92. T. C. Thomas, R. S. Williams (1986) Solid phase equilibria in the Au-Ga-As, Au-Ga-Sb, Au-In-As, and Au-In-Sb ternaries. *J. Mater. Res.* **1**(2), 352–360.
93. F. Font, T. G. Myers (2013) Spherically symmetric nanoparticle melting with a variable phase change temperature. *J. Nanopart.* **15**, 2086.
94. M. B. Panish (1967) Ternary condensed phase systems of gallium and arsenic with group 1B elements. *J. Electrochem. Soc.* **114**, 516–521.
95. M. Ghasemi, J. Johansson (2017) Phase diagrams for understanding gold-seeded growth of GaAs and InAs nanowires. *J. Physics D: Appl. Phys.* 50(13), 134002.
96. J. D. Bryan, D. R. Gamelin (2005) Doped semiconductor nanocrystals: Synthesis, characterization, physical properties, and applications. *Prog. Inorg. Chem.* **54**, 47–126.
97. T. Haggren, J.-P. Kakko, H. Jiang, V. Dhaka, T. Huhtio, H. Lipsanen (2014) Effects of Zn doping on GaAs nanowires. Proc. 14th Intern. Conf. Nanotechnology 18-21 Aug. 2014. ISBN: 978-1-4799-5622-7. *doi:*10.1109/NANO.2014.6968091.
98. N. I. Goktas, E. M. Fiordaliso, R. R. LaPierre (2018) Doping assessment in GaAs nanowires. *Nanotechnology* **29**, 234001.

99. J. A. van Vechten, J. C. Phillips (1970) New set of tetrahedral covalent radii. *Phys. Rev. B* **2**, 2160–2167.

100. I. Teramoto (1972) Calculation of distribution equilibrium of amphoteric silicon in gallium arsenide. *J. Phys. Chem. Solids* **33**(11), 2089–2099.

101. J. Dufouleur, C. Colombo, T. Garma, B. Ketterer, E. Uccelli, M. Nicotra, A. Fontcuberta i Morral (2010) P-doping mechanisms in catalyst-free gallium arsenide nanowires. *Nano Lett.* **10**, 1734.

102. B. Ketterer, E. Mikheev, E. Uccelli, A. Fontcuberta i Morral (2010) Compensation effects in silicon-doped gallium arsenide nanowires. *Appl. Phys. Lett.* **97**, 223103.

103. I. Regolin, V. Khorenko, W. Prost, F. J. Tegude, D. Sudfeld, J. Kästner, G. Dumpich, K. Hitzbleck, H. Wiggers (2007) GaAs whiskers grown by metal-organic vapor-phase epitaxy using Fe nanoparticles. *J. Appl. Phys.* **101**, 054318.

104. B. Chen, X. Fu, J. Tang, M. Lysevych, H. H. Tan, C. Jagadish, A. H. Zewail (2017) Dynamics and control of gold-encapped gallium arsenide nanowires imaged by 4D electron microscopy. *PNAS* **114**, 12876–12881.

105. S. A. Fortuna, J. Wen, I. S. Chun, X. Li (2008) Planar GaAs nanowires on GaAs (100) substrates: Self-aligned, nearly twin-defect free, and transfer-printable. *Nano Lett.* **8**(12), 4421–4427.

106. H. Küpers, R. B. Lewis, A. Tahraoui, M. Matalla, O. Krüger, F. Bastiman, H. Riechert, L. Geelhaar (2018) Diameter evolution of selective area grown Ga-assisted GaAs nanowires. *Nano Res.* **11**(5), 2885–2893.

107. L. Ahtapodov, J. Todorovic, P. Olk, T. Mjåland, P. Slåttnes, D. L. Dheeraj, A. T. J. van Helvoort, B-O. Fimland, H. Weman (2012) A story told by a single nanowire: Optical properties of wurtzite gaas. *Nano Lett.* **12**, 6090–6095.

108. A. Fasoli, A. Colli, F. Martelli, S. Pisana, P- H. Tan, A. C. Ferrari (2011) Photoluminescence of CdSe nanowires grown with and without metal catalyst. *Nano Res.* **4**(4), 343–359.

109. G. R. Amiri, S. Fatahian, S. Mahmoudi (2013) Preparation and optical properties assessment of CdSe quantum dots. *Mater. Sci. Applic.* **4**, 134–137.

110. T. D. T. Ung, T. K. C. Tran, T. N. Pham, D. N. Nguyen, D. K- Din, Q. L. Nguyen (2012) CdTe and CdSe quantum dots: Synthesis, characterizations and applications in agriculture. *Adv. Nat. Sci.; Nanosci. Nanotechnol.* **3**, 043001.

111. A. Pan, R. Liu, Q. Yang, Y. Zhu, G. Yang, B. Zou, K. Chen (2005) stimulated emissions in aligned CdS nanowires at room temperature. *J. Phys. Chem. B* **109**(51), 24268–24272.

112. R. Chen, M. I. B. Utama, Z. Peng, B. Peng, Q. Xiong, H. Sun 2011) Excitonic properties and near-infrared coherent random lasing in vertically aligned CdSe nanowires. *Adv. Mater.* **23**, 1404–1408.

113. N. Kishore, V. Nagarajan, R. Chandiramouli (2019) Mechanical properties and band structure of CdSe and CdTe nanostructures at high pressure – A first-principles study. *Process. Appl. Ceram.* **13**(2), 124–131.

114. Q. Yang, K. Tang, C. Wang, Y. Qian, S. Zhang (2002) PVA-assisted synthesis and characterization of CdSe and CdTe nanowires. *J. Phys. Chem. B* **106**(36), 9227–9230.

115. Z. Tang, N. A. Kotov, M. Giersig (2002) Spontaneous organization of single cdte nanoparticles into luminescent nanowires. *Science* **297**(5579), 237–240.

116. R. Maity, S. Kundoo, K. K. Chattopadhyay (2006) Synthesis and optical characterization of CdS nanowires by chemical route. *Mater. Manuf. Processes* **21**(7), 644–647.

117. S. Chellammal, S. Manivannan (2014) Determination of quantum confinement effect of nanoparticles. *Adv. Mater. Res.* **1051**, 17–20.

118. A. J. Strauss, J. Steininger (1970) Phase diagram of the CdTe-CdSe pseudobinary system. *J. Electrochem. Soc.* **117**(11), 1420–4426.

119. J. D. Major, K. Durose (2011) Early stage growth mechanisms of CdTe thin films deposited by close space sublimation for solar cells. *Sol. Energy Mater. Sol. Cells* **95**(12), 3165–3170.
120. N. Amin, K. S. Rahman (2017) Close-spaced sublimation: A low-cost, high-yield deposition system for Cadmium telluride (CdTe) thin film solar cells. In: *Modern Technologies for Creating the Thin-film Systems and Coatings*, N. N. Nikitenkov Ed., IntechOpen. doi:10.5772/66040.
121. M. I. B. Utama, Z. Peng, R. Chen, B. Peng, X. Xu, Y. Dong, L. M. Wong, S. J. Wang, H. D. Sun, Q. Xiong (2011) Vertically aligned cadmium chalcogenide nanowire arrays on muscovite mica: A demonstration of epitaxial growth strategy. *Nano Lett.* **11**(8), 3051–3057.
122. A. Koma (1999) Van der Waals epitaxy for highly lattice-mismatched systems. *J. Cryst. Growth.* **201–202**, 236–241.
123. A. Colli, A. Fasoli, C. Ducati, J. Robertson, A. C. Fasoli (2006) Deterministic shape-selective synthesis of nanowires, nanoribbons and nanosaws by steady-state vapour-transport. *Nanotechnology* **17**, 1046–1051.
124. A. Fasoli, A. Colli, S. Kudera, L. Manna, S. Hofmann, C. Ducati, J. Robertson, A. C. Ferrari (2007) Catalytic and seeded shape-selective synthesis of II–IV semiconductor nanowires. *Physica E* **37**, 138–141.

5 Optical Signatures of Defects in Semiconductor Nanostructures*

5.1 OPTOELECTRONIC PROPERTIES OF SILICON NANOCRYSTALS

Photoluminescence and optical absorption measurements on silicon nanocrystals are used to investigate the nature of the core- and surface-states responsible for their optical emissions [1–13] and to investigate the effects of hydrogen passivation [14] and oxygen overcoating. We remind here that silicon nanocrystals, typically, present a well-structured core and a defective surface, and are intrinsically prone to oxidation or to reacting with the organic solvents used to prepare their suspensions.

One of the main questions regarding silicon nanocrystals is about the set-up of quantum confinement (QC) conditions, when their average size is less than 10 nm, or, possibly, close to 1 nm.

As an example, silicon nanocrystals of about 3 nm in diameter prepared in the plasma reactor described in Section 3.2.1, properly coated with octadecene, which should, at least in principle, protect the surface of the particles from oxidation, exhibit a single, broad PL emission band peaked at 790 nm (1.8 eV) (see Figure 5.1) with an experimental quantum yield close to 60% under light excitation at 380 nm [4]. Since, under QC conditions, the PL peak emission energy should increase with the decrease of the mean size of the ensembles taken into consideration, we expect such behavior for these nanoparticles.

In fact, Si-NCs of the same origin, dispersed in methanol, do exhibit (see Figure 5.2) an intense PL emission characterized by a broad band centered at 700 nm (1.77 eV) and 800 nm (1.55 eV) for an average particle size of 3.1 and 3.8 nm, respectively.

Similar results were obtained by Beard et al. [10], who measured the absorption cross sections and the PL spectra of ensembles of silicon NCs prepared by Mangolini et al. [6], having an average diameter of 9.5 and 3.8 nm (average distribution 15%) and dispersed in hexane or tetrachloroethylene. The results displayed in Figure 5.3 show that the Tauc's gap (the energy gap) of the NCs with a mean size of 3.8 and 9.5 nm is 1.68 eV and 1.20 eV, respectively, in good agreement with their peak PL emissions at 1.68 and 1.20 eV.

* https://orcid.org/0000-0002-0542-3219.

223

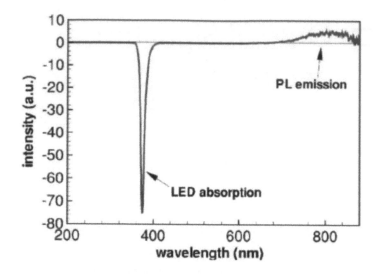

FIGURE 5.1 Room-temperature PL emission of a set of silicon nanocrystals in a colloidal solution. (*After* D. Jurbergs, E. Rogojina, L. Mangolini, and U. Kortshagen (2006) Silicon nanocrystals with ensemble quantum yields exceeding 60% *Appl. Phys. Lett.* **88**, 23311 *Reproduced with permission of AIP publishing, License number 4704870532802 License date Nov 09, 2019.*)

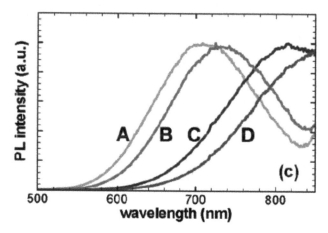

FIGURE 5.2 Room-temperature photoluminescence spectra of silicon nanoparticles dispersed in methanol under laser excitation at 356 nm. (*Reprinted with permission from Nano letters Copyright 2006 American Chemical Society from* L. Mangolini, E. Thimsen, U. Kortshagen (2005) High-yield plasma synthesis of luminescent silicon nanocrystals *Nano Letters* **5** (4), 655–659 *Nov 09, 2019.*)

PL emissions of silicon NCs with a mean size less than 10 nm peaked in this energy range (1.2–1.8 eV), above or well above the energy gap of c-Si (1.14 eV), and, following the size dependence theoretically forecasted, would be qualitatively compatible with quantum confinement conditions. However, the red-orange luminescence typical of these samples calls for oxygen-terminated surfaces, not for

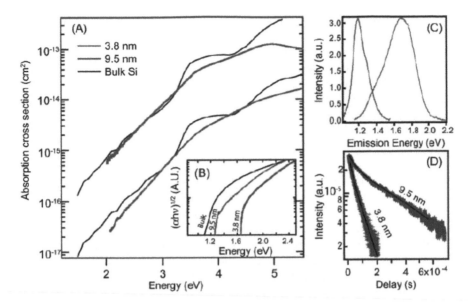

FIGURE 5.3   Optical cross section (A), Tauc's gaps (B) and PL spectra (C and D) of silicon NCs. (*Reproduced with permission of American Chemical Society July 05, 2018 after* M.C. Beard, K.P. Knutsen, P. Yu, J.M. Luther, Q. Song, W.K. Metzger, R.J. Ellingson, and A.J. Nozik (2007) Multiple exciton generation in colloidal silicon nanocrystals *Nanoletters* **7** 2506–2012.)

un-passivated surfaces, as could be deduced from the experimental evidence and by theoretical studies.

In fact, IR spectroscopy and TEM measurements on these materials [4] reveal that the exposure to air activates the presence of surface Si–O–Si bonds and a decrease of crystallinity, due to the formation of amorphous SiO_2.

Moreover, hybrid functional DFT calculations carried out by Zhu et al. [14] predict that oxygen-capped Si-NCs have an *indirect* gap transition in the yellow-red, while hydrogen-passivated NCs have an optically allowed direct gap transition and should luminesce in the blue-violet.

It is, also, experimentally [6] and theoretically known [8, 14, 15] that the visible luminescence of silicon NCs prepared by photoassisted-stain etching of porous silicon* depends on their surface passivation. H-passivated Si-NCs emit in the blue (3.09 eV or 400 nm), while oxygen-passivated NCs emit in the yellow-orange (1.55 eV or 800 nm), with a red-shift from the blue of about 1.5 eV [7].

The experimental evidence [11] of a strong impact of the embedding (ligand) medium, that shifts to the blue the luminescence bands of Si nanoparticles dispersed in colloidal solutions of various chemicals,† indicates, however, that also ligands

---

* Porous silicon prepared by anodic oxidation of c-Si in hydrofluoric solution might be used as the parent phase of Si-NCs.
† Water, methanol, glutamic acid, glycine.

might be involved in the luminescence emission wavelength and bandwidth of Si-NCs.

Therefore, not only the electronic states of the core of Si-NCs involved in the PL emission, but also surface defect states of various origin impede a secure attribution of the luminescence phenomena to QC, unless, as shown by Ledoux et al. [16, 17], the variation of the PL emission energy could be ascribed *solely* to the variation of a single parameter, the particle size.

The aim of the following of this section is to demonstrate that the presence of luminescence phenomena solely depending on the emitting particle size is a rarely fulfilled condition, and that chemical interactions of various origin do influence, in most cases, the optical emissions of Si-NCs.

This situation is well-illustrated by the results of a statistical analysis on 70 different size-selected, porous silicon particles, carried out by Sychugov et al. [12], who show that the variation of the PL emission energy between 1.6 and 2.1 eV is substantially due to surface passivation effects, arising from solvent–NCs interactions.

They show also, see Figure 5.4 (left panel), that while organic ligands broaden the room-temperature PL bandwidth peaked at 700 nm (1.77 eV) of CVD-grown porous Si and of silicon NC ensembles, SiO_2 encapsulation and oxygen-induced passivation of a single silicon NCs lead to a narrow emission linewidth. PL measurements carried out at 10 K on individual NCs capped with SiO_2 additionally show, see Figure 5.4 (right panel) the onset of a very narrow emission at 1.84 eV (673 nm), that well corresponds to the PL emission of a thermal oxide. Here the PL emission of oxygen-passivated Si-NCs seems to be dominated by the SiO_2 cap.

As a further example, Lockwood et al. [18] demonstrate that the room-temperature PL emissions of as-grown porous silicon layers containing spherical Si crystallites with a size within 5 and 3.1 nm in diameter [18], are relatively narrow bands

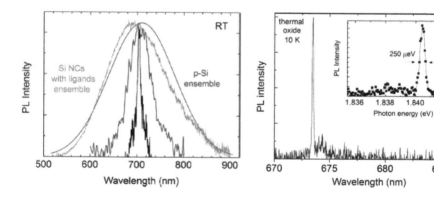

FIGURE 5.4   *Left panel:* Room-temperature PL spectra of an ensemble of size-selected porous silicon particles (blue) and of ligand-passivated silicon NCs (green), of a single silicon NC in an SiO_2 shell (red), with only a thin passivating layer (black). *Right panel:* Low-temperature PL spectra of thermal oxide and of individual NCs capped with SiO_2. (*After* I. Sychugov, A. Fucikova, F. Pevere, Z. Yang, J.G.C. Veinot, and J. Linnros (2014) Ultranarrow luminescence linewidth of silicon nanocrystals and influence of matrix I *ACS Photonics* **1** 998–1005 *ACS* Open access article.)

peaked within 1.42 and 1.54 eV. Since the Raman spectra of the same material are peaked within 515.7 and 510.9 cm^{-1}, there is the indication of rigid phonon confinement conditions [19], and one could also suggest that the PL emissions are quantum confined. The emission range, however, is again typical of an oxygen-passivated material, showing that oxidation of Si dots is a spontaneous thermodynamic process, due to the very large value of the Gibbs free energy of formation of SiO$_2$ (–754.5 KJ/ mole at 298 K).

Since toluene can break porous Si in small particles and hydrosilate their surfaces, leading to a stable protection against oxidation, nc-silicon films prepared by refluxing in toluene H-terminated porous silicon are a precious source of information on the properties of passivated silicon NCs. These films could also be used as vacuum sublimation sources, and the deposits collected on various substrates are shown to consist of clusters of nanoparticles smaller than those present in the parent phase [20].

The Raman spectra of the vacuum-deposited nanoparticles from these hydrosilated films show, as in the previous cases, a peak at 515 cm^{-1}, below the standard signature of c-Si at 520 cm^{-1}, and a PL emission peaked at 2.1 eV, for a measured particle size of 2.5 nm, suggesting that toluene does effectively passivate the Si-NCs and allow the set-up of quantum confinement conditions.

A "solvent effect" of a physical nature could be, however, also involved in the luminescence of suspensions of silicon nanoparticles in various solvents, when the surface structure between the adsorbed organic solvent and the semiconductor NCs provides efficient charge separation for neutral excited states produced by light absorption either on the organic solvent or on the nanocrystal [22]. This would allow the electrons excited from the HOMO to the LUMO states of the surface structure to be efficiently transferred to the conduction band of nc-Si, where these electrons, together with those photoexcited from the valence band of the Si-NCs, would radiatively recombine with the production of the observed PL.

A solvent effect has been supposed by Qiu et al. [21] to be involved in the photoluminescence of suspensions of silicon NCs with an average grain size of 2 nm, in water, chloroform, benzene, and toluene. In fact, the PL emission of these suspensions, excited using the 325 nm (3.814 eV) emission of a He–Ne laser, consists of a broad emission at 390 nm (3.18 eV), whose intensity depends on the nature of the solvent used, and increases from water to chloroform, benzene, and toluene, i.e. with the decrease of the HOMO-LUMO gaps from liquid water [6.83 ± 0.05 eV [25]] chloroform [(5 eV) [24]], benzene [(4.7–4.9 eV) [23], 4.37 eV [24]], to toluene [(4.7 eV) [23] 4.31 eV [24]].

Considering, however, that the PL emission of these samples was excited with a laser beam of 325 nm (3.814 eV), none of these solvents could be involved in the PL emission with their excited LUMO states. The intensity of the PL emissions should, instead, depend on the different extinction coefficients of the solvents [24], and, possibly, on specific solvent–Si-NCs chemical interactions.

From the excitation light wavelength dependence of the optical absorption coefficient α of suspensions of SiNPs in ethanol and toluene, Al Mohaimeed et al. [24] were, instead, able to measure the energy gap of Si NPs in suspension, using a continuous UV source, showing that the energy gap E_G of Si-NCs in ethanol is 3.52 eV,

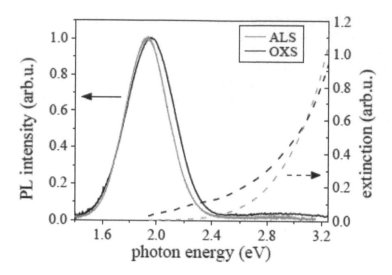

FIGURE 5.5 Photoluminescence and extinction (dashed lines) spectra of alkylated and oxygen-passivated samples of Si dots obtained from porous silicon. (*After* K. Žídek, F. Trojánek, P. Malý, L. Ondič, I. Pelant, K. Dohnalová, L. Šiller, R. Little, and B.R. Horrocks (2010) Femtosecond luminescence spectroscopy of core states in silicon *Optics Express* **18** 2524. Open Access Journal.)

but 3.61 eV in toluene. By a comparison with the experimental values of E_g (3.18 eV) of un-passivated Si-NCs of similar size, it could be concluded that a significant solvent effect could be expected when the solvent is suitably excited in its LUMO state.

The works of Zidek [13], Weeks et al. [26], and Veprek and Veprek-Heijman [27] add final evidence on surface effects on the PL emission of nc-Si samples.

Silicon dots prepared from porous silicon after submission to various chemistries were studied by Žídek et al. [13]. One sample (ALS, size ≈2–2.6 nm) prepared by alkylation of porous silicon and one oxidized sample (OXS, size ≈3 nm) prepared by room-temperature ageing in ambient conditions, after complete structural and optical characterization, were submitted to ultrafast PL measurements to infer the electron confinement properties and the effects of surface passivation on carrier trapping at interface states [13]. Both PL spectra, see Figure 5.5, are peaked at ≈1.9 eV, indicating that alkylation does not protect the sample from spontaneous oxidation.

Weeks et al. [26] demonstrated, instead, that PCVD-deposited Si nanocrystals with a mean size of 4 nm, submitted to oxidation at 100°C in water vapor (10 torr), present an intense PL emission at 1.65 eV, while the same samples submitted to oxidation in oxygen (10 torr) do not present PL activity. The author's conclusion is that the intense PL emission of samples annealed in water vapor might be attributed to the formation of surface –Si–OH (silanolic) groups, while the quenching of luminescence of samples oxidized in oxygen could be attributed to the presence of surface Si–O–Si groups.

This conclusion is well-supported by the work of Veprek and Veprek-Heijman [27] who prepared Si–SiO$_2$ nanocomposites by controlled oxidation of plasma-grown nc-Si films at 870–1200°C with pure oxygen. These nanocomposites were then submitted either to a passivation treatment with H_2 at 350–450°C or were further submitted to a long treatment in boiling water and then passivated with H_2.*

TEM measurements on these samples show that the material consists of Si nanocrystals with a size between 2 and 10 nm well-distributed within a SiO_2 matrix.[†] While the PL emission of H_2-passivated NCs consists of only one band peaked at 1.6 eV, that of the samples exposed to boiling water present also a second PL emission band peaked at 2.8 eV. Since a band peaked at 2.8 eV is exhibited also by Si and Al with surface silanolic –Si–OH and alanolic –Al–OH groups, this band is not associated with the nc-Si core, but with silanolic groups at the surfaces of NCs.

The results of Žídek [13], Weeks [26] and Veprek [27] do prove, therefore, that the surface properties of Si dots strongly influence the PL emission wavelength and intensity. In fact, a uniform capping layer of SiO_2, obtained by low-temperature oxidation with water vapor or by high-temperature oxidation followed by hydrogen passivation, enables the establishment of favorable PL emission conditions of the nanocrystalline core. The presence of a silanolic group at the surface of the nanocrystals enhances the PL intensity which is, instead, quenched in the presence of Si–O–Si– groups.

Apparently, the high-temperature oxidation treatment of nc-Si followed by passivation induces not only the partial oxidation of Si to SiO_2, which eventually fully embeds the un-oxidized silicon core, but affects also the Si–SiO_2 interface properties by decoration with silanolic –Si–OH groups.

Similar results were obtained by Valenta et al. [28], who show that the PL emission of silicon nanopillars (consisting of a silicon core and of an oxide overcoating),[‡] submitted to a forming gas (N_2+H_2) annealing to passivate the surface states, is peaked between 1.58 and 1.88 eV (660–785 nm), with the peak energy increasing with the decrease of the diameter.

According to Delerue et al. [29], nanometric Si-NCs from porous silicon have a complex structure, since they consist of three-dimensional networks of quantum segments, intermediate between dots and wires. They were able to show that LCAO calculations carried out on these objects, with their dangling bonds at the surface fully saturated with hydrogen, lead to an $E_g \approx d^{-1.39}$ law, where d is the nanoparticle diameter, for the size dependence of the optical gap energy.

They could also demonstrate that the experimental values of optical gap measured on porous silicon NCs, as well as those reported by Furukawa and Miyasato [30] relative to nanocrystalline Si:H samples, prepared at low temperatures (100 K), fit extremely well the theoretical behavior, as is shown in Figure 5.6, and could be explained as being due to quantum confinement. Apparently, the highest quantum confinement is obtained for fully hydrogen-passivated nanocrystals having a size close to 1 nm.

Dangling bonds associated with hydrogen desorption and surface oxide layers are, instead, the main cause of luminescence intensity degradation.

---

* A treatment with 4–10% H_2 in nitrogen (forming gas) is a standard process in microelectronics used to reduce defects in gate oxides.

† The authors observed that when the dot inter-distance was less than 1 nm, the PL intensity was quenched due to the overlap of the wave functions of neighbouring nanocrystals with the loss of quantum confinement.

‡ The authors do not give information about the size of the silicon core.

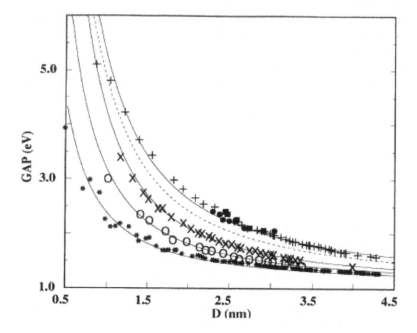

FIGURE 5.6 Calculated and experimental size dependence of optical gap energies for silicon crystallites (+) or wires. (*After* C. Delerue, G. Allan, and M. Lannoo (1993) Theoretical aspects of the luminescence of porous silicon *Phys.Rev.B* **48** 11024–11036 *Reproduced* with *permission of APS, License Number: RNP/19/NOV/020170 License date: Nov 09, 2019.*)

From Table 5.1, which displays a summary of the results discussed in this section concerning the role of size and surface chemistry in the optical emission properties of nanometric silicon crystals, one can eventually conclude that quantum confinement, with optical emissions in the blue, is only observed in the case of nanocrystals fully passivated with hydrogen. For these crystals the excited state is a volume excitation.

Oxygen-capped NCs emit in the yellow-red, and this behavior is associated, according to Zhou [14], with indirect gap properties. For silicon core-oxide shell nanocrystals the excited state is a surface excited state directly involving the capping layer, rather than a volume excited state [14]. We will discuss again this issue in the following section.

5.2 OPTOELECTRONIC PROPERTIES OF NC-Si IN Si–SiO₂ SUPERLATTICES

It was shown in the previous section that the photoluminescence emission energy of silicon NCs depends on their size. Therefore, their PL emission is potentially tunable, provided a suitably passivated material with constant size features could be synthetized.

The optimization of the surface passivation and of the size distribution is quantitatively obtained by the superlattice technology described in Section 3.2, which does

TABLE 5.1

Influence of the Size and Preparation Chemistry on the Spectral Properties of Silicon Nanocrystals (NC=Single Nanocrystals)

Sample	Diameter (nm)	T_{meas}	PL Emission (eV)	Growth Technique/Parent Phase	Post-Growth Chemistry	References
NC	3.9	RT	1.77	SiH_4 plasma		[6]
NC	9.5–3.8	RT	1.20–1.68	SiH_4 plasma		[10]
NC	nd	10 K	1.84	CVD		[12]
NC	nd	70 K	1.6–2.1	CVD		[12]
NC	2.5–8	RT	1.6–2	Laser pyrolysis		[16]
NC	2–3		1.2–2.4	$\mu cSi{:}H$	H-passivated	[30]
NC	<2	RT	3.09	Porous Si	H-passivated	[7]
NC	>4	RT	1.5	Porous Si	O-passivated	[7]
NC		RT	1.84	Porous Si	Oxide-encapsulated	[12]
NC	2.5	RT	2.1	Hydrosilated-porous Si	Vacuum sublimation	[20]
NC	2.2–2.6	RT	1.9	Porous Si	Alkylated	[13]
NC	3.5	RT	>1.9	Porous Si	Oxidized	[13]
NC	1.1–1.4		1.2–1.7	Calculated	Oxidized	[13]
NC	1.1–1.4		2.3	Calculated	H_2-passivated	[14]
NC	4	RT	1.65	SiH_4–Ar plasma	Oxidized in H_2O at 100°C	[26]
NC	2–10		1.6	SiH_4–H_2 plasma	Hydrogen-passivated	[27]
NC	2–10		1.7 and 2.8	SiH_4–H_2 plasma	Oxidized at high temperatures	[27]
Si nanopillars	3–6	RT	1.75	RIE etching followed by oxidation	Hydrogen -passivated	[28]

allow the control of the size of the Si nanocrystals by the thickness of each SiO_{2-x} layer.

The emission intensity depends, instead, on many factors, among which the size distribution [31], the decomposition temperature of the SiO_{2-x} matrix, and the initial composition of the sample [31], the inter-distance among single nanocrystals, the presence of a passivated surface [32–34], and the suitable recovery from defects which could act as recombination centers.

Among them, the P_b defects, which are deep non-radiative defects lying at the Si–SiO_2 interface [35], could severely influence the PL efficiency, but, luckily, they could be fully removed after a hydrogen treatment carried- out at relatively low temperatures (400°C), to avoid any influence on the size and structure of the nanocrystals [34]. In turn, surface passivation would inhibit also the radiative recombination of light-generated carrier at surface defects and at grain boundaries [27, 33, 34], leading to an overall cleaning of the PL spectrum.

Eventually, the nanocrystal inter-distance should be kept sufficiently large (>1 nm) by a careful choice of the growth conditions, in order to avoid the overlap of the wave functions of neighboring nanocrystals, that would induce electron tunneling.

The results obtained by Zacharias et al. [32, 33] using this technology show that the PL emission might be tuned within 950 and 750 nm (1.30–1.8 eV) by changing the dot size from 4 nm to 2 nm; see Figure 5.7. Hydrogen passivation is very effective on the photoluminescence intensity of the main band, which increases by more than a factor of 2 [35]. Apparently, the removal of the P_b defects by passivation does not change, instead, the PL emission energy.

Similar results were obtained by Del Negro et al. [31], see Figure 5.8, on samples prepared by thermal annealing (1000-1300°C) of SiO_x thin films, which present a strong PL emission between 910 and 770 nm, whose peak energy depend on the initial composition of the sample, and by Photopoulos et al. [36] on samples prepared by the direct oxidation of silicon multilayers deposited by low-pressure PECVD. The resulting structure is, here, a Si–SiO_2 superlattice, where the SiO_2 layer thickness is roughly controlled by the oxidation time. This structure exhibits an intense PL emission at 800 nm, with a blue- or red-shift depending on the size of the crystallites.

Features close to the previous systems are also exhibited by Si nanocrystals in films prepared by pulsed plasma ablation in an oxygen atmosphere by Dey and Khare [37]. Here, the nanocrystals are embedded in a matrix consisting of SiO_2 and a-Si, and the PL emission of a sample grown at 0.5 mbar of oxygen, with an NC size of 2–5 nm (see Figure 5.9), is characterized by a main bands peaked at 1.50 eV and a satellite band at 2.36 eV. Following the decrease of the nanocrystal size, the Pl intensities of the band at 1.53 eV and at 2.36 eV increase.

The presence of the band at 1.50 eV, deconvoluted here in three bands peaked at 1.38, 1.53, and 1.73 eV, is typical of Si-NCs embedded in an SiO_2 matrix and consistent with the results of Godefroo [35], though in the case of the Dey and Khare samples [37] the nanocrystal size is not taken constant by the superlattice widths, and ranges within 2 and 4 nm. We will see in the next section that the band at 2.3 eV is, instead, typical of Si-NCs embedded in an a-Si matrix.

The strong correlation observed between luminescence energy and nanocrystal size would suggest that the origin of the luminescence of silicon nanocrystals

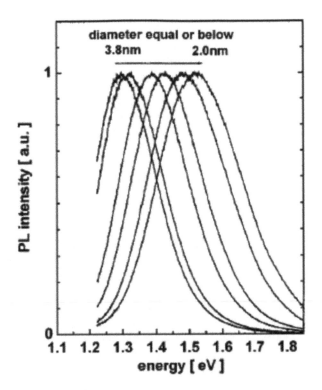

FIGURE 5.7   Size-induced blue-shift of the photoluminescence of Si-NCs prepared with the superlattice approach. (*Reproduced from* M. Zacharias, J. Heitmann, R. Scholz, U. Kahler, M. Schmidt, and J. Bläsing (2002) Size-controlled highly luminescent silicon nanocrystals: A superlattice approach *Appl. Phys. Lett.* **80** 661 *with permission of AIP publishing, Nov 11, 2019.*)

segregated from non-stoichiometric SiO_{2-x} is due to quantum confinement, although the comparison with the results of theoretical calculations (see Figure 5.6) and of experimental results of as-grown and H-passivated nanocrystals show that the emission energy is remarkably shifted toward lower energies [31].

Since the range of the optical emission of silicon nanocrystals prepared with the superlattice approach coincides with that of the oxide-capped crystals described in the former section, it is reasonable to conclude that also in this case the radiative recombination involves SiO_2–Si interfacial defect states [35] with energy levels inside the band gap [31], but it is also necessary to suppose, according to Godefroo [35], that a size-dependent increase of the band gap does occur as a consequence of quantum confinement, to account for the size dependence of the PL energies in the experimental 1.2–1.8 eV range .

It could be suggested, therefore [35], see Figure 5.10, that the luminescence emission is due to a radiative recombination of excited carriers from the energy level E_D of the SiO_2–Si interfacial defect state, lying inside the quantum-confined energy gap of Si-NCs, to the top of the valence band. It could be suggested, as well, that the

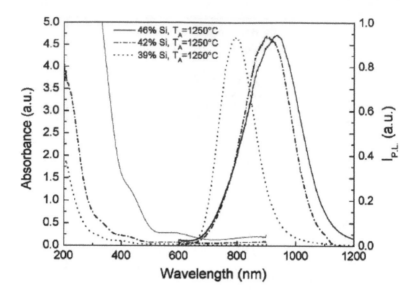

FIGURE 5.8 Absorbance and photoluminescence spectra of nc-Si films. (*Reproduced from* L. Del Negro, M. Cazzanelli, N. Daldosso, Z. Gaburro, L. Pavesi, F. Priolo, D. Pacifici, G. Franzò, F. Iacona (2003) Stimulated emission in plasma-enhanced chemical vapour deposited silicon nanocrystals Physica E **16** 297–308 *http://rss.sciencedirect.com/publication/scie nce/13869477_OA/Open-access.*)

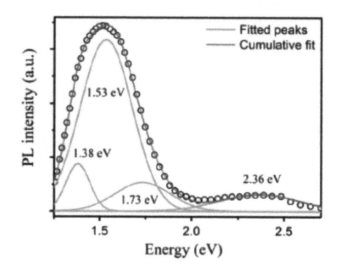

FIGURE 5.9 Photoluminescence of nc-Si/SiO_2 films prepared by laser ablation. (*After* P.P. Dey, A. Khare (2017) Fabrication of photoluminescent nc-Si:SiO_2 thin films prepared by PLD *Phys. Chem. Chem. Phys.*, **19**, 21436–21445 *Permission granted by the Royal Society of Chemistry for academic requestors.*)

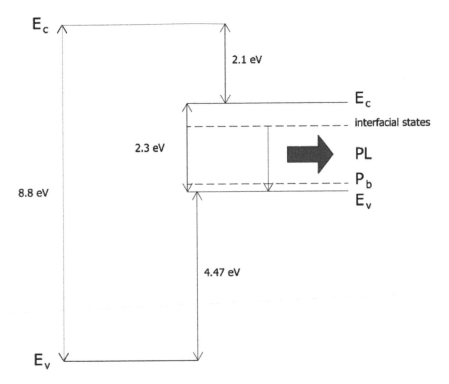

FIGURE 5.10 Schematic diagram of the energy levels involved in the photoluminescence of Si–SiO$_2$ superlattices [35].

position of the SiO$_2$–Si interfacial defect states is invariant with respect to the bottom of the conduction band, in order to follow an $E_D=f(d^{-1})$ law.

5.3 OPTOELECTRONIC PROPERTIES OF NC-SI FILMS

5.3.1 Optoelectronic Properties of Nanocrystalline Silicon Films PECVD-Deposited from SiF$_4$ and SiH$_4$ Plasmas

As already shown in Chapter 3, silicon films grown from SiF$_4$ plasmas at high temperatures are single phase* and consist of an array of Si-NCs with an XRD grain size in the range of 10–20 nm [38–40]. A typical PL spectrum of nanocrystalline Si films deposited from at 450°C from SiF$_4$ plasmas is displayed in Figure 5.11, which shows the presence of a main emission peaked at 1.62 eV, with a shoulder at 1.95 eV, far away from the emission of a-Si that is peaked at 1.2 eV, confirming the absence of a-Si in these films.

The main PL band at 1.62 eV is attributed by the authors [38–40] to the radiative recombination of carriers at deep defect levels, while the energy of the band at 1.95

---

* That is, amorphous silicon is not the embedding medium, like in the case of PECVD-grown films from SiH$_4$–H$_2$.

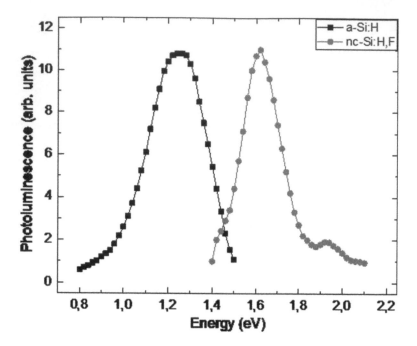

FIGURE 5.11 Room temperature PL spectra of a-Si-H (left) and of nc-Si (right) films deposited at 450°C from SiF₄ plasmas. (*Courtesy of Prof. Giovanni Bruno, CNR Bari.*)

eV corresponds to the value of the optical gap of this material, and therefore to its energy gap, compatible with the average grain size of this material. We will see in the next section that the PL spectrum of Figure 5.11 compares well with the high-energy PL spectrum and with the surface photovoltage spectroscopy (SPS) spectrum of nc-Si-H films LEPECVD-grown from silane–hydrogen mixtures.

These results are, also, comparable with those of Ali [41, 42], who reports a series of PL spectra obtained from nc-Si samples PECVD-grown at 95–250°C using SiF₄, SiH₄, and H₂ mixtures. All of its spectra present two well-pronounced PL bands peaked at 1.75 eV and ≈2.3 eV, whose intensity strongly increases with the increase of the H₂ content in the feeding gas mixture, with the decrease of the grain size, and with the decrease of the growth temperature down to 95°C, while the peak emission energies remain almost constant.* The average grain size depends on the SiH₄ and H₂ content in the mixture, and ranges from 14 to 21 nm, from XRD measurements.

If the PL spectra are, instead, measured on samples grown at increasing SiH₄ fluxes (i.e. with a decrease of the H₂ content) [42], the average grain size decreases, and the intensity of the PL band at 1.7 eV decreases and eventually disappears with the increase of the silane flux, while that of the band at ≈2.2 eV increases, with a systematic blue-shift of the maximum associated with the decrease of the XRD grain size.

---

* The grain size, measured with Raman spectroscopy, varies between 17 and 21 nm with the decrease of the growth temperature from 250 to 95°C at constant H₂ flux of 25 sccm and from 14.3 nm to 18.2 at constant H₂ flux of 46 sccm.

The hypothesis is made that the band at 1.7 eV is due to surface defect states associated with Si–H bonds, whose intensity decreases with the decrease of the available hydrogen in the plasma, while the band at 2.3 eV is the true optical gap of the material, associated to the presence of nanometric, hydrogen-passivated, silicon crystals, whose size ranges around 2–3 nm, in rather good agreement with the theoretical results of energy gap as a function of size, reported in Figure 5.6.

The optical properties of nc-Si films with a grain size ranging from 2 to 15 nm, PECVD-deposited from SiH_4, H_2, and Ar, by Edelberg et al. [43] at 230°C, are very close to those grown in an SiF_4 plasma, and consist of two broad peaks at 1.82 and 2.34 eV.

Similar conclusions hold for the nc-Si films prepared from anodically etched μc–Si, PECVD-grown from SiH_4, B_2H_6, and H_2 mixtures, by Toyama et al. [44, 45], who claim that the PL emissions of these films consist of two gaussian components at 1.75 and 2.28 eV *whose intensity depends on the crystalline volume fraction*, leading to a maximum at $\chi = 60\%$.

A summary of the optical properties of nc-Si films discussed in this section is reported in Table 5.2.

5.3.2 OPTOELECTRONIC PROPERTIES OF NANOCRYSTALLINE SILICON FILMS LEPECVD-DEPOSITED FROM SiH_4–H_2 PLASMAS

We have seen in Chapter 3 that nc-Si:H films LEPECVD-grown from silane plasmas are heterogeneous mixtures of hydrogenated amorphous and nanocrystalline silicon, where a cap of a-Si entirely embeds the Si grains.

Photoluminescence and surface photovoltage spectroscopy measurements carried out by Pizzini et al. [46, 47] on these films show that their optical properties strongly depend on growth temperature and crystallinity χ, that measures the ratio of the mass of c-Si vs. that of a-Si in the heterogeneous mixture.

The PL spectra of nc-Si samples LEPECVD-grown from SiH_4–H_2 plasmas present different features from those grown in SiF_4 plasmas, since the PL emissions, in addition to two broad bands peaked at 1.4 eV and 1.9 eV, include at least three bands in the 0.5 and 0.8 eV range, see Figure 5.12, whose emission intensities strongly depend on the growth temperatures and on the crystallinity.

Apparently, these bands are the convolution of the PL emission from Si crystallites, from the embedding a-Si:H matrix, from the interface states, and from deep defect states [47, 48] in a-Si and c-Si.

The effect of crystallinity χ on the PL spectra of LEPECVD-grown nc-Si samples is illustrated in Figure 5.13, left panel, which shows that both of the two low-crystallinity (5–16%) samples exhibit a single, broad PL band with a maximum at ≈1.2 eV, although the maximum of the PL band of the sample at 16% is at 1.15 eV, and shows also the emergence of a shoulder at 0.8 eV, in good agreement with the low-energy PL spectrum of Figure 5.12.

The PL spectra of Figure 5.13 left panel show a close resemblance with the spectra of Bagolini et al. [48], see Figure 5.13 right panel, measured on NC-Si films grown like the former ones, on which the average grain size was found to lie within

TABLE 5.2
Photoluminescence Properties of Nanocrystalline Silicon Materials (*) from Optical Gap Measurements (**) Porosity

Sample	size (nm)	χ%	T dep °C	Process chemistry	PL emission bands (eV)	References
a-Si:H			160	PECVD	1.85 (*)	[50]
a-Si:H			>200	RF plasma (SiH₄)	0.8, 1.05, 1.2	[50]
a-Si:H				Glow discharge	1.3	[51]
a-Si:H			300–400	PECVD and CCP	1.34	[54]
a-Si:H			200–350	LEPECVD SiH₄/H₂	1.9 (*)	[55]
a-Si:H			160	PECVD	1.8 (*)	[57]
nc-Si	20–2		400	PLD inO₂	1.53, 2.36	[37]
nc-Si			450	PECVD SiF₄/SiH₄/H₂	1.4, 1.62, 1.95, 2.1 (*)	[38–40]
nc-Si	14–21		95–250	PECVD SiF₄/SiH₄/H₂	1.7, 2.3	[41, 42]
nc-Si	2–15		230	PECVD SiH₄/H₂	1.82, 2.34	[43]
nc-Si	2–3	60 (**)	180–500	RF plasma + annealing + anodization	1.7	[44, 45]
nc-Si		72	210–280	LEPECVD SiH₄/H₂	0.8, 1, 1.2, 1.9	[46, 47]
		76			1.4	
nc-Si	20×6	50–60	180–300	VHF-CVD	0.8, 1, 1.3	[49]
nc-Si		87	280	LEPECVD SiH₄/H₂	1.14, 1.18, 1.28, 1.36, 1.52 (*)	[55]
nc-Si	3.6–4.9		250	PECVD SiH₄/H₂	1.77–1.9	[61]

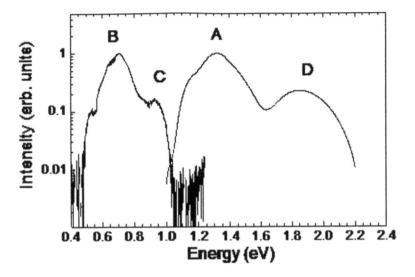

FIGURE 5.12   Typical PL spectrum of a nc-Si film PECVD grown from a SiH_4–H_2 plasma. (*Final Technical Report Nanophoto Project 013944, July 2008.*)

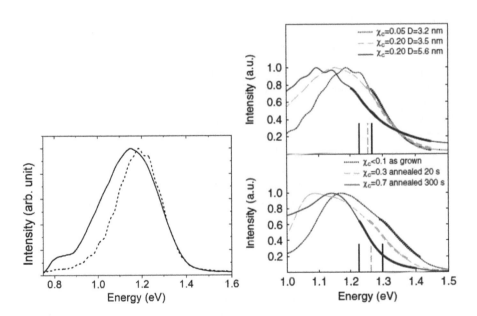

FIGURE 5.13   *Left panel:* PL spectra at 14 K (λ_{exc} 405 nm) of two LEPECVD-grown samples. Dashed line $\chi=5\%$; continuous line $\chi=16\%$. (*After* Pizzini et al. *Final Technical Report Nanophoto Project 013944, July 2008.*) *Right panel:* Effect of the grain size and crystallinity on the peak energies. (*Reproduced with permission of American Physical Society, License Number: RNP/19/NOV/020198 License date: 10-Nov-2019. After* L. Bagolini, A. Mattoni, G. Fugallo, L. Colombo, E. Poliani, S. Sanguinetti, E. Grilli (2010) Quantum confinement by an order-disorder boundary in nanocrystalline silicon *Phys. Rev. Lett.* **104** 176803.)

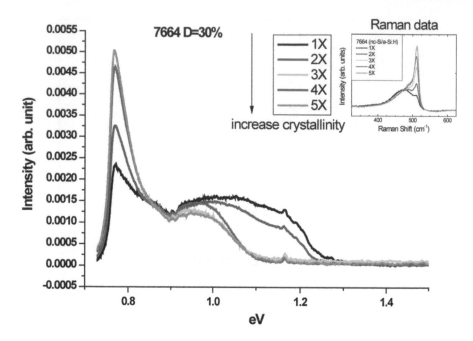

FIGURE 5.14 PL spectra at 15 K of five nc-Si samples presenting increasing crystallinity [1x χ% = 8.7, 2x χ% = 22.9, 3x χ% = 37.9, 4x χ% = 44.8, 5x χ% = 48.3]. (*After* Pizzini et al. *Final Technical Report Nanophoto Project 013944, July 2008.*)

3.2 and 5.6 nm.* One can observe that with the decrease of the average grain size a sizable blue-shift could be observed, while the effect of crystallinity is marginal.

These spectra are also close to those measured by Fields et al. [49] on intentionally oxygen- contaminated NC Si:H films, which show the presence of a main band at 0.95 eV with a shoulder at 0.8 eV and of another weak band at 1.2 eV, whose intensities depend on the crystallinity. In fact, with the increase of the crystallinity[†] the band peaked at 1.2 eV disappears, while those at 0.8 and 1 eV become prominent. In agreement with earlier literature results, Fields et al. [48] attribute the band at 0.95 eV to band tail–band tail transitions at the nc-Si grain boundaries and the band at 1.2 eV to band-tail band-tail transitions in the amorphous phase.

It should be noted that, according to Engemann and Fischer [50], Street [51], Schubert et al. [52], Oheda [53], and Nguyen et al. [54], the bands at 0.8 and 0.95–1.0 eV are defect bands of a-Si (see Table 5.2), of them that peaked at $\approx$0.8 eV is supposedly due to the oxygen contamination of the material.

The role of crystallinity is more evident in Figure 5.14 which displays the PL spectra of five nc-Si samples presenting increasing crystallinity from 8.7 to 48%. It is possible to observe that the intensity of the band at 0.8 eV increases with increase of the crystallinity, together with the decrease of the intensity of a broad band peaked

---

* The discussion of these spectra was carried out assuming localization of only the valence band states of Si-NCs embedded in the amorphous Si matrix: see ref. [47] for details.
[†] Only relative data are given in the paper.

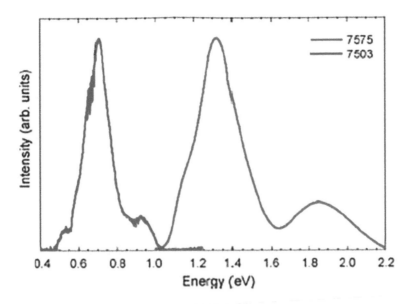

FIGURE 5.15   PL spectra (T = 12K) of a nc-Si sample (7503) grown at 210°C and of a nc-Si sample (7575) grown at 280°C. (*After* Pizzini et al. *Final Technical Report Nanophoto Project 013944, July 2008.*)

at 1.1 eV and with the increase of the intensity of a band peaked at 0.95 eV. From the Raman data in the inset of Figure 5.14, it is, also, possible to observe the systematic presence of an almost constant amorphous phase contribution in the whole set of samples.

The effect of the growth temperature is, also, well-apparent in Figure 5.15, which displays the PL spectra of a sample grown at 210°C (sample 7503, $\chi = 76\%$, $\delta = 22$ nm) and of a sample grown at 280°C (sample 7575, $\chi = 72\%$, $\delta = 17$ nm)

From this figure it is possible to remark that the sample prepared at low temperatures presents only the PL emissions at ≈ 0.8 eV and at ≈ 0.95 eV, already observed in the samples of Figures 5.12 and 5.13, while the sample prepared at 280°C presents the emission at ≈ 1.35 eV and at ≈ 1.9 eV,* while the band at 1.2 eV, typical of low-crystallinity materials, see Figure 5.12, is here only present as a shoulder of the main band.

Since the average crystallinity of both samples is very close, it has been suggested [46] that the low-temperature sample consists of a distribution of optically inactive c-Si nuclei in an a-Si matrix, while the onset of the PL emissions at 1.35 and 1.9 eV in the sample grown at 280°C calls for a kind of "surface passivation" and optical activation of Si nanocrystallites, now well-embedded in the amorphous silicon matrix.

Optical absorption measurements and surface photovoltage spectroscopy (SPS) carried out by Cavallini et al. [55] on similar samples add further information on the optical properties of mixed nc-Si/a-Si phases in nc-Si films.

---

* The energy of this emission ranges from 1.9 to 2eV, depending on morphological factors which are not always well-controllable.

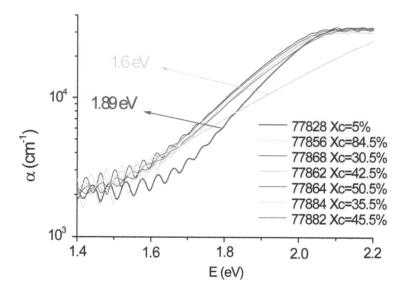

FIGURE 5.16 Experimental dependence of the optical absorption coefficients α of nc-Si films as a function of the crystallinity (in red the α values of the almost fully amorphous sample and its Tauc's gap value, in grey the α values of a highly crystalline (χ=84%) sample and its Tauc's gap value). (*Final Technical Report Nanophoto Project 013944, July 2008.*)

In fact, the optical absorption coefficients (α) spectra of a series of low-, medium-, and high-crystallinity nc-Si films, displayed in Figure 5.16, do allow the calculation of the values of the optical gap for a quasi-amorphous material (χ=5%) and for a high-crystallinity (χ=84%) film, which hold 1.89 eV and 1.6 eV, respectively.

The (Tauc's) optical gap of the entire series of film, calculated using the following equation

$$(\alpha h \nu)^{1/2} \sim \left(h\nu - E_{\text{Tauc}} \right) \tag{5.1}$$

might be, instead, deduced from the SPS spectra displayed in Figure 5.17 left panel, whose values are plotted in Figure 5.17, right panel, as a function of the crystallinity.

The experimental values of the optical gap as a function of the crystallinity are well-fitted by the solid curve drawn in Figure 5.17, calculated from the weighted average of the energy gap $E_{G(a)}$=1.87 eV of the amorphous and $E_{G(c)}$=1.12 eV of the crystalline phase, with a weighting factor x_c for the crystalline phase and $x_a = (1 - x_c)$ for the amorphous phase

$$E_G = \frac{E_{G(a)} x_a + E_{G(ac)} x_c}{x_a + x_c} \tag{5.2}$$

It is apparent that the optical gap increases almost linearly with the decrease of the crystallinity, leading to a value of 1.9 eV for the quasi-amorphous sample, which

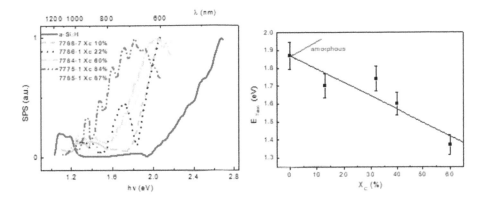

FIGURE 5.17 *Left panel:* SPS spectra of a series of nc-Si films at decreasing crystallinity. *Right panel:* Calculated dependence of the optical gap on the crystallinity. (*Final Technical Report Nanophoto Project 013944, July 2008.*)

corresponds to the optical gap of a-Si deduced from its SPS spectrum displayed in Figure 5.17, left panel.

This value of the optical gap of a-Si fits well also with the SPV values of 1.66–1.7 eV measured by Fefer et al. [56], and with a value of 1.8 eV measured by Abdulraheem et al. [57] on ultra-thin a-Si:H films.

Apparently, nc-Si behaves as a pseudo-single phase, optically hybrid material [58], where a-Si and c-Si work as the (two) pseudo-components of the material, not as separate phases.

This conclusion provides, also, a reasonable hypothesis about the nature of the PL emissions typical of the majority of PECVD-grown nc-Si films reported in Table 5.2, which consist, in fact, of a single PL band at energies ranging between 1.4 and 1.9 eV.

A detailed analysis of the SPS spectra of three LEPECVD-grown low-crystallinity samples and of a fully amorphous sample, see Figure 5.18 left panel, and of the SPS spectrum of a high-crystallinity sample, see Figure 5.18 right panel [55], leads to a further improvement of our knowledge about the optical properties of nc-Si.

The analysis of SPS spectra could be carried out following the suggestions of Kronik and Shapira [59], who show that the band gap energy could be evaluated by a knee or a peak energy in the SPS spectrum, while defect levels come as slope changes in the below-gap range of the SPS spectrum.*

One can observe that the SPS spectra of the low-crystallinity ($9\% \geq \chi \leq 55\%$) films present only a peak at ≈ 1.2 eV, and a knee at ≈ 1.4–1.75 eV, whose energies decrease with the increase of the crystallinity. The peak at ≈ 1.2 eV corresponds to the band at 1.2 eV in Figure 5.12, which we assign according to Fields et al. [48], to band tail to band tail transitions of a-Si, while the knees at ≈ 1.4–1.75 eV correspond to the Tauc's (optical) gaps of nc-Si films with these crystallinities, obtained from the optical absorption spectra reported in Figure 5.16.

---

* Other details can be found in the Kronik and Shapira paper [59], to which the interested reader is addressed.

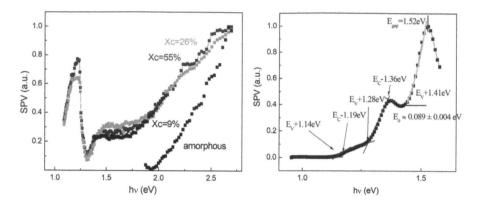

FIGURE 5.18 *Left panel:* SPS spectra of three low-crystallinity samples and of an amorphous silicon sample. *Right panel:* SPS spectrum of a high-crystallinity ($\chi=75\%$) sample. (*Final Technical Report Nanophoto Project 013944, July 2008.*)

The SPS spectrum of a high-crystallinity material displayed in Figure 5.18, right panel, is rather different from those of low-crystallinity samples. In fact, not only is the peak at 1.2 eV absent, but the spectrum is well-structured, with a sub-band gap defect structure consisting of three transitions at E_v+ 1.14 eV, Ec − 1.19 eV, and E_v+ 1.28 eV, of a peak at 1.36 eV, and of a second peak at 1.52 eV. According to Cavallini et al. [55] the transitions at E_v+ 1.14 eV, Ec − 1.19 eV, and E_v+ 1.28 eV do correspond to electron or hole transitions from or to defect levels, the first of which corresponding to the ionization of a c-Si level, with c-Si present in the matrix as a solute, not as a separate phase, and the other two to transitions from or to defect levels of a-Si, as could be deduced from the a-Si data reported in Table 5.2 and from Figure 5.18. Eventually, the peak at 1.36 eV fits closely with the value of the optical gap of an nc-Si:H film of a crystallinity larger than 60% deduced from Figure 5.17, right panel, while the peak at 1.52 eV, like the PL band at 1.9 eV, is presumably associated with surface defect states, common to most of the materials prepared with PECVD processes.

From a comparison of the experimental values of the PL emissions of PECVD and LEPECVD nc-Si films reported in Table 5.2, which displays also the data of a-Si, one could immediately note the large dispersion of the energies of the PL bands of films with very close crystallinity values. In fact, as is shown in the Annex I at the end of the chapter, the XRD or Raman crystallinities are not the sole parameters, at constant growth temperatures, that determine the optical properties of nc-Si, but the size dispersion also plays a critical role.

A nice example of the influence of the size dispersion on both the mean PL emission energy and intensity is given by Chen et al. [61], who demonstrate that n-doped nc-Si films, with a grain diameter L in the range of 3.8–4.9 nm, present a single PL band whose energy increases from 1.77 eV to 1.9 eV with the decrease of the size dispersion σ/L, not with the absolute decrease of the size L.

The conclusions that could be deduced from the results of PL, optical absorption, and SPS measurements on PECVD- and LEPECVD-grown nc-Si films could be summarized as follows:

- For nc-Si film samples, which structurally consist of a heterogenous mixture of crystalline silicon and amorphous silicon, with a-Si which fully embeds Si nanocrystals, we would expect to observe the gap signatures of both materials. Gap energies would differ from those of the bulk phases in the presence of surface strain and quantum confinement
- Most of the PECVD-grown nc-Si films present, instead, only a single PL emission, at energies ranging between 1.4 and 1.8 eV, although a few of them present a second emission at 2–2.3 eV, that should be associated with surface states.
- We observe, also, that the optical gap of LEPECVD-grown nc-Si films depends on the average content of c-Si and a-Si, showing that nc-Si behaves as a pseudo-single phase, hybrid material.
- nc-Si films which have been grown in the full absence of a-Si contamination present, instead, only the emission associated with c-Si, with QC effects when the grain size is appropriate [39–42, 60].

5.4   OPTOELECTRONIC PROPERTIES OF SILICON NANOWIRES

The main interest in silicon nanowires comes from their compatibility with the current microelectronic processes and from the possibility that their optoelectronic properties could drift toward direct gap conditions, provided the NW diameter is at or below the length scale of quantum confinement [1]. There are however some critical factors which should be overcome to develop the full potential of Si NWs. As an example, their surfaces are chemically reactive, leading to the presence of interface defects which could degrade their optical properties. Furthermore, different surface terminations could influence the diameter dependence of the energy gap of the nanowire, as is shown in Figure 5.19 [1], which displays the results of DFT calculation for –H, –OH, and –NH terminated surfaces. It is apparent that only passivated, H-terminated surfaces do allow the establishment of quantum confinement conditions.

It will be seen that the experimental evidence of similar effects has been given by Legesse et al. [62] in the case of Ge nanowires with different surface terminating chemical groups.

As a further example, only a few of the growth processes discussed in Chapter 3 allow a satisfactory control of the diameter of the Si NWs, which, in fact, requires the use of very sophisticated wire nucleation processes, which were solved only very recently by Irrera et al. [63] for MACE-grown NWs.

Eventually, a phase transition to the hexagonal silicon structure, induced by surface strain, and impurity contamination, induced by high-temperature diffusion of the nucleation metals, lead to unwanted electronic and optical properties [64]. For

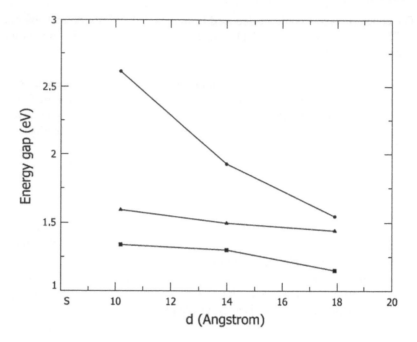

FIGURE 5.19 DFT-calculated effects of surface termination of (100) silicon nanowires on their energy gaps as the function of wire diameter ● H-terminated △ NH₃-terminated ■ OH-terminated. (*After* M. Hasan, Md F. Huq, Z Hasan-Mahmood (2013) A review on electronic and optical properties of silicon nanowire and its different growth techniques *SpringerPlus* **2** 151 Open Access Journal.)

this reason, the discussion of the optical properties of Si NWs will be carried out for individual growth processes.

5.4.1 OPTICAL PROPERTIES OF SI NWS DEPOSITED BY METAL-ASSISTED CVD

Despite the fact that it is well-known (as shown in Chapter 3) that wurtzite-type silicon is thermodynamically stable at the nanosize, as theoretically demonstrated by Zhang et al. [65] also for the case of Si nanowires, for most of the work carried out before 2014 on the optical properties of metal-assisted CVD-grown Si NWs, the crystalline structure of the NWs grown with metal-assisted CVD processes was not taken into consideration.

As a first example, the optical properties of 80 μm long and 200 nm thick, straight silicon nanowires, CVD-grown at 800°C using copper as the catalyst were investigated by Demichel et al. [66] using an Nd-YAG laser (355 nm) as the excitation source. The PL spectrum of a mat (500 μm in diameter) of as-grown NWs displayed in Figure 5.20, left panel, is the tail of a broad band, which turns out to a relatively narrow band peaked at 1.04 eV after a thermal oxidation carried out at 960°C, lying below the PL energy of crystalline silicon.

Although the authors do not explain the nature of the broad band emission from the as-grown NWs, a comparison with the results of Fabbri et al. [64] displayed

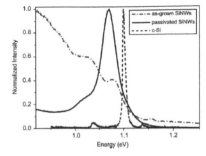
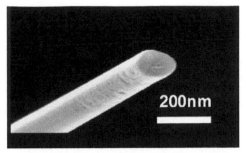

FIGURE 5.20 *Left panel:* Comparison of the PL spectra of as grown and passivated Si NWs with that of crystalline silicon. *Right panel:* TEM image of an oxidized Si NW. (*After* O. Demichel, F. Oehler, V. Calvo, P. Noé, N. Pauc, P. Gentile, P. Ferret, T. Baron, N. Magnea (2009) Photoluminescence of silicon nanowires obtained by epitaxial chemical vapor deposition *Physica E: Low-dimensional Systems and Nanostructures* **41**, 963–965 Open Access Article.)

in Figures 5.21 and 5.22, which we will discuss below, shows that it probably corresponds to the tail of a band peaked at 0.9 eV, typical of wurtzite silicon. The origin of the PL band at 1.08 eV from the oxygen-passivated silicon NWs is, instead, attributed to the radiative recombination of an e-h plasma localized at the Si–SiO_2 interface, a phenomenon already observed in 100 nm thick SOI thin films by Tajima and Ibuka [67].

It could be suggested, also, that the different behavior of oxygen-passivated silicon NWs and oxygen-capped Si-NCs, where the latter exhibit a PL band at 1.8 eV (see Table 5.1), is due to the oxidation temperature of Si NWS (960°C), high enough to create a several nm-thick oxide layer on the wire, displayed in Figure 5.20, right panel.

The absorption- and photoluminescence-spectra of SiO_2-coated straight Si NWs with a 38 nm core of unknown crystalline structure, CVD-grown with Au as the catalyst and SiH_4 as the precursor, were, instead, studied by King et al. [68]. The absorption spectrum exhibits a maximum of absorbance at 3.4 eV, in correspondence with the direct transition of c-Si, and a Tauc's gap at ≈2 eV. The PL of these samples is only observed under excitation at 200–230 nm, and the Pl spectrum consists of a broad band extending from 300 to 680 nm, with several shoulders and large peaks between 450 and 650 nm (2.7–1.8 eV), most of which are above the experimental Tauc's gap value. Like in the case of oxygen-encapsulated Si-NCs, an emission at 1.8 eV is present, while the main emissions lie at higher energies.

The effect of Au contamination on B- and P-doped silicon nanowires grown at 600°C from mixtures of silane and diborane using Au as the catalyst was investigated by K. Sato et al. [69]. B-doped and P-doped NWs displayed cylindrical structures with an average diameter of 400 nm and 40 nm, respectively, significant of a relevant role of the dopant in the morphology of the wires, but large enough to exclude quantum confinement effects. DLTS measurements on B- doped NWs show the presence

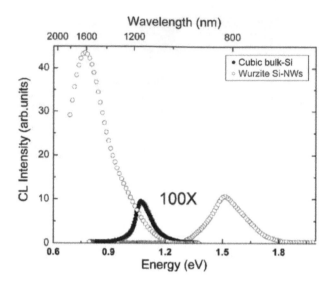

FIGURE 5.21 Comparison of the cathodoluminescence spectrum of cubic bulk silicon (in black) with that of B-doped wurtzite Si NWs. (*After* F. Fabbri, E. Rotunno, L. Lazzarini, N. Fukata, G. Salviati (2014) Boron-doped wurtzite silicon nanowires *Scientific Reports* 4:3603 1–7 Open Access Journal.)

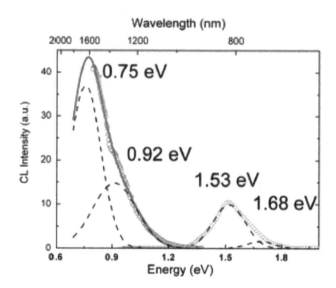

FIGURE 5.22 Cathodoluminescence spectra of B-doped wurtzite Si NWs. (*After* F. Fabbri, E. Rotunno, L. Lazzarini, N. Fukata, G. Salviati (2014) Boron-doped wurtzite silicon nanowires *Scientific Reports* 4:3603 1–7 Open Access Journal.)

of three trap levels at 0.055 eV, 0.1 eV, and 0.36 eV, of which the first corresponds to the shallow level of B in substitutional position, and the third corresponds to that of an active Au–H complex. Photoinduced current transient spectroscopy (PICTS) was, instead, used in the case of P-doped NWs, which allowed the detection of a trap level at 0.38 eV, that present the same features of the level at 0.36 eV in B-doped NWs. These results confirm the hypothesis that Au contaminates the silicon nanowires, with a negative impact on their use in photonic devices.

Fabbri et al. [64] were, apparently, the first to explicitly demonstrate that Si NWs grown from a gas phase of silane and diborane are a mixture of cubic, wurtzite (hexagonal), and defective (flawed) nanowires. These last have a cubic core embedded in a thick, amorphous Si shell, with a large amount of stacking faults at the interface.

They, also, demonstrated that with the increase of the diborane content in the gas phase, the amount of hexagonal silicon NWs decreases, while it takes a maximum in undoped samples, as is shown in Figure 5.23.

The same authors [64, 70, 71] investigated the optical properties of Si NWs grown with the wurtzite structure using cathodoluminescence (CL) and surface photovoltage measurements.

Some results of this study are displayed in Figure 5.21, which displays the CL spectrum of bulk c-Si and the CL spectrum of wurtzite Si NWs, in Figure 5.22, which displays the CL spectra of a wurtzite Si NW deconvolved by gaussian peaks, and in Figure 5.24, which displays the CL spectra of doped and undoped Si NWs.

One could observe in Figure 5.21 that the CL emission consists of two main bands peaked at 0.77 and 1.53 eV, and in Figure 5.22 that the band at 0.77 eV could be deconvolved in two bands peaked at 0.75 and 0.92 eV, and the band at 1.53 eV has a shoulder at 1.68 eV.

The emissions at 0.92 eV and 1.53 eV compare reasonably well with results of DFT calculations on the band structure of wurtzite Si carried out by Persson and Janzen [72], who assign an indirect band gap of 0.99 eV to hexagonal-Si and a direct

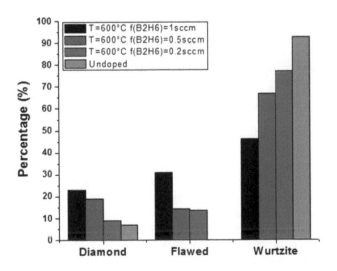

FIGURE 5.23 Effect of the diborane flux on the structure of Si NWs. (*After* F. Fabbri, E. Rotunno, L. Lazzarini, N. Fukata, G. Salviati (2014) Boron-doped wurtzite silicon nanowires *Scientific Reports* **4**:3603 1–7 Open Access Journal.)

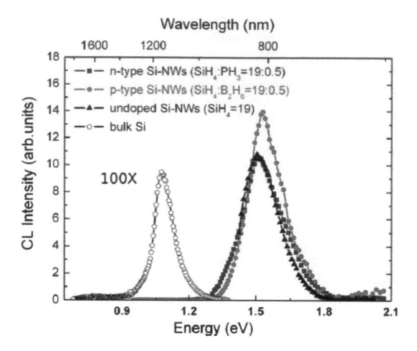

FIGURE 5.24 Cathodoluminescence spectra of doped and undoped Si NWs. The intensity of the CL spectrum of bulk silicon is 100× magnified with respect to those of Si NWs. (*Reproduced with permission from* F. Fabbri, E. Rotunno, L. Lazzarini, D. Cavalcoli, A. Castaldini, N. Fukata, K. Sato, G. Salviati, A. Cavallini (2013) Preparing the way for doping wurtzite silicon nanowires while retaining the phase Nano Letters **13** (12) 5900–5906 *Copyright 2013 American Chemical Society, Dec 1, 2019.*)

transition at the Γ point at 1.4 eV. Tentatively, the band at 0.75 eV is, instead, assigned to the radiative recombination at a boron shallow state, in view of the dependence of the peak intensity on the diborane flux.

5.4.2 Optical Properties of MACE-Grown Silicon Nanowires

The MACE-growth of silicon nanowires has been the subject of a large number of studies until recently, but the technique developed by Irrera et al. [63, 73], using the maskless etching technique discussed in detail in Chapter 3, represents, probably, the best tool available today, which grants, as well, the most representative results.

It enables, in fact, the growth of truly nanometric, monocrystalline, and defect-free nanowires, with diameters ranging within 3 nm and 9 nm. TEM measurements on these wires do evidence their core-shell structure, with a crystalline core and an amorphous shell, consisting of a 2 nm thick layer of SiO_2, due to the reaction of surface silicon with atmospheric oxygen.

The Raman spectra of these wires are red-shifted from the bulk c-Si position, and the red-shift increases with the decrease of the wire diameter. Some typical room-temperature PL spectra of 2.5 um long wires are displayed in Figure 5.25,

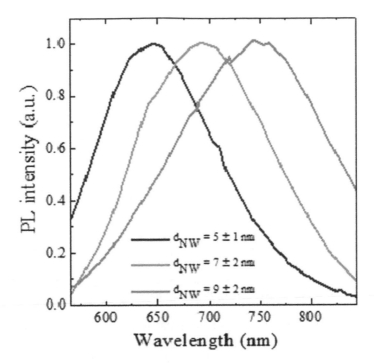

FIGURE 5.25 Photoluminescence spectra of maskless MACE-grown silicon nanowires excited with the 488 nm line of an Ar laser. *(Courtesy of Alessia Irrera, CNR - Istituto per i Processi Chimico-Fisici Viale F. Stagno D'Alcontres, 37 I-98158, Messina, Italy.)*

which show a PL maximum at 640 nm (1.93 eV) for a wire diameter of 5 nm, a PL maximum at 690 nm (1.77 eV) for a wire diameter of 7 nm, and a PL maximum at 750 nm (1.65 eV) for a wire diameter of 9 nm.

Incidentally, the PL intensity of the signal at 690 nm increases by a factor of four with the increase of the wire length from 1.3 to 7.8 μm, due to the increase of the volume in which e-h recombination events take place.

It is, also, apparent that the confined band gaps deduced from the PL energy maxima for the 5–9 nm diameter wires are very close to those predicted for quantum confinement in Figure 5.6 [29], thus showing that here a SiO_2-shell does not rule the recombination process.

MACE grown nanowires, using Ag as the catalyst, could also present efficient up-conversion features, as demonstrated by Li et al. [74], measuring the PL spectra of Si NWs with a diameter in the range of 30–10 nm, above the range of quantum confinement conditions, and 6 μm in length.

These wires present a Raman peak at 519.9 cm⁻¹, that fits well with the experimental and theoretical value of pure crystalline silicon, while the PL emission, excited with a 980 nm (1.265 eV) continuous wave (CW) laser, of a suspension of wires scraped from the substrate and dispersed in ethyl alcohol, consists (see Figure 5.26, left panel) of three phonon-assisted, up-converted emissions, of which the band main band (P1) is peaked at 733 nm (1.692 eV), and the two additional bands P2 and P3

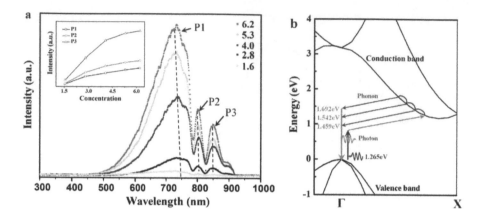

FIGURE 5.26 *Left panel:* PL bands of MACE-grown Si NWs. *Right panel:* Schematic electronic diagram illustrating the origin of the photoluminescence and the role of phonon assistance. (*After* Z. Li, X. Wang, S. Liu, J. Yang, K. Shi, H. Wang, D. Zhu, X. Xing (2018) Broadband photoluminescence of silicon nanowires excited by near-infrared continuous wave lasers *Optics & Laser Technology* **99** 81–85 Open Access Journal.)

are peaked at 804 (1.542 eV) and 850 nm (1.459 eV), whose intensities increase with the increase of the density* of wires in the suspension.

It is already known [75] that anti-Stokes photoluminescence in semiconductors requires defect states within the band gap and excitation by an Auger process or by two-photon absorption (for direct gap semiconductors). Alternatively, for both direct- and indirect-gap semiconductors the simultaneous absorption of photons and phonons can lead as well to up-conversion.

According to Li et al. [74] (see Figure 5.26 right panel) the up-conversion process does involve one-photon excitation near the Γ point towards a defect level in the band gap lying at an energy level of 1.265 eV and the absorption of three phonons, without mentioning, however, the nature of this defect level.

As expected, and in agreement with the arguments presented in the introduction of the section, the optical properties of MACE-grown Si NWs strongly depend either on the process conditions adopted or on post-growth treatments.

As a first example, Saxena et al. [76] studied the optical properties of Si NWs fabricated on a single crystal Si substrate by metal-induced (Ag) chemical etching (MIE), after a 5-hr-long oxidation step at room temperature in air. They found that the optical gap, measured by diffuse reflectance analysis [77], is substantially independent (2.35–2.45 eV) of the etching time, while the intensity of the photoluminescence spectra, which consist of a single broad PL band at 1.91 eV, ≃0.5 eV below the optical gap, decreases with the etching time.

While the author's suggestion that the effect of the etching-time on the decrease of PL intensity relies on a decrease of the NW diameter, and on an increase of quantum confinement, is not tenable, their suggestion that the PL emission energy accounts for the presence of a sub-band level associated with Si–O covalent bonds, formed

---

* Called "concentration" in the paper and in the figure.

during the post-growth oxidation, is, instead, quite appropriate. This case, in fact, is exactly the homologue of that concerning oxidized Si-NCs, which systematically presents a PL band at $\simeq 1.8$ eV, associated with the presence of a sub-band gap level due to Si–SiO$_2$ interfacial defects, see Figure 5.10, and an optical gap at $\simeq 2.3$ eV.

The effect of an SiO$_2$ shell over a crystalline core has been studied by Oda et al. [78], who analyzed the structural and optical properties of mesoscopic silicon nanowires synthesized by metal-assisted chemical etching, using Ag nanoparticles with a size ranging within 50 and 200 nm as the catalysts. Unlike previous works, the final step is a HNO$_3$-etch to remove the Ag nanoparticles, that leads to the surface oxidation of the NWs. Due to the microstructure of the catalyst, the diameter of the wires, that present core-shell structures, with a shell consisting of SiO$_2$, ranges between 200 nm and 1 μm. The photoluminescence of mats of these wires consists of a broad band which could be deconvolved in two gaussian components peaked at 1.95 eV and at 1.76 eV. The study of the PL of single nanowires shows, however, that the PL emission, a broad band with the same components, occurs only at the very top of the nanowires, which present a porous structure consisting of an array of silicon nanocrystals. We could, therefore, understand why the PL emission of these mesoscopic wires mimics that of oxidized Si-NCs, discussed in Section 5.1, and assign the emission at 1.76 eV to defect states located at the core-shell interface of the Si-NCs and that at 1.96 eV to a core excitonic transition.*

5.4.3 Optical Properties of Si NWs Grown with Other Methods

The optical properties of silicon NWs prepared by the laser ablation of a powder consisting of a mixture of Si (99%) and Au (1%) have been studied by Barsotti et al. [79] and by Bhattacharya et al. [80].

The material thus prepared consists of a random distribution of wires, grown with an Au-assisted process, some of which are nanochains of silicon nanowires connected by many nanoparticles. The nanowires present a core-shell structure with a silicon core with an average diameter of 6.7 ± 2.9 nm and a SiO$_x$ shell of a few nm in thickness. The average length of the wires is around 10 μm.

The PL spectrum of selected wires consists of a single band peaked at 1.5 eV for samples with an average diameter of 3.65 nm ($\sigma = 0.11$ nm) and at 1.4 eV for samples with an average diameter of 5 nm ($\sigma = 0.15$ nm). The corresponding Tauc's optical gap obtained from optical absorption measurements is 1.63 eV.

These results are compatible with those of Fabbri et al. [64] relative to Au-assisted grown Si NWs (see Figures 5.21 and 5.22) who found two PL maxima at 1.56 and 1.68 eV, arising from excitation at defects states.

5.4.4 Effect of Morphology and Structural Defects on
the Optical Properties of Silicon Nanowires

The effect of the surface morphology of mesoporous silicon NWs, with diameters ranging between 50 and 350 nm, on their optical properties was studied by Gosh

---

* This is our conclusion, not that of the authors.

et al. [81]. The surface of these wires is decorated by silicon nanocrystals with a nanometric size ($\simeq 6$ nm), whose size actually depends on the etching conditions, which are at the origin* of the visible photoluminescence of the wires, consisting of a single broad band peaked within 1.6–1.8 eV, with a long tail and a second band at 2.3–2.5 eV, in reasonable agreement with the hypothesis that the oxidized silicon nanocrystals are responsible for the PL emission.

Different structural, morphological, and optical behavior is, instead, presented by samples of Si NWs prepared with the laser ablation process, using as catalysts a mixture of 97% of Si containing 3% of Fe impurities and Ni and Co nano-powders, by Yu et al. [83, 84]. The hot-pressed powder was used as the target that was ablated at 1200°C. The ablated species were recovered from the inner surface of the quartz tube of the furnace in which the process is carried out, and consist of a sponge-like mesh of pure Si nanowires, as results from EDAX measurements. The RX structure of the wires corresponds to that of cubic silicon, with a slight lattice expansion indicating a distortion of the structure, is demonstrated to be single crystalline from HRTEM measurements on individual nanowires.

Samples prepared with the same method were studied by Zhang et al. [85] who could, instead, demonstrate by careful TEM measurements that the crystalline structure of the wires is, actually, broken in several nanocrystals with different shapes and orientations.

This kind of nanostructure is compatible with the PL features of these NWs, consisting of a single broad band peak at 2.5 eV, and with conditions of effective quantum confinement.

The effect of twin boundaries and stacking faults on the PL emission of Si NWs has been studied by Li et al. [86] on Si NWs CVD-grown at 550°C in a tube furnace using Au as the catalyst and SiH_4 as the silicon source, working at a pressure of 100 m torr. Unlike typical CVD-grown NWs, with which we dealt in Chapter 3, Section 3.4.1, these wires, see Figure 5.27, present a chain-like morphology, consisting of wire segments separated by twin boundaries, possibly induced by the relatively high growth pressure.

TEM measurements on cross sections of the wires put in evidence regions with high densities of stacking faults where the structure is wurtzite, while the structure is cubic across twin interfaces.

The visible and IR PL features of these NWs are displayed in Figure 5.28. The visible PL consists of a broad band peaked at 650 nm (1.91 eV), split in two maxima at 600 and 700 nm (2.06 and 1.77 eV), while the IR PL consists of a broad band peaked at 1400 nm (0.88 eV). It is interesting to note that the visible PL corresponds to the PL of MACE-grown cubic NWs (see Figure 5.25) with a diameter of 5 nm, while the IR emission, which could be never observed in MACE-grown Si NWs, probably arises from the presence of cubic/wurtzite heterostructures.

Eventually, the effect of porosity and surface oxidation of Si NWS grown with the MACE process, carried out with the use of high concentrations of H_2O_2, on the optical emissions of these wires has been studied by Zhang et al. [87]. It was shown that with a high concentration of H_2O_2 in solution the wires, which have a diameter around

---

* According the authors' opinion and that of the author of this book.

FIGURE 5.27   SEM images of silicon nanowires CVD-grown at 550°C. (*Reproduced with permission from Y.Li, Z.Liu, X.Lu, Z.Su, Y.Wang, R.Liu, D.Wang, J.Jan, J.H.Lee, H.Wang, Q.Yiu, J.Bao (2015) Broadband infrared luminescence in silicon nanowires with high density of stacking faults Nanoscale 7 160110. Copyright 2015 Royal Society of Chemistry(UK) Order Number: 1003228 Order Date: Nov 11, 2019 Order license ID 1003228-1; ISSN 2040-3372.*)*

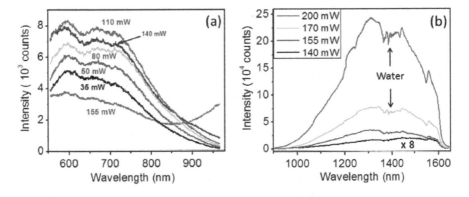

FIGURE 5.28   Photoluminescence spectra of chain-like Si NWs. (*Reproduced with permission from Y. Li, Z. Liu, X. Lu, Z. Su, Y. Wang, R. Liu, D. Wang, J. Jan, J.H. Lee, H. Wang, Q. Yiu, J. Bao (2015) Broadband infrared luminescence in silicon nanowires with high density of stacking faults Nanoscale 7 160110. Copyright 2015 Royal Society of Chemistry (UK) Order Number: 1003228. Order Date: Nov 11, 2019. Order license ID 1003228-1; ISSN 2040-3372.*)

20 nm, become nanoporous, and that the intensity of their optical emission at 750 nm (1.65 eV), typical of oxidized Si-NC (see Table 5.1) (see Figure 5.29 left panel), but also common to MACE-grown NWs, does increase with the increase of their porosity.

A red-shift toward 800 nm (1.54 eV), associated with a decrease of the emission intensity, is instead observed (see Figure 5.29 left panel) when the wires are oxidized at 1000°C, due to defect levels at Si–SiO$_x$ interfaces. As expected, see Figure 5.29, right panel, the emission intensity increases with the decrease of the measurement temperature.

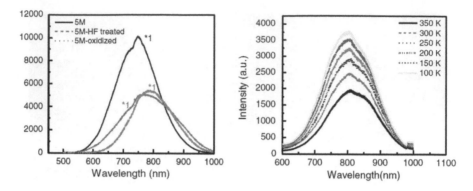

FIGURE 5.29 Effect of porosity and oxidation (left panel) and temperature (right panel) on the luminescence of MACE-grow Si NWs. (*After* C. Zhang, C. Li, Z. Liu, J. Zheng, C. Xue, Y. Zuo, B. Cheng, and Q. Wang (2013) Enhanced photoluminescence from porous silicon nanowire arrays *Nanoscale Research Letters* **8** 277 Open access article distributed under the terms of the Creative Commons Attribution License (http://creativecommons.org/licenses/by/2.0).)

It is, therefore, quite possible that the PL emission at 650–700 nm, typical of MACE-grown Si NWs with diameters within 5 and 9 nm, and of thick, nanoporous Si NWs, would have a common origin in the nanoporosity of their surfaces.

5.5 OPTICAL PROPERTIES OF GE NWs

The small difference (136 meV) between the direct Γ and the indirect L valley stimulated recent interest in Ge nanowires, looking at the opportunity to tune its optical properties and its band gap by strain [88] or by surface termination [62]. Greil et al. [88] succeeded in demonstrating that applying a mechanical uniaxial strain of 2.5% to a single Ge NW caused a Raman downshift to 290 cm^{-1} from the 300.5 cm^{-1} typical of unstrained wires and bulk Ge. From photocurrent measurements a decrease of the direct gap with the increase of the strain was also shown, but the lowering is smaller than that predicted by the theory.

It is known that that the band gap of H-terminated Ge NWs increases with the decrease of the size by QC effect [89], but that decreases when their surfaces are F- or Cl-terminated [90].

Starting from this experimental evidence, Legesse et al. [62] carried out a DFT study on Ge nanowires with different diameters (1, 1.4, and 1.7 nm) with three different surface-terminating chemical groups, –H, –OH, and –NH$_2$, in order to show the respective role of size and surface termination on the decrease of the band gap. They succeeded in demonstrating that the band gap of Ge nanowires can be shifted by more than 1.0 eV by changing the surface termination from –H to –OH, but that this effect is most evident with small-diameter nanowires. The decrease of the surface-terminating effect with the increase of the diameter of the wires is the most obvious effect of the smaller surface to volume ratio.

Also the corresponding shift of the absorption edge to longer wavelength with the –OH and –NH$_2$ surface terminations, by maintaining constant the diameter of the wires, apparently depends on a pure chemical effect.

5.6   OPTICAL PROPERTIES vs. DEFECTIVITY OF COMPOUND NANOCRYSTALLINE SEMICONDUCTORS

5.6.1   Silicon Carbide Nanowires

The optical properties of aligned SiC nanowires grown on a silicon wafer by heating at 1200°C a mixture of ZnS powder, which works as the catalyst, and carbon were studied by Niu et al. [91], using argon as the cover gas. The Field Emission scanning electron microscopy (FESEM) images, at different magnification, of a forest of vertically aligned wires are displayed in Figure 5.30. XRD measurements carried out on these wires indicate that their structure is that of β–SiC, and that they are vertically

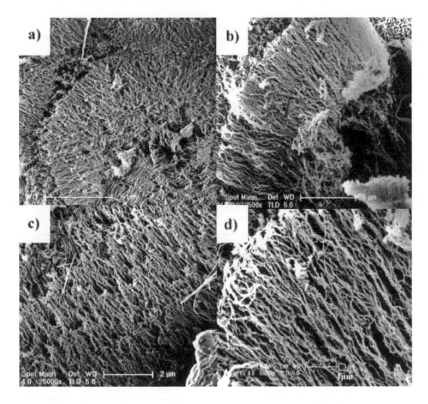

FIGURE 5.30   Field Emission scanning electron microscopy images of arrays of aligned SiC NWs (a), (b), and (c) low magnification, (d) high magnification image. (*After* J.J. Niu, J.N. Wang (2007) A simple route to synthesize scales of aligned single-crystalline SiC nanowires arrays with very small diameter and optical properties *J. Chem. Phys. B* **111** 4368-4373 *Reproduced with permission of American Chemical Society Nov 14, 2019.*)

aligned along the [111] direction. TEM measurement confirm the structure of the wires and show that their mean diameter is ≈10 nm, larger than the Bohr radius of 3C-SiC (≈2.7 nm), though NWs with a diameter of 20 nm are also present. HRTEM measurements show in addition that stacking faults are almost absent, but a very thin amorphous SiO_x shell is present.

A typical room-temperature PL spectrum of these NWs is displayed in Figure 5.31, left panel, which presents a main emission in the green at 510 nm (2.43 eV) and a feebler one at 430 nm (2.89 eV).

The main strong emission at 2.43 eV fits well with the band energy of cubic SiC at 300 K (2.36 eV) [92] or with the value of 2.417 eV reported in Chapter 4, with a very small blue-shift. The emission at 2.89 eV might be, instead, attributed to an interface state arising from the presence of a shell of SiO_x.

The Raman spectrum of a sample of these nanowires, see Figure 5.31, right panel, is the convolution of two bands peaked at ≈787 and ≈936 cm^{-1}, of which the first corresponds well to the signature of the TO Raman band of bulk cubic SiC at 300 K [93], Figure 5.32.

Similar results were obtained by Tabassum et al. [94] concerning the array of ultra-thin (10–40 nm) nanocrystalline SiC films, CVD-deposited at 800°C on an HSQ ribbon to form a self-aligned SiC nanowire array, see Figure 4.16, and further thermally annealed at 1100°C in 5% H_2–Ar mixtures on which details were reported in Chapter 4, Section 4.1.2.

From FTIR measurements, the intensity of the absorption corresponding to the Si-C stretching mode is shown (see Figure 5.33) to be greatly enhanced by increasing the annealing temperature in 5% H_2–Ar mixtures, indicating the beneficial effect of H_2 passivation. Eventually, from optical absorption measurements the Tauc's gap could be determined, which approaches the energy gap of cubic SiC (2.4 eV) for high-temperature annealed nanowire arrays. Therefore, in agreement with the anticipations given in Chapter 4, Section 4.1, quantum confinement is absent in this kind of

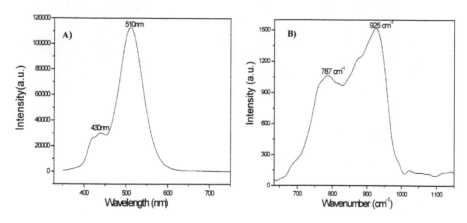

FIGURE 5.31 Photoluminescence (A) and Raman (B) spectra of aligned SiC NWs. (*After* J.J. Niu, J.N. Wang (2007) A simple route to synthesize scales of aligned single-crystalline SiC nanowires arrays with very small diameter and optical properties *J.Chem.Phys.B* **111** 4368–4373 *Reproduced with permission of American Chemical Society Nov 14, 2019.*)

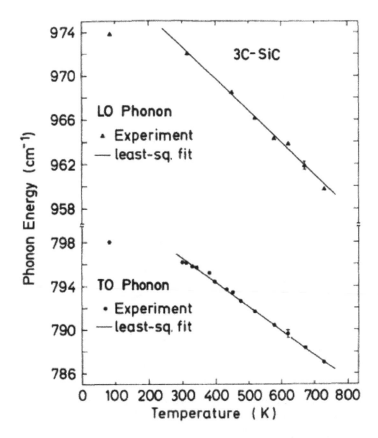

FIGURE 5.32 Temperature dependence of the peak positions of the LO and TO Raman lines of cubic (3C) SiC. (*Reproduced with permission from* J.J. Niu, J.N. Wang (2007) A simple route to synthesize scales of aligned single-crystalline SiC nanowires arrays with very small diameter and optical properties *J.Chem.Phys.B* **111** 4368–4373 *Copyright American Physical Society, License Number: RNP/19/NOV/020360 License date: Nov 14, 2019.*)

SiC nanowire, since clean grain boundaries do not behave as quantum confinement barriers.

Different is the case of the periodically twinned SiC nanowires studied by Wang et al. [95], already discussed in Chapter 4, Section 4.1.1, where twin boundaries create a periodic sequence of nanocells, with a diameter of 50 nm and a length of about 15 nm, which could favor quantum confinement effects, as shown by the photoluminescence spectra of these nanowires, with three peaks at 404 nm (3.07 eV), 434 nm (2.86 eV), and 547 nm (2.27 eV), none of which is consistent with the PL emissions of bulk SiC.

Instead, the first at 3.07 eV shows a blue-shift of about 0.6 eV vs. the main emission of bulk 3C-SiC at 2.417 eV, due to effective quantum confinement, the second at 2.86 eV corresponds to emission from a defect at the SiC–SiO$_x$ interface and the third should, instead, have its origin in a transition from a deep level within the band gap. Apparently, the 3D cells are of a size suitable for the set-up of quantum confinement.

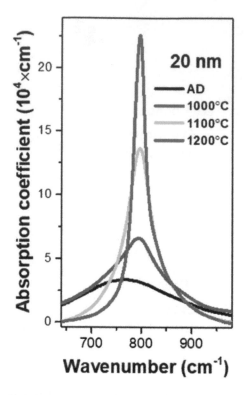

FIGURE 5.33 Annealing temperature dependence of the Si-C stretching mode. (*After* B.N. Tabassum, M. Kotha, V. Kaushik, B. Ford, S. Dey, E. Crawford, V. Nikas, S. Gallis (2018) On-demand CMOS-compatible fabrication of ultrathin self-aligned SiC nanowire arrays *Nanomaterials* **8** 906. This article is an open access article distributed under the terms and conditions of the Creative Commons Attribution (CC BY) license (http://creativecommons.o rg/licenses/by/4.0/).)

It could be interesting to compare the optical behavior of nanocrystalline and multi-twinned NWs with that of a dispersion in ethylic alcohol of mechanically interconnected 6H-SiC individual* nanocrystal prepared by electrochemical etching by Botsoa et al. [96], to see the role of size and solvent effects in radiative and non-radiative recombination.

It was observed that the room-temperature photoluminescence of a wet dispersion of the nanocrystals consists of a broad band peaked at 2.65 eV, see Figure 5.34, due to carrier recombination between a couple of donor–acceptor pairs (Al and N impurity centers) well-known in bulk 6H-SiC, and that a shoulder at 3 eV, corresponding to the band gap energy of bulk 6H-SiC, emerges in the PL spectrum. Apparently, the average size of the nanocrystals is too large to allow QC, given their size distribution centered at 100 nm.

The effect of centrifugation of the powder, which does allow an efficient separation of small particles from the coarse ones, leading to an average size of 1.9 nm,

---

* Which could simulate the behaviour of porous SiC nanowires.

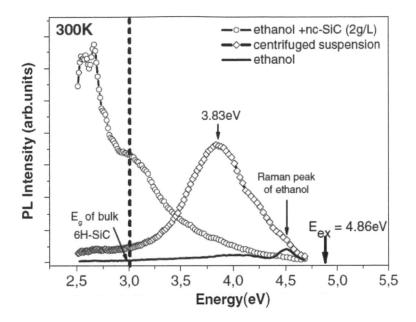

FIGURE 5.34 Effect of physical and chemical treatment on the photoluminescence spectra of 6H-SiC interconnected nanopowders. (E_{ex} is the energy of higher excited state for nanoparticles of 1.9 nm). (*Reproduced with permission from* J. Botsoa, J.M. Bluet, V. Lysenko, L. Sfaxi, Y. Zakharko, O. Marty, G. Guillot (2009) Luminescence mechanisms in 6H-SiC nanocrystals *Phys. Rev. B* **80** 155317 *Copyright 2009 American Physical Society License Number: RNP/19/NOV/020374 License date: Nov 15, 2019.*)

could be observed in Figure 5.34, which shows that the PL spectrum is a band peaked at 3.83 eV, blue-shifted by 0.8 eV above the energy gap of bulk 6H-SiC.

The emission at 3.83 eV could be explained as due both to a size selection and to a solvent effect, this last leading to the displacement of the levels of surface defect centers responsible for non-radiative transitions toward the band edges or to their full disappearance from the band gap, leaving undisturbed the blue-shifted transition due to quantum confinement.

The doubt, however, remains that the emission at 3.83 eV would be due to the ethylated surface of SiC nanocrystals, as already seen for the case of Si-NC in Section 5.1.

5.6.2 Optical Properties of Nanocrystalline GaAs Nanowires

It has been shown in Chapter 4 that bistability should be a typical character of GaAs NWs, due to the very small difference of the formation energy ΔE_{W-ZB} of the zinc-blende (ZB) phase, which is the thermodynamically stable phase, and that of wurtzite (WZ), with a calculated value of $\Delta E_{W-ZB} \approx 24$ meV/pair, or of 2.5 KJ/mol [97].

It has been, also, seen in Chapter 4 that GaAs nanowires could be grown with a variety of Au-assisted and self-assisted processes, that Au-assisted NWs mostly crystallize with the wurtzite (WZ) structure while the self-assisted ones crystallize with the zincblende (ZB) structure.

This structural difference associated with small ΔE_{W-ZB} values leads to very small differences in their energy gaps, since the room-temperature energy gap E_G^{ZB} of bulk ZB GaAs is 1.42 eV and that of WZ GaAs E_G^{WZ} is 1.46 eV [98], leading to an experimental difference $\Delta E_G = E_G^{WZ} - E_G^{ZB}$ of 56 meV measured with RT cathodoluminescence measurement [99].

A comparison of the temperature dependence of the optical emission energy of WZ NWs with that of ZB GaAs epilayers and of ZB bulk GaAs shows, also, that the emission energy at 10K of ZB and WZ coincides, while diverges at higher temperatures, leading to a room-temperature emission energy of WZ nanowires of 1.45 eV, while that of ZB bulk GaAs [100] and ZB epilayer materials lies at 1.42 eV.

Due to these small ΔE_{W-ZB} and ΔE_G values, the morphology and the optical properties of GaAs NWs depend by minute differences on the growth process conditions (temperature, nature of the metal catalyst and of the gaseous precursors, impurity contamination), which lead, eventually, to the presence of ZB and WZ segments in the same wire, and of extended defects (twins and stacking faults) with severe influence on their optical properties. As claimed by Glas et al. [101], nucleation processes should also be taken into account, as is the case of Au-assisted GaAs NWs that crystallize with the WZ structure. They, in fact, demonstrate that the interface energy at the Au–GaAs interface favors the nucleation of a WZ phase at high liquid supersaturation.

Relatively large nanometric diameters, comparable with the Bohr exciton radius of GaAs (11.6 nm) are expected to set up quantum confinement (QC) conditions, which are, however seldomly observed, with the significant exception of the single crystal GaAs NWs, studied by Duan et al. [102].

These wires, prepared by laser ablation, using Au as the catalyst, at furnace growth temperatures within 800 and 1030°C, present a uniform diameter of 3–5 nm along their entire lengths of 10 μm. Contrary to expectations, these wires crystallize with the zincblende structure. No detectable Au contamination was observed by energy-dispersive X-ray (EDX) spectroscopy,* and TEM measurements indicated the absence of extended defects. The room-temperature photoluminescence spectra of these wires consist of three PL bands peaked at 752 nm (1.65 eV), 794 nm (1.56 eV), and 836 nm (1.48 eV), of which the first two are significantly different to the emission of bulk zincblende GaAs, consisting of a single emission band at 870 nm (1.42 eV). The blue-shift of these bands from the bulk GaAs emission could be, therefore, assigned to size-induced quantum confinement.

The problem of degradation of the optical properties of GaAs NWs by process-induced Au-contamination was, instead, discussed by Breuer et al. [103] who, as already mentioned in Chapter 4, carried out PL measurements on core-shell GaAs/AlGaAs† nanowires MBE-grown either in the self-assisted mode (which did grow

---

* That could not be taken as an indication of a total absence of contamination, since EDX measurements are not trace measurements.
† The AlGaAs shell is used to reduce the surface recombination of minority carriers, as already shown in Chapter 4.

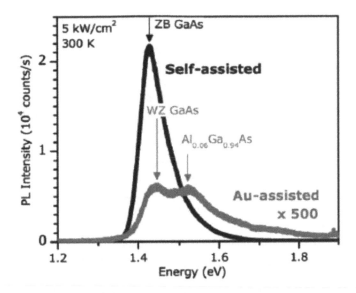

FIGURE 5.35   Photoluminescence of GaAs NWs grown with Au-assistance and with self-assistance. (*Reproduced with permission of American Chemical Society from S.* Breuer, C. Pfu'ller, T. Flissikowski, O. Brandt, H.T. Grahn, L. Geelhaar, H. Riechert (2011) Suitability of Au- and self-assisted GaAs nanowires for optoelectronic applications *Nano Lett.* **11** (3) 1276–1279 *Dec 02, 2019*.)

with the ZB structure), or in the Au-assisted mode (which did grow with the wurtz-ite structure), at considerably lower temperatures* than in the previous case. These wires are, also, much thicker (the core of the Au-assisted wire is 42 nm in diameter, that of self-assisted wire is 106 nm) than the wires grown by Duan et al. [102] and for both their room-temperature PL emission consists of a single peak at ≈1.45 eV, see Figure 5.35, though the wire grown with the Au-assistance shows a second band associated with the emission from the AlGaAs sheet.

It appears, however, that the intensity of the PL band measured on Au-assisted wires is more than one order of magnitude lower than that of the self-assisted nanow-ires, due, according to the authors, to Au contamination.

Different behavior is exhibited by an individual, wurtzite, nearly defect-free, 40 nm diameter, core-shell GaAs/AlGaAs nanowire, MBE-grown with Au-assistance at a substrate temperature of 540°C by Ahtapodov et al. [104], which presented along its entire length only 17 SFs, as shown in Chapter 4. The optical emission measured in correspondence with defect-free segments of the wire consists of a single band peaked at 1.516 eV at 15 K and at 1.444 eV ± 1 meV at 294 K, almost coinciding with the bulk ZB GaAs, free exciton emission energy

Thicker[†] core-shell GaAs/AlGaAs NWs, VLS-grown on Au-patterned silicon <111> substrates studied by Bailon-Somintac et al. [105] confirm the previous

---

* 500°C for the Au-assisted process and 580°C for the self-assisted process.
[†] The diameters ranged between 105 ± 44 nm and 176 ± 42 nm.

results. The room-temperature photoluminescence spectra of these wires generally consist of a peak at 1.426 eV ± 0.004 meV and of a higher energy shoulder due to the emission of the AlGaAs sheet, while the low-temperature PL (10 K) consists of a main peak at 1.5 eV and a second peak at 1.64 eV, due to AlGaAs.

The analysis of the optical properties of two core-shell GaAs/AlGaAs NWs samples, MBE -grown with Ga-assistance on single crystal <111> Si substrates at 580°C, of which the first (sample #1) was grown with a stoichiometric flux of Ga and As (F_{As}/F_{Ga} = 1), and the second (sample #2) doubling the As flux (F_{As}/F_{Ga} = 2) was, instead, carried out by Jahn et al. [99] to give deeper insight into the effect of the stoichiometry of the growth atmosphere on the properties of GaAs NWs.

The effect of the process gas stoichiometry is reflected by the diameter of the NWs: in fact, the sample #1 wires grew with an average diameter of 150 nm, while the sample #2 wires grew with an average diameter of 89 nm, over the entire length (9–10 μm) of both.

Moreover, XRD measurements indicate the presence of segments with the zinc-blende and wurtzite structure in both the wires, as confirmed also by their Raman spectra.

Eventually, μ-luminescence measurements carried out at 10 K on these samples display broad, spike-like spectra with a long tail in the low-energy part, see Figure 5.36, where the peak energy of the PL emission band of the sample #1 corresponds to that of a reference sample grown with Au assistance. Both peak energies exceed the free exciton energy of ZB GaAs at 10 K of more than 10 meV. Instead,

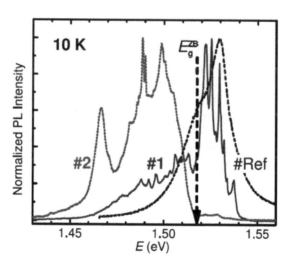

FIGURE 5.36 Micro-luminescence spectra at 10 K of two GaAs NWs samples grown without metal assistance (#1 and #2) and of a reference sample grown with Au-assistance. (The vertical dotted line indicates the energy gap of a bulk ZB GaAs sample.) (*Reproduced with permission from* U. Jahn, J. Lähnemann, C. Pfüller, O. Brandt, S. Breuer, B. Jenrichen, M. Ramsteiner, L. Geelhaar, and H. Riechert (2012) Luminescence of GaAs nanowires consisting of wurtzite and zincblende segments *Phys. Rev. B* **85**, 045323 *Copyright (2012) of American Physical Society License Number: RNP/19/NOV/020511 License date: Nov 21, 2019.*)

the peak of the μ-luminescence spectrum of the sample #2 is red-shifted by about 30 meV with respect to that of the sample #1.

The slight blue-shift of the emission of the sample #1 vs. the free exciton energy of ZB GaAs could not be ascribed to quantum confinement, considering that the diameter of the wire is largely in excess of the Bohr exciton radius of GaAs. Instead, it is well possible that the different blue-shifts observed in both samples could arise from different local arrangements of heterostructures consisting of ZB and WZ segments along the wires, leading to different band alignments. This view is consistent with results of scanning μ-luminescence measurements carried out along a single wire, which did manifest a change of the spectral peak energy with the scanning position.

It is, therefore, apparent that GaAsNWs could behave as mixed-phase system, with key consequences for their optical behavior.

Similar results were obtained by A. Fontcuberta i Morral [106] while studying the effect of three different As-fluxes by maintaining the Ga rate and the substrate temperature constant at 630°C on the MBE growth of GaAs nanowires. It was observed that by decreasing the As-flux and the As-supersaturation, the growth rate decreased. At the higher As-fluxes the sample α has the ZB structure, while the other two samples (β and γ) present increasing amounts of the wurtzite phase, associated with 180° rotational twins, each of which creates a WZ stacking unit and a quantum well. The PL spectra at 4 K of these samples are displayed in Figure 5.37, which shows that the emission of the sample α is a single band peaked at 1.516 eV, but those

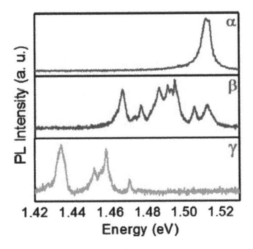

FIGURE 5.37 Comparison of the photoluminescence spectrum of a pure ZB GaAs nanowire (α) and of mixed-phase ZB-WZ GaAs nanowires (β and γ). (*Reprinted with permission from* D. Spirkoska, J. Arbiol, A. Gustafsson, S. Conesa-Boj, F. Glas, I. Zardo, M. Heigoldt, M.H. Gass, A.L. Bleloch, S. Estrade, M. Kaniber, J. Rossler, F. Peiro, J.R. Morante, G. Abstreiter, L. Samuelson, and A. Fontcuberta i Morral (2009) Structural and optical properties of high quality zinc-blende/wurtzite GaAs nanowire heterostructures *Phys. Rev. B* 80 245325 *Copyright 2009 American Physical Society License Number: RNP/19/NOV/020553 License date: Nov 23, 2019.*)

of the samples β and γ consist of bands with two main peaks, which shift down to 1.42–1.46 eV for the sample γ which contains the major amounts of WZ segments. A different band alignment of these WZ/ZB heterostructures is claimed at the origin of their optical emissions.

This conclusion is, also, supported by the results of a work of Hoang et al. [107], who studied the effect of 40 nm long, ZB GaAsSb (20% Sb) inserts in the 40 nm-core diameter of individual core-shell WZ GaAs/AlGaAs nanowire, MBE-grown with Au-assistance.

The PL emissions, measured at 10 K, from the GaAsSb core inserts and from the GaAs NW consist of two broad bands, whose peak energies depend on the excitation power.* At low excitation power (10 μW) they are peaked, respectively, at 1.28 and 1.49 eV, of which the first could be attributed to the GaAsSb inserts and the second to the GaAs core, whereas at high power (100 μW) they are peaked at 1.36 eV and at 1.515 eV, of which the second is the ZB GaAs free exciton energy.

The fact that the emission energy of the GaAs core at low excitation energies is below the free exciton energy of the ZB GaAs is considered by the authors to be due to stacking faults in the WZ GaAs phase, where one stacking fault in the wurtzite GaAs phase is a 1 nm thick segment of a ZB GaAs phase, and, consequently, to the presence of a ZB/WZ heterostructure. The emission process, in this case, is supposed to be due to localized electron–hole recombination, with electrons localized in the nm-thick ZB GaAs segments and holes in the WZ GaAs core.

5.6.3 II–VI Semiconductors

Unlike the case of GaAs nanowires, polytypism does not affect so much the optical properties of CdTe, CdSe, and CdS, the three II–VI compound semiconductors on which a detailed discussion will be carried out in this section. In fact, the wires are always single phase, and almost free of twins and stacking faults. Therefore their optical properties are dictated by their size and by the presence of metallic impurities used as catalysts more than by structural defects, with some exceptions that will be discussed here.

The optical properties of CdSe nanowires prepared with Au-assistance and of self-assisted grown NWs (see details in Chapter 4, Section 4.3) were studied among others by Fasoli et al. [108, 109] in two successive works, which help in understanding the role of morphology and Au-contamination in the optical properties of CdSe nanowires.

The effect of Au-assistance on the optical emission CdSe nanorods[†] is well-evident by comparing the PL spectra of Au-assisted grown NRDs with that of self-seeded NRDs (see Figure 5.38). In fact, while the first is a broad spectrum with two peaks at 1.8 eV and 1.65 eV, the second is a shallow band peaked at 1.8 eV, corresponding to

---

* Polarization effects observed with these samples are not discussed here, being outside the general interest of the book.

[†] A nanorod is a nanowire with a large diameter with respect to the length.

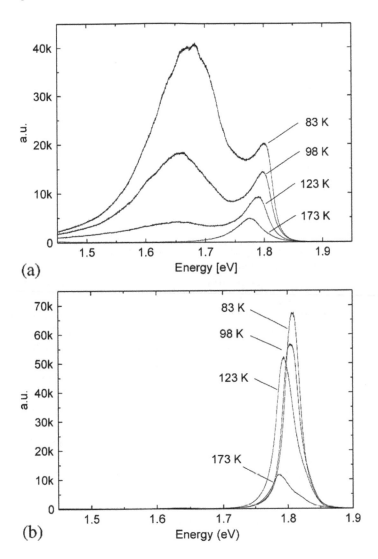

FIGURE 5.38 Photoluminescence spectra of Au-assisted CdSe nanorods (a) and of self-seeded CdSe nanorods (b). (*Reproduced with permission of Elsevier after* A. Fasoli, A. Colli, S. Kudera, L. Manna, S. Hofmann, C. Ducati, J. Robertson, A.C. Ferrari (2007) Catalytic and seeded shape-selective synthesis of II–IV semiconductor nanowires *Physica E* **37** 138–141 *License number 4720820732437, License date Dec 02, 2019.*)

the recombination of a A-type* excitons of bulk CdSe. QC effects are not expected to be displayed by these wires, given that the diameter of both Au-seeded and self-seeded NRDs is well above the exciton Bohr radius (5.4 nm) of CdSe.

---

* An exciton in the Cd sublattice.

In their further work, they confirm a near band edge emission at 1.8 eV at 70 K, due to a free A-type exciton recombination, while they associate the emission of the band at ≈1.7 eV with defect states due to morphological defects of the wires, well-detected by TEM measurements on different nanowires.

The optical properties of WZ CdTe nanowires grown by spontaneous aggregation of cubic ZB CdTe nanocrystals (see Chapter 4 for the details) were instead studied by Tang et al. [110]. These wires present the same diameter as the original nanocrystals (2.5–5.4 nm), despite the phase change accompanying the agglomeration process. These wires present an RT broad emission band peaked at 650 nm (1.91 eV), shifted well above the band gap of bulk CdTe (1.513 eV), measured by Fonthal [111] with PL measurements. The blue-shifted emission of these wires could be explained not by 2D QC along the length of the wires, which have a diameter smaller than the Bohr exciton radius of CdTe (7.3 nm), but to a 3D QC inside the wurtzite segments of the wire, separated presumably by twin boundaries.

CdS nanowires 1.5 μm long and with an average diameter of 25–30 nm (but a few of them well below these figures) distributed at the surface in 300 nm thick CdS films prepared with a room-temperature wet process were, instead, studied by Maity et al. [112]. The wires crystallize with the ZB structure and present RT Tauc's gaps (from optical absorption measurements) of 3.07 eV in the case of as-grown films and 3.00, 2.89, and 2.86 eV for films annealed at 100, 150, and 200°C, respectively. An RT broad photoluminescence band extended between 510 and 625 nm, peaked at 566 nm (2.19 eV) is, instead, obtained on un-annealed films. Though the analysis of these results is definitely biased by the heterogeneous features of the matrix, still, considering the thickness of the films, the large blue-shift of their Tauc's gap from the energy gap of CdS (2.40 eV) could be attributed to QC effects in correspondence with the nanowires, despite their diameter being much larger than the Bohr radius (5.8 nm) of CdS. Instead, the photoluminescence at 2.19 eV is attributed to minority carrier recombination at the surface of the films.

Quantum confinement and lasing capability is, eventually, observed as a property of vertically aligned CdSe nanowires [113], a property already observed in CdS NWs [114].

These nanowires have a diameter around 70 nm, present a quasi-conical tip, and are free of structural defects. Strong excitonic features dominate their low-temperature (10 K) absorption spectra, with a peak at 1.8248 eV, that corresponds to the free exciton absorption of CdSe, and a second peak at 1.8515 eV. The energy difference corresponds to the energy of a longitudinal optical phonon. Under pulsed laser excitation a broad emission peaked at 720 nm could be observed, but a sharp peak, with a linewidth of 0.4 nm, emerges at an excitation threshold of 0.296 MW cm^{-2}, which could be taken as an indication of the appearance of a lasing action.

APPENDIX 1

A purely phenomenological analysis is unable to lead to a comprehensive understanding of the physics of the luminescence of nc-Si films, as shown by Islam and Kumar [115], who demonstrated that a quantitative simulation of the PL spectra of a nanocrystallite needs to account not only for the crystallite size L, but for its dispersion σ (standard deviation for a gaussian distribution) as well as for surface states induced by compositional (i.e. chemical) and structural surface disorder.

They start by considering that for the PL emission energy E^* of a nanocrystallite the following equation holds

$$E^* = E_G + \Delta E - E_s - E_b \tag{5.3}$$

where E_G is the energy gap of the bulk material, ΔE is the energy upshift due to quantum confinement (QC), E_s is the localization energy of surface states, and E_b is the exciton-binding energy.

Then, for a log-normal particle size distribution, the PL intensity profile for an ensemble of nanocrystallites should be given by the following equation

$$I(\Delta E) \propto \frac{\left(C/\Delta E\right)^{(5-\alpha+n)/n}}{nC\sigma} \exp-\frac{\left\{\ln\left(C/\Delta E\right)^{1/n} - \ln\left(L_o\right)\right\}^2}{2\sigma^2} \tag{5.4}$$

where $\Delta E = \dfrac{C}{L^n}$ is the band-gap up-shift due to QC, C and n are constants, and L_o is the mean average radius.

The critical effect of the size distribution is reported, as an example, in Figure 5.39. It displays in the left panel the calculated PL spectra for three nanocrystal ensembles having a mean size of 2.2, 3.2, and 4.3 nm, taking at 10% the standard deviation L_o of the mean average radius. Apparently, nanocrystals with a mean size of 2.2 eV present a calculated PL emission at 2.3 eV, which is also, but seldomly, experimentally observed (see Table 5.2).

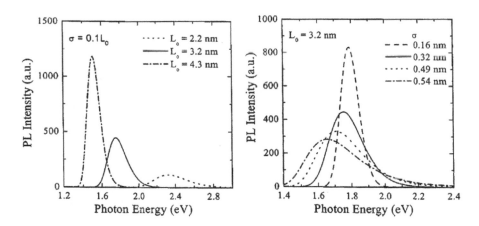

FIGURE 5.39 *Left panel:* Computed PL spectra for three nanocrystal ensembles having a mean size of 2.2, 3.2, and 4.3 nm (the standard deviation L_o of the mean average radius is 10%). *Right panel:* Computed PL spectra for silicon nanocrystallites having a mean diameter $L_o = 3.2$ nm with different normal-log distributions σ. (*Reproduced with permission of AIP Publishing, from* Md.N. Islam and S. Kumar (2003) Influence of surface states on the photoluminescence from silicon nanostructures *J.Appl. Phys.* **93**, 1753–1759 *License Number 4725811182783, License date Dec 11, 2019.*)

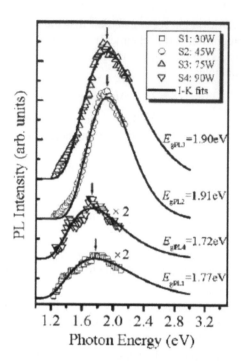

FIGURE 5.40 RF-power and size dependence of the PL emission energy of nc-Si thin films □ 3.79 nm; ○ 4.25 nm; V 4.09 nm; Δ 4.39 nm. (*Reproduced with permission of AIP publishing after* H. Chen and W.Z. Shen (2006) Structural order effect in visible photoluminescence properties of nanocrystalline Si:H thin films *Appl. Phys. Lett.* 88 121921 *License Number 4715990465822 License date Nov 25, 2019.*)

It is possible to see in the right panel the strong effect of different normal-log distributions σ of the ensemble of nanocrystallites having a mean diameter $L_o = 3.2$ nm displayed in the left panel on the intensity. The shape and peak energies of their PL spectra seem to pertain to different materials.

Islam and Kumar [115] were able to fit well with their model a variety of experimental data reported in the literature [60, 116], but the work of Chen and Shen [61] is particularly interesting.

They measured the PL spectra of nanocrystalline Si-films, PECVD-grown in silane/hydrogen at 250°C on c-Si substrates at increasing RF power. The spectra consist of a single PL band as is shown in Figure 5.40. It is immediately apparent from the figure that the peak energies, which range between 1.77 eV and 1.9 eV, do not follow the expected QC up-shift with the decrease of the nanocrystal size. Instead, the spectra (solid curves in the figure) could be finely fitted by applying the Islam and Kumar model accounting for both surface disorder and size distribution.

REFERENCES

1. M. Hasan, Md F. Huq, Z. Hasan Mahmood (2013) A review on electronic and optical properties of silicon nanowire and its different growth techniques. *Springerplus* **2**, 151. doi:10.1186/2193-1801-2-151.

2. W. Lu, C. M. Lieber (2006) Semiconductor nanowires. *J. Phys. D. Appl. Phys.* **39**, R387–R406.
3. A. Fojtik, A. Henglein (1994) Luminescent colloidal silicon particles. *Chem. Phys. Lett.* **221**, 363–367.
4. D. Jurbergs, E. Rogojina, L. Mangolini, U. Kortshagen (2006) Silicon nanocrystals with ensemble quantum yields exceeding 60%. *Appl. Phys. Lett.* **88**, 233116.
5. D. Engemann, R. Fischer (1977) Photoluminescence in amorphous silicon. *Phys. Status Solidi B* **79**, 195–202.
6. L. Mangolini, E. Thimsen, U. Kortshagen (2005) High-yield plasma synthesis of luminescent silicon nanocrystals. *Nano Lett.* **5**, 655–659.
7. M. V. Wolkin, J. Jorne, P. M. Fauchet, G. Allan, C. Delerue (1999) Electronic states and luminescence in porous silicon quantum dots: The role of oxygen. *Phys. Rev. Lett.* **82**, 197.
8. A. Puzder, A. J. Williamson, J. C. Grossman, G. Galli (2002) Surface chemistry of silicon nanoclusters. *Phys. Rev. Lett.* **88**, 097401–097404.
9. A. Fojtik, A. Henglein (2006) Surface chemistry of luminescent colloidal silicon nanoparticles. *J. Phys. Chem. B.* **110**, 1994–1998.
10. M. C. Beard, K. P. Knutsen, P. Yu, J. M. Luther, Q. Song, W. K. Metzger, R. J. Ellingson, A. J. Nozik (2007) Multiple exciton generation in colloidal silicon nanocrystals. *Nanoletters* **7**, 2506–2012.
11. K. Herynkova, C. Vorkotter, P. Simakova, J. Benedikt, O. Cibulka (2016) Structural and luminescence properties of silicon nanocrystals in colloidal solutions for bio applications. *Phys. Status Solidi A* **213**, 2873–2878.
12. I. Sychugov, A. Fucikova, F. Pevere, Z. Yang, J. G. C. Veinot, J. Linnros (2014) Ultranarrow luminescence linewidth of silicon nanocrystals and influence of matrix. *ACS Photonics* **1**, 998–1005.
13. K. Žídek, F. Trojánek, P. Malý, L. Ondič, I. Pelant, K. Dohnalová, L. Šiller, R. Little, B. R. Horrocks (2010) Femtosecond luminescence spectroscopy of core states in silicon. *Opt. Express* **18**, 25242.
14. Z. Zhou, L. Brus, R. Friesner (2003) Electronic structure and luminescence of 1.1 and 1.4 nm silicon nanocrystals: Oxide shall versus hydrogen passivation. *Nano Lett.* **3**, 163–167.
15. I. Vasiliev, J. R. Chelikowsky, R. M. Martin (2002) Surface oxidation effects on the optical properties of silicon nanocrystals. *Phys. Rev. B: Condens. Matter* **65**, 121302.
16. G. Ledoux, J. Gong, F. Huisken O. Guillois, C. Reynaud (2002) Photoluminescence of size-separated silicon nanocrystals: Confirmation of quantum confinement. *Appl. Phys. Lett.* **80**, 4834.
17. G. Ledoux, O. Guillois, D. Porterat, C. Reynaud, F. Huiskens, B. Kohn, V. Paillard (2000) Photoluminescence properties of silicon nanocrystals as a function of their size. *Phys. Rev. B* **62**, 15942–15951.
18. D. J. Lockwood, A. Wang, B. Bryskiewicz (1994) Optical absorption evidence for quantum confinement effects in porous silicon. *Solid State Commun.* **89**, 587–589.
19. H. Campbell ,P. M. Fauchet (1986) The effects of microcrystal size and shape on the one phonon Raman spectra of crystalline semiconductors. *Solid State Commun.* **58**, 739–741.
20. Y. Chao, L. Šiller, S. Krishnamurthy, P. R. Coxon, U. Bangert, M. Gass, L. Kjeldgaard, S. N. Patole, L. H. Lie, N. O'Farrell, T. A. Alsop, A. Houlton, B. R. Horrocks (2007) Evaporation and deposition of alkyl-capped silicon nanocrystals in ultrahigh vacuum. *Nat. Nanotechnol.* **2**, 486.
21. T. Qiu, X. L. Wu, F. Kong, H. B. Ma. P. K. Chu (2005) Solvent effect on light -emitting property of Si nanocrystals. *Phys. Lett. A* **334**, 447–445.
22. N. C. Greenham, X. Peng, A. P. Alivisatos (1996) Charge separation and transport in conjugated-polymer/semiconductor-nanocrystal composites studied by photoluminescence quenching and photoconductivity. *Phys. Rev. B* **54**, 17628–17637.

23. J. B. Lambert, H. F. Shurvell, D. A. Liqhtner (1998) *Organic Stuctural Spectroscopy*, Prentice-Hall, Upper Saddle River, New Jersey.
24. R. M. Al Mohaimeed, A. A. Ansari, A. Aldwayyan (2018) The role of solvent environment on the optical behavior of chemically synthesized silicon nanoparticles. *J. Spectrosc.* (Hindawi) **2018**, 9. Article ID 6870645.
25. P. Cabral do Couto, S. G. Estácio, B. J. Costa Cabral (2005) The Kohn-Sham density of states and band gap of water: From small clusters to liquid water. *J. Chem. Phys.* **123**, 054510 D.
26. S. L. Weeks, R. P. Chaukulkar, P. Stradins, S. Agarwal (2014) Photoluminescence behavior of plasma synthesized Si nanocrystals oxidized at low temperature in pure O2 and H2O. *J. Vac. Sci. Technol. A* **32**, 050604.
27. S. Veprek, M. G. J. Veprek-Heijman (2015) Photoluminescence from nanocrystalline silicon, nc-Si/SiO$_2$ nanocomposites, and nc-Si oxidized in O$_2$ and treated in H$_2$O. *J. Vac. Sci. Technol. A* **33**, 043001-4.
28. J. Valenta, R. Juhasz, J. Linnros (2002) Photoluminescence spectroscopy of single silicon quantum dots. *Appl. Phys. Lett.* **80**, 1070.
29. C. Delerue, G. Allan, M. Lannoo (1993) Theoretical aspects of the luminescence of porous silicon. *Phys. Rev. B* **48**, 11024–11036.
30. S. Furukawa, T. Miyasato (1988) Quantum size effects on the optical band gap of microcrystalline Si:H. *Phys. Rev. B* **38**, 5726–5729.
31. L. Del Negro, M. Cazzanelli, N. Daldosso, Z. Gaburro, L. Pavesi, F. Priolo, D. Pacifici, G. Franzò, F. Iacona (2003) Stimulated emission in plasma-enhanced chemical vapour deposited silicon nanocrystals. *Physica E* **16**, 297–308.
32. M. Zacharias, J. Heitmann, R. Scholz, U. Kahler, M. Schmidt, J. Bläsing (2002) Size-controlled highly luminescent silicon nanocrystals: A superlattice approach. *Appl. Phys. Lett.* **80**, 661.
33. J. Heitmann, F. Muller, M. Zacharias, U. Gosele (2005) Silicon nanocystals: Size matter. *Adv. Mater.* **17**, 795–803.
34. Zh. Yuan, A. Anopchenko, L. Pavesi (2012) Innovative quantum effects in Silicon for photovoltaic applications. In: *Advanced Silicon Materials for Photovoltaic Applications*, S. Pizzini Ed., Wiley, New York, NY.
35. S. Godefroo, M. Heyne, M. Jivanescu, A. Stesmans, M. Zacharias, O. I. Lebedev, G. Van Tendeloo, V. V. Moshchalkov (2008) Classification and control of the origin of photoluminescence from Si nanocrystals. *Nat. Nanotechnol.* **3**, 174–178.
36. P. Photopoulos, A. G. Nassiopoulou, D. N. Kouvatsos, A. Travlos (2000) Photoluminescence from nanocrystalline silicon in Si/SiO$_2$Si superlattices. *Appl. Phys. Lett.* **76**, 3588–3590.
37. P. P. Dey, A. Khare (2017) Fabrication of photoluminescent nc-Si:SiO$_2$ thin films prepared by PLD. *Phys. Chem. Chem. Phys.* **19**, 21436–21445.
38. G. Cicala, P. Capezzuto, G. Bruno, L. Schiavulli, G. Perna, V. Capozzi (1996) Plasma deposition and characterization of photoluminescent fluorinated nanocrystalline silicon films. *J. Appl. Phys.* **80**, 6564–6566.
39. G. Cicala, P. Capezzuto, G. Bruno (2005) From amorphous to microcrystalline silicon deposition in SiF$_4$, H$_2$, He plasmas: In situ control by optical emission spectroscopy. *Thin Solid Films* **383**, 203–205.
40. G. Cicala, G. Bruno, P. Capezzuto, L. Schiavulli, V. Capozzi, G. Perna (2011) Deposition of photoluminescent nanocrystalline silicon films by SiF$_4$-SiH$_4$-H$_2$ plasmas. *MRS Online Proceedings Library*.
41. M. Ali (2006) Origin of photoluminescence in nanocrystalline Si:H films. *J. Lumin.* **127**, 614–622.
42. A. M. Ali, H. Kobayashi, T. Inokuma, A. Al-Hajry (2013) Morphological, luminescence and structural properties of nanocrystalline silicon thin films. *Mater. Res. Bull.* **48**, 1027–1033.

43. E. Edelberg, S. Bergh, R. Naone, M. Hall, E. S. Ayrdil (1997) Luminescence from plasma deposited silicon films. *J. Appl. Phys.* **81**, 2410.
44. T. Toyama, T. Matsui, Y. Kurokawa, H. Okamoto, Y. Hamakawa (1996) Visible photo- and electroluminescence from electrochemically formed nanocrystalline Si thin film. *Appl. Phys. Lett.* **69**, 1261.
45. T. Toyama, Y. Kotami, A. Shimode, H. Okamoto (1999) Direct transition at the fundamental gap in light-emitting nanocrystalline Si thin films. *Appl. Phys Lett.* **74**, 3323.
46. S. Pizzini, M. Acciarri, S. Binetti, D. Cavalcoli, A. Cavallini, D. Chrastina, L. Colombo, E. Grilli, G. Isella, M. Lancin, A. Le Donnne, A. Mattoni, K. Peter, B. Pichaud, E. Poliani, M. Rossi, S. Sanguinetti, M. Texier, H. von Känel (2006) Nanocrystalline silicon films as multifunctional material for optoelectronic and photovoltaic applications. *Mat. Sci. Eng. B* **134**, 118–124.
47. S. Pizzini Final technical report of the nanophoto project 013944, July 2008.
48. L. Bagolini, A. Mattoni, G. Fugallo, L. Colombo, E. Poliani, S. Sanguinetti, E. Grilli (2010) Quantum confinement by an order-disorder boundary in nanocrystalline silicon. *Phys. Rev. Lett.* 104, 176803.
49. J. D. Fields, B. Gorman, T. Merdzhanova, B. Yan, T. Su, P. C. Taylor (2013) On the origin of deep oxygen defects in hydrogenated nanocrystalline silicon thin films used in photovoltaic applications. *Sol. Energy Mater. Sol. Cells* **113**, 61–70.
50. D. Engemann, R. Fischer (1977) Photoluminescence in amorphous silicon. *Phys. Stat. Solidi B* **79**, 192–202.
51. R. A. Street (1981) Luminescence and recombination in hydrogenated amorphous silicon. *Adv. Phys.* **30**, 593–676.
52. M. Schubert, R. Stachowitz, W. Fuchs (1996) Geminate and non-geminate recombination in amorphous silicon (a-Si:H). *J. Non-Cryst. Solids* **198–200**, 251–254.
53. H. Oheda (2007) Two characteristic photoluminescence states and their metastability in hydrogenated amorphous silicon and its alloys. *J. Appl. Phys.* **101**, 053711.
54. H. Nguyen, F. E. Rougieux, D. Yan, Y. Wan, S. Mokkapati, S. Martin de Nicolas, J. P. Seif, S. De Wolf, D. Macdonald (2016) Characterizing amorphous silicon, silicon nitride, and diffused layers in crystalline silicon solar cells using micro-photoluminescence spectroscopy. *Sol. Energy Mater. Sol. Cells* **145**, 403–411.
55. A. Cavallini, D. Cavalcoli (2008) Nanostructures in silicon investigated by atomic force microscopy and Surface photovoltage spectroscopy. *Scanning* **30**, 358–363.
56. E. Fefer, Y. Shapira, I. Balberg (1995) Direct determination of the band-gap states in hydrogenated amorphous silicon using surface photovoltage spectroscopy. *Appl. Phys. Lett.* **67**, 371.
57. Y. Abdulraheem, I. Gordon, T. Beard, H. Meddeb, J. Poortmans (2014) Optical band-gap of ultra-thin amorphous silicon films deposited on crystalline silicon by PECVD. *AIP Adv.* **4**, 057122. doi:10.1063/1.4879807.
58. M. Kato, T. Fujiseki, T. Miyadera, T. Sugita, S. Fujimoto, M. Tamakoshi, M. Chikamatsu, H. Fujiwara (2017) Universal rules for visible-light absorption in hybrid perovskite materials. *J. Appl. Phys.* **121**, 115501.
59. L. Kronik, L. Shapira (1999) Surface photovoltage phenomena: Theory, experiment, and applications. *Surf. Sci. Rep.* **37**, 1–206.
60. M. Ehrbrecht, B. Kohn, F. Huisken, M. A. Laguna, V. Paillard (1997) Photoluminescence and resonant Raman spectra of silicon films produced by size-selected cluster beam deposition. *Phys. Rev. B* **56**, 6958–6964.
61. H. Chen, W. Z. Shen, W. S. Wei (2006) Structural order effect in visible photoluminescence properties of nanocrystalline Si:H thin films. *Appl. Phys. Lett.* **88**, 121921.
62. M. Legesse, G. Fagas, M. Nolan (2017) Modifying the band gap and optical properties of Germanium nanowires by surface termination. *Appl. Surf. Sci.* **396**, 1155–1163.

63. A. Irrera, P. Artoni, F. Iacona, E. F. Pecora, G. Franzò, M. Galli, B. Fazio, S. Boninelli, F. Priolo (2012) Quantum confinement and electroluminescence in ultrathin silicon nanowires fabricated by a maskless etching technique. *Nanotechnology* **23**, 075204.

64. F. Fabbri, E. Rotunno, L. Lazzarini, N. Fukata, G. Salviati (2014) Visible and infra-red light emission in boron-doped wurtzite silicon nanowires. *Sci. Rep.* **4**, 3603 1–7.

65. D. B. Zhang, M. Hua, T. Dumitrică (2008) Stability of polycrystalline and wurtzite Si nanowires via symmetry-adapted tight-binding objective molecular dynamics. *J. Chem. Phys.* **128**, 084104.

66. O. Demichel, F. Oehler, V. Calvo, P. Noé, N. Pauc, P. Gentile, P. Ferret, T. Baron, N. Magnea (2009) Photoluminescence of silicon nano wires obtained by epitaxial chemical vapor deposition. *Physica E* **41**, 963–965.

67. M. Tajima, S. Ibuka (1998) Luminescence due to electron-hole condensation in silicon-on-insulator. *J. Appl. Phys.* **84**, 2224–2228.

68. M. King, S. Chaure, S. Krishnamurthy, W. J. Blau, A. Colli, A. C. Ferrari (2008) Optical characterization of oxide encapsulated silicon nanowires of various morphologies. *J. Nanosci. Nanotechnol.* **8**, 4202–4206.

69. K. Sato, A. Castaldini, N. Fukata, A. Cavallini (2012) Electronic level scheme in boron- and phosphorus-doped silicon nanowires. *Nano Lett.*, **12**, 3012–3017.

70. F.Fabbri, E. Rotunno, L. Lazzarini, D. Cavalcoli, A. Castaldini, N. Fukata, K. Sato, G. Salviati, A. Cavallini (2013) Preparing the way for doping wurtzite silicon nanowires while retaining the phase. *Nano Lett.* **13**(12), 5900–5906.

71. F. Fabbri, Y.-T. Lin, G. Bertoni, F. Rossi, M. J. Smith, S. Gradečak, E. Mazur, G. Salviati (2015) Origin of the visible emission of black silicon microstructures. *Appl. Phys. Lett.* **107**, 021907.

72. C. Persson, E. Janzen (1998) Electronic band structure in hexagonal close-packed Si polytypes. *J. Phys-Condens. Matter* **10**, 10549–10555.

73. A. Irrera, M. J. Lo Faro, C. D'Andrea, A. A. Leonardi, P. Artoni, B. Fazio, R. A. Picca, N. Cioffi, S. Trusso, G. Franzò, P. Musumeci, F. Priolo, F. Iacona (2017) Light emitting silicon nanowires obtained by metal-assisted chemical etching. *Semicond. Sci. Techn.* **32**, 043004 (20pp).

74. Z. Li, X. Wang, S. Liu, J. Yang, K. Shi, H. Wang, D. Zhu, X. Xing (2018) Broadband photoluminescence of silicon nanowires excited by near-infrared continuous wave lasers. *Opt. Laser Technol.* **99**, 81–85.

75. K. Mergenthaler, N. Anttu, N. Vainorius, M. Aghaeipour, S. Lehmann, M. T. Borgström, L. Samuelson, M.-E. Pistol (2014) Anti-Stokes photoluminescence probing k-conservation and thermalization of minority carriers in degenerately doped semiconductors. *Nat. Commun.* **8**, 1634.

76. S. K. Saxena, H. M. Rai, R. Late, P. R. Sagdeo, R. Kumar (2015) Origin of photoluminescence from silicon nanowires prepared by metal induced etching (MIE). *AIP Conf. Proc.* **1661**, 080027.

77. A. B. Murphy (2007) Band-gap determination from diffuse reflectance measurements of semiconductor films, and application to photoelectrochemical water-splitting. *Sol. Energy Mater. Sol. Cells* **91**, 1326–1337.

78. K. Oda, Y. Nanai, T. Sato, S. Kimura, T. Okuno (2014) Correlation between photoluminescence and structure in silicon nanowires fabricated by metal-assisted etching. *Phys. Status Solidi (A)* **211**, 848–855.

79. R. J. Barsotti, Jr., J. E. Fischer, C. H. Lee, J. Mahmood, C. K. W. Adu, P. C. Eklund (2002) Imaging, structural, and chemical analysis of silicon nanowires. *Appl. Phys. Lett.* **81**, 2866.

80. S. Bhattacharya, D. Banerjee (2004) Confinement in silicon nanowires: Optical properties. *Appl. Phys. Lett.* **85**, 2008.

81. R. Ghosh, P K Giri, K. Imakita, Minoru Fujii (2014) Origin of visible and near-infrared photoluminescence from chemically etched Si nanowires decorated with arbitrarily shaped Si nanocrystals. *Nanotechnology* 25(2014), 045703 (13pp).
82. K. Oda, Y. Nanai, T. Sato, S. Kimura, T. Okuno (2014) Correlation between photoluminescence and structure in silicon nanowires fabricated by metal-assisted etching. *Phys. Status Solidi (A)* 211(4), 848–855.
83. D. Yu, C. S. Lee, I. Bello (1998) Synthesis of nano-scale silicon wires by excimer laser ablation. *Solid State Commun.* 105, 403.
84. C. S. Lee, I. Bello, X. S. Sun, Y. H. Tang, G. W. Zhou, Z. G. Bai, Z. Zhang, S. Q. Feng (1998) Synthesis of nanoscale silicon wires by excimer laser ablation at high temperature. *Solid State Commun.* 105, 403–407.
85. L. Zhang, W. Ding, Y. Yan, J. Qu, B. Li, L-Yu Li, K. T. Yue, D. Yu (2002) Variation of the Raman feature on excitation wavelength of silicon nanowires. *Appl. Phys. Lett.* 81, 4446.
86. Y. Li, Z. Liu, X. Lu, Z. Su, Y. Wang, R. Liu, D. Wang, J. Jan, J. H. Lee, H. Wang, Q. Yiu, J. Bao (2015) Broadband infrared luminescence in silicon nanowires with high density of stacking faults. *Nanoscale* 7, 160110.
87. C. Zhang, C. Li, Z. Liu, J. Zheng, C. Xue, Y Zuo, B. Cheng, Q Wang (2013) Enhanced photoluminescence from porous silicon nanowire arrays. *Nanoscale Res. Lett.* 8, 277.
88. J. Greil, A. Lugstein, C. Zeiner, G. Strasser, E. Bertagnolli (2012) Tuning the electro-optical properties of germanium nanowires by tensile strain. *Nano Lett.* 12, 6230–6234.
89. G. Collins, J. D. Holmes (2011) Chemical functionalisation of silicon and germanium nanowires. *J. Mater. Chem.* 21(2011), 11052–11069.
90. H. Adhikari, P. C. McIntyre, S. Y. Sun, P. Pianetta, C. E. D. Chidsey (2005) Photoemission studies of passivation of germanium nanowires. *Appl. Phys. Lett.* 87(2005), 263109.
91. J. J. Niu, J. N. Wang (2007) A simple route to synthesize scales of aligned single-crystalline SiC Nanowires arrays with very small diameter and optical properties. *J. Chem. Phys. B* 111, 4368–4373.
92. Yu. Goldberg Yu., M. E. Levinshtein, S. L. Rumyantsev (2001) *Properties of Advanced Semiconductor Materials GaN, AlN, SiC, BN, SiC, SiGe*, M. E. Levinshtein, S. L. Rumyantsev, M. S. Shur, Eds., John Wiley & Sons, Inc., New York, pp. 93–148.
93. D. Olego, M. Cardona (1982) Temperature dependence of the optical phonons and transverse effective charge in 3C-SiC. *Phys. Rev. B* 25, 3889.
94. B. N. Tabassum, M. Kotha, V. Kaushik, B. Ford, S. Dey, E. Crawford, V. Nikas, S. Gallis (2018) On-demand CMOS-compatible fabrication of ultrathin self-aligned SiC nanowire arrays. *Nanomaterials* 8, 906.
95. D-H. Wang, D. Xu, Q. Wang, Y-J- Hao, G-Q. Jin, X-Y. Guo, K. N. Tu (2008) Periodically twinned SiC nanowires. *Nanotechnology* 19, 215602.
96. J. Botsoa, J. M. Bluet, V. Lysenko, L. Sfaxi, Y. Zakharko, O. Marty, G. Guillot (2009) Luminescence mechanisms in 6H-SiC nanocrystals. *Phys. Rev. B* 80, 155317.
97. Y. Yeh, Z. W. Lu, S. Froyen, A. Zunger (1992) Zinkblende-wurtzite polytypism in semiconductors. *Phys. Rev. B* 46(16), 10086–1009.
98. P. Kusch, S. Breuer, M. Ramsteiner, L. Geelhaar, H. Riechert, S. Reich (2012) Band gap of wurtzite GaAs: A resonant Raman study. *Phys. Rev. B* 86, 075317.
99. U. Jahn, J. Lähnemann, C. Pfüller, O. Brandt, S. Breuer, B. Jenichen, M. Ramsteiner, L. Geelhaar, H. Riechert (2012) Luminescence of GaAs nanowires consisting of wurtzite and zincblende segments. *Phys. Rev. B* 85, 045323.
100. P. Lautenschlager, M. Garriga, S. Logothetidis, M. Cardona (1987) Interband critical points of GaAs and their temperature dependence. *Phys. Rev. B* 35(17), 9174–9189.
101. F. Glas, J.-C. Harmand, G. Patriarche (2007) Why does wurtzite form in nanowires of III-V zinc blende semiconductors? *Phys. Rev. Lett.* 99, 146101.

102. X. Duan, J. Wang, C. M. Lieber (2000) Synthesis and optical properties of gallium arsenide nanowires. *Appl. Phys. Lett.* **76**, 1116.
103. S. Breuer, C. Pfüller, T. Flissikowski, O. Brandt, H. T. Grahn, L. Geelhaar, H. Riechert (2011) Suitability of au- and self-assisted GaAs nanowires for optoelectronic applications. *Nano Lett.* **11**(3), 1276–1279.
104. L. Ahtapodov, J. Todorovic, P. Olk, T. Mjåland, P. Slåttnes, D. L. Dheeraj, A. T. J. van Helvoort, B-O. Fimland, H. Weman (2012) A story told by a single nanowire: Optical properties of wurtzite GaAs. *Nano Lett.* **12**, 6090–6095.
105. M. F. Bailon-Somintac, J. J. Iban̄ez, R. B. Jaculbia, R. A. Loberternos, M. J. Defensor, A. A. Salvador, A. S. Somintac (2011) Low temperature photoluminescence and Raman phonon modes of Au-catalyzed MBE-grown GaAs–AlGaAs core–shell nanowires grown on a pre-patterned Si (111) substrate. *J. Cryst. Growth* **314**, 268–273.
106. D. Spirkoska, J. Arbiol, A. Gustafsson, S. Conesa-Boj, F. Glas, I. Zardo, M. Heigoldt, M. H. Gass, A. L. Bleloch, S. Estrade, M. Kaniber, J. Rossler, F. Peiro, J. R. Morante, G. Abstreiter, L. Samuelson, A. Fontcuberta i Morral (2009) Structural and optical properties of high quality zinc-blende/wurtzite GaAs nanowire heterostructures. *Phys. Rev. B* **80**, 245325.
107. T. B. Hoang, A. F. Moses, L. Ahtapodov, H. Zhou, D. L. Dheeraj, A. T. J. van Helvoort, B.-O. Fimland, H. Weman (2010) Engineering parallel and perpendicular polarized photoluminescence from a single semiconductor nanowire by crystal phase control. *Nano Lett.* **10**, 2927–293.
108. A. Fasoli, A. Colli, S. Kudera, L. Manna, S. Hofmann, C. Ducati, J. Robertson, A. C. Ferrari (2007) Catalytic and seeded shape-selective synthesis of II–IV semiconductor nanowires. *Physica E* **37**, 138–141.
109. A. Fasoli, A. Colli, F. Martelli, S. Pisana, P- H. Tan, A. C. Ferrari (2011) Photoluminescence of CdSe nanowires grown with and without metal catalyst. *Nano Res.* **4**(4), 343–359.
110. Z. Tang, N. A. Kotov, M. Giersig (2002) Spontaneous organization of single CdTe nanoparticles into luminescent nanowires. *Science* **297**(5579), 237–240.
111. G. Fonthal, L. Tirado-Mejía, J. I. Marín-Hurtado, H. Ariza-Calderón, J. G. Mendoza-Alvarez (2000) Temperature dependence of the band gap energy of crystalline CdTe. *J. Phys. Chem. Solids* **61**, 579–583.
112. R. Maity, S. Kundoo, K. K. Chattopadhyay (2006) Synthesis and optical characterization of CdS nanowires by chemical route. *Mater. Manuf. Process.* **21**(7), 644–647.
113. R. Chen, M. I. B. Utama, Z. Peng, B. Peng, Q. Xiong, H. Sun 2011) Excitonic properties and near-infrared coherent random lasing in vertically aligned CdSe nanowires. *Adv. Mater.* **23**, 1404–1408.
114. A. Pan, R. Liu, Q. Yang, Y. Zhu, G. Yang, B. Zou, K. Chen (2005) Stimulated emissions in aligned CdS nanowires at room temperature. *J. Phys. Chem. B* **109**(51), 24268–24272.
115. Md. N. Islam, S. Kumar (2003) Influence of surface states on the photoluminescence from silicon nanostructures. *J. Appl. Phys.* **93**, 1753–1759.
116. M. Binder, T. Edelmann, T. H. Metzger, G. Mauckner, G. Goerigk, J. Peisl (1996) Bimodal size distribution in p- porous silicon studied by small angle X-ray scattering. *Thin Solid Films* **276**, 65–68.
117. S. Pizzini, E. Leoni, S. Binetti, M. Acciarri, A. Le Donne, B. Pichaud (2004) Luminescence of dislocations and oxide precipitates in Si. *Solid State Phenom.* **95–96**, 273–282.
118. Y. Kanzawa, T. Kageyama, S. Takeoka, M. Fujii, S. Hayashi, K. Yamamoto (1997) Size-dependent near-infrared photoluminescence spectra of Si nanocrystals embedded in SiO_2 matrices. *Solid State Commun.* **102**, 533–537.

Index

Milton Keynes UK
Ingram Content Group UK Ltd.
UKHW051014071024
449327UK00012B/247

9 780367 489076